MONOGRAPHS ON STATISTICS AND

General Editors

V. Isham, N. Keiding, T. Louis, N. Reid, R. Tibshirani, and H. Tong

Statistical
Inference
Based on the likelihood

ADELCHI AZZALINI

Professor of Statistics
University of Padua
Italy

CRC Press
Taylor & Francis Group
Boca Raton London New York

CRC Press is an imprint of the
Taylor & Francis Group, an **informa** business

A CHAPMAN & HALL BOOK

Originally published by Chapman & Hall
First edition 1996

Published 2002 by CRC Press
Taylor & Francis Group
6000 Broken Sound Parkway NW, Suite 300
Boca Raton, FL 33487-2742

First issued in paperback 2022

© 1996 by Adelchi Azzalini
CRC Press is an imprint of Taylor & Francis Group, an Informa business

No claim to original U.S. Government works

ISBN 13: 978-1-03-247801-2 (pbk)
ISBN 13: 978-0-412-60650-2 (hbk)

DOI: 10.1201/9780203738627

**Visit the Taylor & Francis Web site at
http://www.taylorandfrancis.com**

**and the CRC Press Web site at
http://www.crcpress.com**

Library of Congress Cataloging-in-Publication Data

Catalog record is available from the Library of Congress

Contents

Preface

This book is an account of statistics at a level above the elementary. More precisely, it covers a selection of concepts and methods which appear to me to form the core of the discipline in its present state of development. The unifying element which pervades the whole book is the likelihood function. The aim is to show how the main body of currently used statistical techniques can be generated from a few key concepts, in particular the likelihood.

No attempt is made to cover the whole discipline Some people may well express surprise, if not irritation, that certain ideas are completely missing, notably the Bayesian approach. The reason for this choice is not only that these areas are more distant from my own statistical interests, but also that the treatment of certain topics would have been in conflict with the aim of presenting the material in a homogeneous form. As a result, I have not even included certain topics within my own research interests, such as nonparametrics.

The prime origin of this book lies in notes prepared several years ago for a second course in statistics at the University of Padua. Subsequent additions and amendments have raised the technical level of the text somewhat, but not changed substantially its original profile.

The book will thus be of interest primarily to junior graduate students, or advanced undergraduates, in statistics. It is assumed that the reader is well acquainted with probability theory, but no use of measure theory is involved. To provide any necessary revision of the probability theory background, the appendix briefly recalls some standard material, and describes in greater detail other concepts less commonly found in probability courses. Some sporadic examples in the text involve elementary notions of stochastic processes, but it is not essential to understand these examples in order to follow the bulk of the material. It is advisable that the reader is already familiar with statistical ideas, at least at an elementary level. However, the book starts, in principle, from scratch, and thus is suitable for a mathematician who wants to gain an understanding of statistics

The mathematical style is not abstract, and 'proofs' are presented in such a way as to show the underlying thread of reasoning, rather than aiming at absolute mathematical correctness or weakening conditions as much as possible.

The following people have read either portions or the entire draft of the book, and made very useful comments: M. Chiogna, D. R. Cox, D. Firth, G. Masarotto, A. Salvan, S. Weisberg. The final version has been edited with great competence by R. Leigh. To all of them goes my sincere appreciation.

Adelchi Azzalini
University of Padua

May 1995

CHAPTER 1

Introduction and overview

1.1 Statistical inference

This book is about the theory of statistical inference, or at least about a considerable part of it. To explain the meaning of the term 'statistical inference', let us start with what an encyclopedia says about it.

Statistics
This can be defined as the body of methods concerned with the abstraction of synthetic information from observed data, with the purpose of characterizing those aspects of a phenomenon of interest which are relevant to particular aims Thus, statistics has a wide range of applications in studying all phenomena where it is supposed that some erratic factors are present in addition to some systematic factors whose effects are to be highlighted; so it turns out that the outcome of the latter factors, which could have been described by some 'mathematical law', is overlapped by a component which transforms such a law into 'statistical regularity'
To achieve the aim of examining the essential characteristics of a phenomenon, statistical theory makes wide use of probability theory techniques, especially so when the available observations do not cover all possible instances of the phenomenon under considerations In such a case, 'statistical inference' problems arise, where the aim is to infer the characteristics of the whole from an observed portion of it

(A Naddeo, 1963)*

Therefore, in order to properly carry out a statistical investigation, it is necessary to establish exactly what is the phenomenon under consideration, and to state explicitly which are the observable characteristics, called **variables**, relevant for our purposes. We must also specify the subset of instances which we want to focus on. For instance, suppose that the variable of interest is the height of people. We might consider it in total generality (for the entire human population) or focus on a certain geographic context (such as a nation) and/or on an certain time. This set of instances is called the **population**, and the single instances are called **individuals** or **subjects**; sometimes the terms **cases** and **units** are

* Translated by the author

used. This derivation of this terminology is historical, statistics having originally developed alongside and in interaction with demography; nowadays, the terms 'individual' and 'population' do not necessarily refer to human beings, when they are used in a technical sense.

In some investigations, the entire population is examined, and this is called **census**. In other cases, only a **sample**, i e a subset of the population, is examined, but the target is still to investigate the properties of the whole population. **Statistical inference** constitutes the operation through which information provided by the sample is used to draw conclusions about the characteristics of the population. This sort of procedure is a step forward with respect to a purely **descriptive** statistical approach, which simply attempts to synthesize the more relevant features of the *observed* data The present text is about the theory and the methods which direct the statistical inference operation.

To illustrate the concepts just introduced, let us introduce a very simple practical example. An industry which produces hydraulic pumps purchases from different suppliers many components necessary for its production. In particular, plastic gaskets used to join mechanical elements are supplied by a chemical company in batches of 5000

Obviously, the buyer needs to evaluate the quality of the gaskets supplied in order to eliminate, or at least substantially reduce, the possibility that a faulty gasket is used. Since the cost of repairing a pump found to be faulty is far higher than the cost of the gasket itself, it would be desirable to test the gaskets by putting them to work at appropriate water pressure for a short time, before adopting them for production. On the other hand, also the time required to set up and perform the test of the gaskets represents a cost

One way out is to examine not all the gaskets supplied but only a few, 50 say, and use the information provided by these tests to evaluate the number of faulty gaskets in the entire batch, and to decide about possibly returning the batch to the supplier if it is unsatisfactory. In doing so, we have to take into account that the subset tested will not in general contain faulty gaskets in exactly the same ratio as the batch.

In this example, we can regard the batch of 5000 gaskets as the population under investigation, and each gasket as an individual. At this point in time, our interest in the individual is confined to one specific aspect, namely whether it is 'conforming' or 'not conforming' to the specifications The observed characteristics of the sample elements are not of direct interest, except as a means of making inferences about the characteristics of the population as a whole

Let us now discuss in detail why we often examine only a sample instead of the whole population, which is apparently a preferable approach since it would avoid any indeterminacy in our conclusions (when carried out with full accuracy).

- The inspection cost of the entire population may be excessive, either because the number of subjects is large or because the individual inspection cost is high. Even when economic resources would allow a census of the population, the damage caused by a sample-type study may well not exceed the census cost.

- Since a complete investigation of the population often takes a long time, this may easily conflict with promptness requirements. For instance, the general census of the human population of a nation is performed at very long intervals (usually every ten years), and the results of these investigations are published much later. On the other hand, there are many economic and social problems, such as those related to cost of living or unemployment, for which substantially delayed information is not acceptable, since governments and agencies must take their decisions far more promptly.

- In very many cases, there is a virtual population which is effectively infinite; this situation occurs when instances of the phenomenon under study can be replicated as many times as we like. Suppose, for instance, that we want to study the capacity of a drug in lowering blood pressure in human beings; the relevant population is then formed by all people to whom the drug could possibly be given, i e the entire human population, present and future. Clearly, only a sample study is feasible here. Note that most scientific and technological experiments fall in this situation.

- In some cases, the inspection of the sample units destroys the units themselves. For example, the reliability analysis of a batch of light bulbs involves keeping them on for a rather long time, possibly until they wear out. Therefore, at the end of the inspection, the light bulbs are of degraded quality, if not dead. If the population is finite, clearly a census of the population is ruled out, unless the existence of the population after testing is irrelevant.

1.2 Sampling

It is evident that the sample has to represent, as far as possible, the population characteristics, in order to allow extension of the features of the sample to the population. This requirement is sometimes referred to by saying that the sample must be **representative**.

To illustrate this point, suppose for example that members of an amateur scientific society wish to carry out a survey of the attitudes and behaviour of the inhabitants of their city concerning racial problems. Initially, the members of the society may think of resorting to their personal acquaintances, and administer to these people a suitable questionnaire. It is, however, easy to see why this conduct is grossly unsuitable. Society members form a group of people sharing, to some extent, relevant features such as culture, social status, and possibly also political inclinations. These common elements can be expected to be found among their relatives, friends, neighbours and colleagues, even if to a weaker extent Therefore, this procedure for selecting the sample elements tends to select subjects sharing certain characteristics, which in addition are not unrelated to the topic of the questionnaire. In short, this method would produce an unrepresentative sample.

Therefore, to obtain an adequate sample, the subjects should be selected for inclusion in the sample on the basis of characteristics which are independent of the characteristics under study. On the other hand, it is often difficult, and in some cases virtually impossible, to state which characteristics enjoy this property.

One solution to the problem is to select the sample elements *randomly*, hence by definition independently of any characteristics. To visualize a random selection process, it is convenient to imagine an urn of numbered balls, each of which is associated with a subject of the population Balls are drawn from the urn, and the subjects associated with each ball will constitute the sample.

Even this very simple scheme lends itself to two variants: with or without replacement of the balls. Moreover, 'randomly' does not mean drawing 'with constant probability'; therefore, additional variants arise by assigning different selection probabilities to the balls. All these variants give rise to different sampling schemes, of which only the simplest (but also the most important) will be discussed here, with special attention to the case of an 'infinite' population. We anticipate that the sampling procedure adopted has an effect on the inferential procedures.

1.3 Statistics and probability

We have already pointed out that inference is subject to a certain unavoidable degree of uncertainty, since it is completely unrealistic, except in exceptional circumstances, to imagine that the sample represents exactly all the characteristics of the population.

The degree of closeness between the sample and the population characteristics is a variable entity which can be studied with the tools of probability theory. This area of mathematics plays, then, an essential role in the development of statistical theory, to such an extent that it becomes sometimes difficult to establish the border between these two disciplines.

Notice that it is the very fact of selecting the sample according to a random scheme that allows a mathematical analysis of the degree of correspondence between sample and population. Such an analysis would not be possible using different selection criteria, such as subjective selection of the sample units decided by the experimenter, or self-selection of the sample, i.e. a situation where the units put themselves forward for interview.

To begin to explain how probability theory enters into statistical inference theory, let us go back to our sample of 50 gaskets from a batch of 5000. It is convenient to think of the batch as an urn with 5000 numbered balls, of which an unknown proportion θ are black (i e defective), and the remaining proportion white. Of 50 balls drawn, a certain number y are black; suppose $y = 4$. To estimate the overall number x of black balls in the urn, or equivalently to estimate the proportion θ, it is natural to consider the equality

$$4 \div 50 = x \div 5000 \qquad (1.1)$$

which leads to

$$\theta = 4/50.$$

However, various questions arise.

- Are there reasonable alternatives to this reasoning, hence to the choice of θ?

- How accurate is the present estimate of θ compared to its true value? Can we obtain an interval of plausible values of θ?

- If we had drawn 100 balls, 8 of which were black, the above argument would lead to 8/100, i.e. the same as 4/50. Does the estimate based on a sample of 100 contain more information than that based on 50? It should be so, but the mere proportions 4/50 and 8/100 do not reflect this difference in 'information'.

- If the supplier had guaranteed that the batch contained a proportion of faulty gaskets no greater than 5%, is there sufficient evidence to state that the batch is inadequate, and return it to the supplier? Take into account that the supplier can easily claim that we had simply selected an 'unlucky' sample, that 5% is not so different from 8%, and that a test of the entire batch would certainly reveal that it has less than 5% defectives.

- Equation (1 1) is certainly a reasonable criterion, but it can be used only to evaluate the size of a sub-population, x in this case, or equivalently a proportion, θ in this case. Can we construct a *general* criterion, usable in any practical situation, even one far more complicated than our gasket test? An example of non-trivial statistical problem is the following: in a medical study, a group of patients affected by a certain type of cancer undergo chemotherapy treatment using an appropriate combination of several drugs. Ideally, drug dosages should be sufficiently high to fight the tumour, but not so high as to induce severe toxicity (defined as putting the patient's life at risk); if necessary, treatment cycles are repeated until the tumour vanishes. After all patients have been treated and discharged, data are analysed to study the relationships between severe toxicity and drug dosage, with the aim of identifying the minimal dosage and the optimal mix of the drugs required to remove the tumour, avoiding unnecessary toxicity. If known, concomitant factors such as the sex of the patient, age, tumour stage, histological type of tumour should also be included in the study. Clearly, a problem of this sort cannot be handled simply with the aid of equation (1.1).

Returning to our simple quality control example, the above raised questions demand an analysis, from the viewpoint of probability theory, of the relationship between the unknown proportion, θ, and the observed number of defectives, y. In this context, the observed number y can be regarded as a value sampled from a random variable Y, whose probability distribution depends both on θ and on the sampling scheme adopted. Specifically, the distribution of Y is binomial or hypergeometric, depending on whether we sample with or without replacement, since the population is finite. However, these distributions are quite close in this case, because of the small fraction sampled, 50 out of 5000. Hence, for the sake of simplicity, we shall restrict the discussion to the binomial case.

A priori with respect to the sampling, the probability that Y yields the observed value y is

$$\Pr\{Y = y\} = \binom{50}{y} \theta^y (1 - \theta)^{50-y} \tag{1.2}$$

where y is an integer between 0 and 50. The set of possible values for θ is the set of ratios of type $k/5000$, where $k \in \{0, 1, \ldots, 5000\}$; however, this set can be approximated reasonably well by the set of all numbers in $[0, 1]$ As θ varies in $[0, 1]$, expression (1.2) spans a whole family of probability distributions, although there is only one value of θ which actually controls the data generation.

From this point on, inference can be regarded as an operation dealing with the **parameter** θ *which identifies the true probability distribution of Y within the family of distributions (1.2).*

Therefore, from now on, we shall no longer talk about populations, but rather about random variables and inference about the parameters of their distributions, as the connection between populations and random variables is understood.

1.4 Some typical problems

The aim of this entire text is, in fact, to address the questions raised in section 1.3. Although a more systematic discussion will be presented in the following chapters, it is useful to introduce here some key ideas in a rather informal fashion.

1.4.1 Likelihood and estimates

One of the queries in section 1.3 asked for a general criterion for the construction of parameter estimates, where **estimate** can be informally thought of as a plausible value, $\hat{\theta}$ say, in place of the unknown value θ.

Once the sample data are given, $y = 4$ in our example, (1.2) is

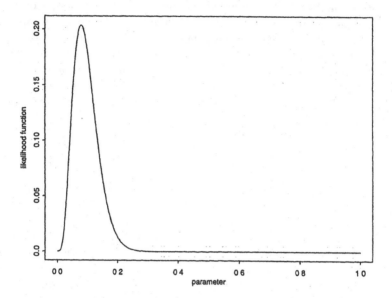

Figure 1.1 *A likelihood function $L(\theta)$ plotted against the parameter θ*

a function of θ only, namely

$$L(\theta) = \binom{50}{4} \theta^4 (1 - \theta)^{46} \qquad (0 \le \theta \le 1).$$

This function is plotted in Figure 1.1.

This expression gives, as a function of θ, the a priori probability of observing what has actually been observed. Conversely, it can be regarded as a measure of agreement between any nominated value of θ and the observations, hence explaining the name **likelihood** given to $L(\theta)$. Notice that, although all ordinates of $L(\theta)$ are probabilities, the function $L(\theta)$ itself is not a probability distribution.

It is then quite natural to select the point with highest likelihood, if an estimate of θ is required. A simple mathematical exercise shows that the maximum of $L(\theta)$ occurs at

$$\hat{\theta} = 4/50,$$

and this is called, for obvious reasons, the **maximum likelihood estimate**.

We can immediately extend the method to a slightly more general situation by considering a generic number n of sampled elements, with an observed number of defectives denoted by y, an integer between 0 and n. Then the random variable Y has binomial distribution with index n and parameter θ, and the likelihood is now

$$L(\theta) = \binom{n}{y}\theta^y(1-\theta)^{n-y} \qquad (0 \le \theta \le 1) \qquad (1.3)$$

with corresponding estimate

$$\hat{\theta} = y/n. \qquad (1.4)$$

Although the criterion of maximization of the likelihood has produced nothing different from what we did before by a simpler, more direct argument, the present criterion has the major advantage that it can also be applied to problems where (1.1) makes no sense. In fact, it can be employed whenever we can write the likelihood, hence covering an enormous number of practical situations, as will be demonstrated in the following chapters.

1.4.2 Interval estimation and hypothesis testing

At first sight, it may seem that estimation is the only issue in data analysis. The discussion in section 1.3 shows the error of this belief, by raising a number of additional questions, whose discussion we now wish to start.

Since the estimate cannot be expected to coincide with the true value of the parameter, it is natural to look for an interval which should presumably contain the parameter. In the above example with $\hat{\theta} = 4/50$, we look for an interval around 4/50 in which θ is likely to lie. Such an interval in called an **interval estimate** of θ.

A different, but related, problem is raised when it is required to compare the estimate, or more generally the observations, with a reference value of the parameter. For instance, in the discussion of section 1.3, the question was raised of deciding whether the data provided evidence for or against the claim that the proportion of defectives in the batch does not exceed 5%. Problems of this sort are called **hypothesis testing**.

Let us explore, arguing rather informally, why interval estimation and hypothesis testing are similar problems, in a sense. If we have selected some interval estimate (θ_1, θ_2) for θ, then any value of θ within this interval is somehow compatible with the empirical evidence contained in the sample. On the other hand, if the hypotized value $\theta = 5\%$ considered in the hypothesis testing problem is within the interval (θ_1, θ_2), then the hypothesis about θ is not in conflict with the data.

To choose the interval (θ_1, θ_2), we again make use of the likelihood function. If $\theta = 4/50$ is the preferred point because it has the highest likelihood, then it is also true that other points in the interval $[0, 1]$ are more or less plausible depending on their likelihood value. This remark implies that the criterion for inclusion of a point in a confidence interval must be based on the value of the likelihood at that point, compared to the maximum value. We are then led to consider the **relative likelihood**

$$\tilde{L}(\theta) = \frac{L(\theta)}{L(\hat{\theta})} \tag{1.5}$$

which varies in $[0, 1]$. Figure 1.2 represents the relative likelihood corresponding to the likelihood of Figure 1.1, discarding the portion of the interval $[0, 1]$ where the likelihood itself is negligible. Since $L(\hat{\theta})$ is constant with respect to θ, the graph is essentially the same as before, except for a scaling factor.

If we had to choose an interval for θ, we could select the set of points satisfying the condition

$$\tilde{L}(\theta) > 1/2$$

or some other similar constraint, for instance

$$\tilde{L}(\theta) > 1/5.$$

For the latter choice, Figure 1.2 shows the corresponding confidence interval.

Consider now the question of deciding about the reference value $\theta = 5\%$. The relative likelihood for this point is 0.668, a reasonably high value, so that 5% does not seem in conflict with the observed data. Inspection of Figure 1.2 illustrates the connection between confidence intervals and hypothesis testing: the plausibility of the value 5% derives from its relative likelihood, and is in direct correspondence with the fact that 5% is within the confidence interval.

It is interesting to examine the effect of increasing the sample size, hence the 'amount of information', without changing the parameter estimate. In our example, suppose that both the sample size and the number of defectives are doubled. Although $\hat{\theta}$ does not change, the likelihood function and the relative likelihood do change; Figure 1.3 shows plots of the relative likelihood for both

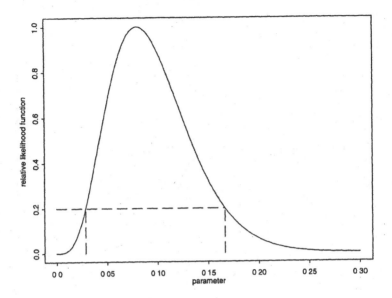

Figure 1.2 *A relative likelihood* $\bar{L}(\theta)$ *plotted against* θ *and the confidence interval for* θ *specified by the condition* $\bar{L}(\theta) > 1/5$

cases, $n = 50$ and $n = 100$. For the new relative likelihood function, values of θ different from $\hat{\theta}$ are now less likely; the new likelihood function is said to be 'more concentrated' around $\hat{\theta}$.

In particular, the relative likelihood at $\theta = 5\%$ drops from 0.668 to 0.446. Although not extremely low, the new value is far less reassuring than before. The question is whether or not this decrease of confidence in the value 5% is sufficient to reject it as a plausible value; this problem is essentially related to the choice of the threshold value for the relative likelihood, 1/5 in the above discussion and in Figure 1.2. An extended discussion of this point will is deferred until Chapter 4; for the moment, we have confined ourselves to laying down some of the basic elements

1.4.3 Repeated sampling principle

So far the motivation of the principles and methods discussed has been entirely intuitive. It is, however, possible to examine these

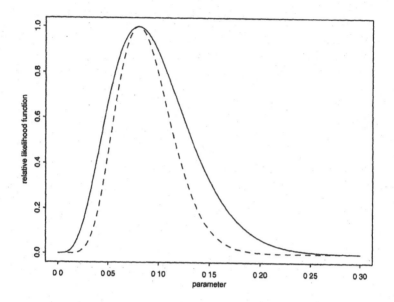

Figure 1.3 *Two relative likelihood functions, associated with $y = 4$, $n = 50$ (continuous curve) and $y = 8$, $n = 100$ (dashed curve)*

methods from a different perspective, by studying their mathematical properties.

In so doing, we adopt the criterion of *regarding estimates and other quantities related to statistical techniques as values sampled from random variables.* For instance, the estimate $\hat{\theta} = y/n$ depends on y, which is a value sampled from a random variable; $\hat{\theta}$ is of the same nature. This viewpoint gives rise to the **repeated sampling principle** which consists in regarding $\hat{\theta}$ as a quantity sampled from a random variable and in studying the properties of this random variable.

The word *repeated* arises from the following consideration. If the sample had been drawn a large number of times and, for each sample, the maximum likelihood estimate had been computed, then each individual value of the estimate could indeed be regarded as sampled from a random variable. In the example discussed so far, the specific batch of gaskets could be one of a regular supply; hence this situation would fall naturally within the repeated sampling principle. In many other cases, this replication of samples does not really take place, but we still argue as if it did.

Adopting the repeated sampling principle is a completely separate choice of the criterion which has produced the statistical method under consideration: the same principle can be applied to techniques not related to the likelihood, as methods can be adopted purely from qualitative considerations, not involving the repeated sampling principle.

To demonstrate how the repeated sampling principle works in practice, consider (1.4), where y is sampled from a random variable Y with binomial distribution with index n and parameter θ. By using elementary results from probability theory, we immediately obtain

$$\mathrm{E}\left[\hat{\theta}\right] = \mathrm{E}\left[\frac{Y}{n}\right] = \frac{\mathrm{E}[Y]}{n} = \theta,$$

which says that *on average* $\hat{\theta}$ is centred at θ, whatever this value is. Clearly, we have no assurance that the observed value of $\hat{\theta}$ is equal to θ; in fact, in most cases this will not be true, but it is reassuring to know that the method has no systematic **bias**.

To assess the degree of accuracy of the estimate, we can consider a measure of variability of $\hat{\theta}$, in particular

$$\mathrm{var}\left[\hat{\theta}\right] = \mathrm{var}\left[\frac{Y}{n}\right] = \frac{\mathrm{var}[Y]}{n^2} = \frac{\theta(1-\theta)}{n}.$$

An important implication of this result is that this variance goes to 0 as $n \to \infty$, an highly desirable feature. If a numerical quantification of the above variance is needed, it is obtained by replacing θ by the observed value of $\hat{\theta}$ in the final expression.

1.5 Statistics and real problems

The theory of statistical inference is about general principles and criteria which motivate certain mathematical constructs, in particular methods which can be used to tackle problems arising in the real world

The logical path from these basic principles to the solution of real problems is extremely long, and crosses diverse territories. The first part is entirely within the realm of mathematical sciences; in the final stage, one has to enter the world of empirical evidence.

This book deals with the first stage only, except for brief comments on the other aspects. When we wish to move on to applications of statistical methods to real problems, important and difficult problems arise, as we have to face the task of matching the two worlds of formal deduction and empirical evidence; this is the field of **applied statistics**.

Over the years, applied statisticians have developed a wide range of techniques to help in this operation Some of these tools are extremely powerful and sophisticated, to the point of involving methods from artificial intelligence. However, these tools cannot solve by themselves the core of the problem, since a meaningful application of statistical inference methods also entails some understanding of the phenomenon under consideration.

In the simplest cases, common knowledge may suffice; in less trivial cases, cooperation with a subject-matter expert will be necessary. The fruitful integration of different academic backgrounds is, in general, not an easy target to pursue It requires from both sides, the statistician and the subject-matter scientist, an open-minded ability to penetrate and accept principles and methods from a different discipline.

Whether the statistician works individually or jointly with a subject-matter scientist, it is an important requirement that statistical methods are applied taking into account the basic principles of the context discipline In some applications, the motivation of the statistical analysis is pragmatic rather than explanatory of the phenomenon, and often no proper 'context discipline' exists; however, at least some form of background information on the phenomenon under study is virtually always present. Whenever some form of context information is available, the statistician must make use of it, avoiding the temptation of blind numerical manipulation of data, since this would produce meaningless results from the point of view of the subject specialist Mills (1965, p. 5) has expressed this concept concisely and effectively:

> If such [statistical] techniques are to be well and wisely employed they must be adapted, with understanding, to the materials under study

Exercises

1.1 Assuming (here and in the following exercises) the setting of section 1.3, obtain the likelihood for x instead of θ.

1.2 Write down the likelihood for the case of sampling without replacement.

1 3 If $n = 100$ and $y = 8$, obtain the interval estimate for θ associated with the condition $\tilde{L}(\theta) > 1/5$ either by an approximate graphical procedure based on Figure 1.3, or more accurately via numerical computation.

1.4 In the repeated sampling principle, obtain $E[R]$ and $\text{var}[R]$, where $R = Y/n$, if sampling is without replacement.

CHAPTER 2

Likelihood

The aim of this chapter is to construct a theoretical framework and some general concepts which will be useful for the development of specific methods in the subsequent chapters.

2.1 Statistical models

2.1.1 Introduction

Our starting point is an empirical study, either experimental or observational, which is supposed to have provided a certain set of data, called **sample values** or even simply a **sample**, and denoted by y.

In the simplest cases, y is a vector of numbers, $y = (y_1, \ldots, y_n)^{\top}$ say, but it can be a far more complex data structure. However, for our present purposes, it is not useful, and may even be misleading, to think of it as a composite entity; for the moment, let us regard it as a unit

The **fundamental assumption** is to consider y as *a value taken on by a random variable Y*, and our goal is to make use of y in drawing conclusions on the unknown distribution $F_*(\cdot)$ of Y, as already mentioned in Chapter 1. Sometimes we shall refer to the physical mechanism which generated y as an **experiment** whatever its actual nature, be it an observational study or an experiment in the proper sense.

Obviously, our conclusions about F_* are subject to uncertainty since the randomness governing Y, which produces y, affects our statements. Our goal is to make sure that:

- the level of uncertainty is the smallest possible, considering the randomness of Y;

- we are able to assess the level of uncertainty in our conclusions.

The physical nature of the phenomenon which generates y, the sampling scheme, and possibly other available information, will place limits on the set of possible choices for F_*. This set, denoted by $\mathcal{F}$, is called the **(statistical) model**. It is intuitive that our inferences will be more accurate if we are able to select the set $\mathcal{F}$ so that it is as small as possible, under the requirement that $F_* \in \mathcal{F}$.

In some cases, it can be assumed that Y is a random variable with independent and identically distributed components, in this case, we say that y is a **simple random sample** (s.r.s.) from Y.

2.1.2 Parametric models

In principle, $\mathcal{F}$ can be any set of distribution functions, but there is a category of such sets which plays a prominent role, from both the theoretical and the applied point of view. This case occurs when all elements of $\mathcal{F}$ are functions of the same mathematical form, identified only by the different specification of θ, which varies within a set $\Theta \in \mathbb{R}^k$ In other words, we have

$$\mathcal{F} = \{F(\cdot\,;\theta) : \theta \in \Theta \subseteq \mathbb{R}^k\}$$

where, for any fixed θ, $F(\cdot\,;\theta)$ is a distribution function whose support (see p.19) is a subset of $\mathbb{R}^n$, with k and n positive integers

In the overwhelming majority of cases (and certainly in all the cases we shall consider), all distribution functions which are members of a given $\mathcal{F}$ refer either to discrete or to continuous random variables Then, $\mathcal{F}$ can be defined using the corresponding density functions or probability functions. In the following, we shall often make use of the expression 'density function' in both cases, which is legitimate in any case when one interprets the phrase 'density function' as in a measure theory context.

Therefore, it is more convenient to define a statistical model $\mathcal{F}$ as a set of density functions, namely

$$\mathcal{F} = \{f(\cdot\,;\theta) : \theta \in \Theta \subseteq \mathbb{R}^k\} \qquad (2\;1)$$

for some density function f The quantity θ is called the **parameter**, the set Θ is called the **parameter space** and class (2.1) is called the **parametric class** or **parametric (statistical) model**.

Thus, the elements of $\mathcal{F}$ are associated to the elements of Θ. In particular, there exists a value $\theta_* \in \Theta$, associated with F_*, called the 'true value' of the parameter, and our inferences will be about θ_*.

Obviously, not all statistical models are of parametric type For instance, the classes

$\mathcal{F}_1 = \{$the set of all density functions in one variable$\}$,

$\mathcal{F}_2 = \mathcal{F}_1 \cap$
$\qquad$ $\{$the set of all +ve functions whose logarithm is concave$\}$,

in fact define **nonparametric models**. Inference for this kind of situation is handled by a different, broad section of statistics, which is correspondingly called **nonparametric statistics**, whose treatment is outside the scope of the present book.

Connected to the concept of the statistical model is the notion of the **sample space**, which is the set $\mathcal{Y}$ of all possible sample outcomes y compatible with a given parametric model. Formally, denoting by $\mathcal{Y}_\theta$ the support* of the density function $f(\cdot\,;\theta)$, the sample space is given by

$$\mathcal{Y} = \bigcup_{\theta \in \Theta} \mathcal{Y}_\theta.$$

Quite often, however, $\mathcal{Y}_\theta$ is the same for all possible choices of θ, and this set coincides with $\mathcal{Y}$.

Example 2.1.1 Assume that $Y \sim Bin(n, \theta)$, where the notation indicates that Y has binomial distribution with index n and parameter $\theta \in [0, 1]$. Then $\mathcal{Y}_\theta$ is the same for all possible θ, and coincides with the sample space

$$\mathcal{Y} = \{0, 1, \ldots, n\}.$$

It is not necessary, however, that $\Theta = [0, 1]$, i e Θ does not need to include all possible values of θ which are mathematically meaningful $\qquad\qquad$ □

Example 2.1.2 If two values are sampled independently from a $N(\theta, 1)$ random variable, then $y = (y_1, y_2)^\top$, where $y_i \in \mathbb{R}$ ($i = 1, 2$),

$$\mathcal{Y} = \mathbb{R} \times \mathbb{R}, \quad Y \sim N_2\left(\begin{pmatrix} \theta \\ \theta \end{pmatrix}, I_2\right),$$

where I_2 is the identity matrix of order 2, and

$$f(y; \theta) = \phi(y_1 - \theta)\,\phi(y_2 - \theta) \qquad (y \in \mathcal{Y})$$

where ϕ is defined by (A.7). If there are no constraints on θ then

---

* That is, the smallest set to which the density function assigns probability 1

$\Theta = \mathbb{R}$, but if it happens, for instance, that we know θ to be positive, then $\Theta = \mathbb{R}^+$.   □

Example 2.1.3 A parametric family of distributions is said to form a location and scale family if the density function of a single component can be written as

$$g(t; \theta_1, \theta_2) = \frac{1}{\theta_2} \, g_0 \left(\frac{t - \theta_1}{\theta_2} \right)$$

for a fixed density $g_0(t)$ and two parameters θ_1, θ_2, called location and scale parameters, respectively ($\theta_2 > 0$). The reason for the terminology is that all distributions of the family can be obtained by applying a linear transformation of the form

$$Y = \theta_1 + \theta_2 Y_0$$

to a variable Y_0 with density $g_0(\ )$

As an example, consider the parametric class of one-dimensional density functions

$$g(t; \omega, \lambda) = \begin{cases} \lambda e^{-\lambda(t-\omega)} & \text{if } t > \omega, \\ 0 & \text{otherwise,} \end{cases}$$

where the parameter is now $\theta = (\omega, \lambda)^\top$ with $\lambda > 0$. This class of distributions is said to form a location and scale family, with

$$\Theta = \{\theta : \theta = (\omega, \lambda), \lambda \in \mathbb{R}^+, \omega \in \mathbb{R}\},$$
$$\mathcal{Y}_\theta = \{t : t \in \mathbb{R}, t > \omega\}$$

and, by the union of the sets $\mathcal{Y}_\theta$, we have $\mathcal{Y} = \mathbb{R}$.

By saying that the above density function is one-dimensional, we are implicitly referring to a single observation from a random variable Y. If n observations are drawn from the same random variable, and Θ remains unchanged, then $\mathcal{Y} = \mathbb{R}^n$ and $\mathcal{F}$ is given by the set of the n-dimensional density functions obtained by n-fold multiplication of the individual density function g.   □

2.1.3 Parametrizations

In specifying $\mathcal{F}$, we can choose various equivalent formulations, i.e. various **parametrizations**. If h is a one-to-one function from Θ to Ψ, we can rewrite (2.1) as

$$\begin{aligned} \mathcal{F} &= \{f(\cdot\,; \psi) : \psi = h(\theta), \theta \in \Theta\} \\ &= \{f(\cdot\,; \psi) : \psi \in \Psi\}, \end{aligned}$$

where
$$\Psi = \{\psi : \psi = h(\theta), \theta \in \Theta\}.$$

In some cases, one specific parametrization is clearly superior to others; the typical reason for preference is that one parametrization has a clearer physical interpretation. In other cases, there is some degree of arbitrariness in the choice of the parametrization.

Since the choice of parametrization is not uniquely determined, it is desirable that inferences are **invariant** with respect to the choice of parametrization. There are, however, various statistical methods which are *not* parametrization-invariant, as we shall see later.

Example 2.1.4 Let Y be a negative exponential random variable whose density function at point t ($t > 0$) can be written as $\lambda e^{-\lambda t}$, where $\lambda > 0$; alternatively, the same density function can be written as $\psi^{-1} e^{-t/\psi}$, where $\psi = 1/\lambda$. If Y represents the inter-arrival time between events in a Poisson process, then both parameters have direct physical meaning, since λ is the mean number of events in a time unit, and ψ is the mean time between events. □

Example 2.1.5 If $Y \sim N(\mu, \sigma^2)$, the parameter μ is expressed directly on the same scale as the variable Y itself, and it will rarely be appropriate to reparametrize this component. If a change of parametrization is introduced for the purposes of statistical analysis, the final conclusions need to be converted back to the original scale; this stresses the requirement of invariance of the methods involved with respect to a change of parametrization.

As indices of variability both σ and σ^2 are candidates. Although σ has the advantage of being measured on the same scale Y, σ^2 is also commonly used. □

To avoid ambiguity, it is essential that distinct values of θ correspond to distinct probability distributions. If this property did not hold, there could exist two values of θ associated with the true distribution F_*, making it impossible to say which one is the true parameter value. Formally, we require that, for any two distinct elements θ_1 and θ_2 in Θ, there exists at least one set B in the sample space such that

$$\Pr\{Y \in B; \theta_1\} \neq \Pr\{Y \in B; \theta_2\},$$

where the notation $\Pr\{A; \theta\}$ means that the probability of the event A is computed for the distribution specified by θ. This property of a model is called **identifiability**.

Remark 2.1.6 In the above definition the term 'set' should really be replaced by 'measurable set'. Here, as elsewhere in this text, we avoid subtle mathematical details. *For the rest of this book,* we adopt the following terminology: 'set' stands for 'measurable set', 'function' for 'measurable function' With these conventions, the statements are correct, although an absolutely rigorous mathematical treatment is not attempted. An expression such as

$$\int_{\mathcal{Y}} g(y)\, d\nu(y)$$

must be interpreted as a Riemann integral or as a summation, depending on whether the underlying random variable Y is continuous or discrete. Obviously, one can interpret the expression in a measure-theoretic context, with ν indicating a measure defined on $\mathcal{Y}$.   □

2.2 Statistical likelihood

The key word in Statistics is information
After all, this is what the subject is all about

(D Basu, 1975)

2.2.1 The likelihood function

Consider a given statistical model of type (2.1). Once a sample value, y say, has been observed, the value of the density function $f(y; \theta)$ depends only on θ. This function gives us the probability (density) which, a priori with respect to the experiment, we had of *observing what we have in fact observed.*

If we need to rank our preferences between two elements of Θ, θ' and θ'' say, then the relevant quantity is the ratio $f(y; \theta')/f(y; \theta'')$, provided the denominator does not vanish. Since this ratio does not change if both terms are multiplied by a positive constant c independent of θ, then for comparing the elements of Θ the relevant quantity is $f(y; \theta)$ up to a multiplicative constant.

Definition 2.2.1 *For the statistical model (2.1), from which a sample $y \in \mathcal{Y}$ has been observed, we use the term* likelihood function, *or simply* likelihood, *for the function from Θ into $\mathbb{R}^+ \cup \{0\}$ written as*

$$L(\theta) = L(\theta; y) = c(y) f(y; \theta) \tag{2.2}$$

where $c(y)$ is a positive constant independent of θ.

Although the likelihood is a function of θ, the notation $L(\theta; y)$ is sometimes used to emphasize that $L(\theta)$ depends on y, in the sense that for a different sample y' we obtain a different likelihood function $L(\theta; y')$. Notice also that it does not matter whether we write c or $c(y)$ in (2.2), since the likelihood is a function of θ.

In fact, (2.2) identifies a family of functions, whose elements differ by a multiplicative constant; more precisely, it is an equivalence class of functions. It follows that two points in the sample space associated with density functions which are proportional determine the *same* likelihood, although we shall commonly use the term **equivalent likelihoods**.

Even if every value of $L(\theta)$ is determined by a probability distribution, and in spite of the graphical appearance of Figure 1.1, the likelihood function is *not* a probability distribution.

Since $L(\theta)$ is a non-negative quantity, and in most cases is positive for all Θ, we can define the **log-likelihood function** as

$$\ell(\theta) = \ln L(\theta) = c + \ln f(y; \theta)$$

with the convention that $\ell(\theta) = -\infty$ if $L(\theta) = 0$. Again, the 'function' is a family of functions, all parallel to each other In Chapter 3 we shall see that it is the log-likelihood (and its derivatives), rather than the likelihood itself, which summarizes the 'information' available in the sample, and determines the properties of the associated statistical methods.

Example 2.2.2 Extending Example 2.1.2 above, consider an s r s $y = (y_1, \ldots, y_n)^{\top}$ from a random variable $N(\mu, \sigma^2)$, where $\theta = (\mu, \sigma^2)$ varies over the entire admissible space, i e $\Theta = \mathbb{R} \times \mathbb{R}^+$. Because of the independence of the components, we have

$$
\begin{aligned}
L(\theta) &= c \prod_{i=1}^{n} \frac{1}{\sigma} \phi\left(\frac{y_i - \mu}{\sigma}\right) \\
&= c \prod_{i=1}^{n} \frac{1}{\sqrt{2\pi}\sigma} \exp\left\{-\frac{1}{2}\left(\frac{y_i - \mu}{\sigma}\right)^2\right\}, \\
&= c\,\sigma^{-1/n} \exp\left\{-\frac{1}{2\sigma^2}\left(\sum_i y_i^2 - 2\mu \sum_i y_i + n\mu^2\right)\right\}
\end{aligned}
$$

where the symbol c is used to denote any constant not depending on θ, so that the c on the third line is not the same as that in the previous two lines. In the following, we shall use the same convention without further comment. The corresponding log-likelihood function is

$$\ell(\theta) = c - \frac{n}{2}\ln\sigma^2 - \frac{1}{2\sigma^2}\left(\sum_i y_i^2 - 2\mu\sum_i y_i + n\mu^2\right).$$

$\square$

Example 2.2.3 Consider an s r s $y = (y_1, \ldots, y_n)^\mathsf{T}$ from a random variable $U(0, \theta)$ where $\theta > 0$. The probability density associated with a generic term y_i is $1/\theta$ if $y_i \in (0, \theta)$, but in multiplying such densities we cannot simply multiply the term $1/\theta$ without considering the condition 'if $y_i \in (0, \theta)$'. Therefore, we write the density function for a single observation as

$$\frac{1}{\theta}I_{(0,\theta)}(t) \qquad (t \in \mathbb{R})$$

where $I_A(\cdot)$ is the indicator function of set A. Obviously, we could also have used $I_{[0,\theta]}$ or $I_{[0,\theta)}$, since this does not affect the probability distribution. Then the likelihood is

$$
\begin{aligned}
L(\theta) &= c\prod_{i=1}^{n}\frac{1}{\theta}I_{(0,\theta)}(y_i) \\
&= \frac{c}{\theta^n}\prod_{i=1}^{n}I_{(0,1)}(y_i/\theta) \\
&= \frac{c}{\theta^n}I_{(0,1)}(y_{(n)}/\theta) \\
&= \frac{c}{\theta^n}I_{(1,\infty)}(\theta/y_{(n)}) \\
&= \frac{c}{\theta^n}I_{(y_{(n)},\infty)}(\theta),
\end{aligned}
$$

where we use the fact that the product of the terms $I_{(0,1)}(y_i/\theta)$ is 1 when θ is greater than all the y_i, i.e. when θ is greater than the greatest sample value, $y_{(n)}$. Then the likelihood function is 0 to the left of $y_{(n)}$, jumps to a positive value at $y_{(n)}$ and then decreases geometrically, as shown in Figure 2.1. The corresponding log-likelihood is

$$\ell(\theta) = \begin{cases} -\infty & \text{if } \theta \leq y_{(n)}, \\ c - n\ln\theta & \text{if } \theta > y_{(n)}, \end{cases}$$

but notice that the value of ℓ at $y_{(n)}$ can be changed to $c - n\ln\theta$,

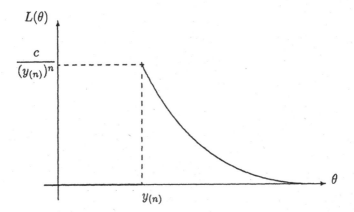

Figure 2.1 *Likelihood for the parameter θ of a $U(0,\theta)$ random variable in the case of an s r s*

since at the beginning we could have used the indicator function $I_{(0,\theta]}$ in the density function of a single observation.   □

Example 2.2.4 Consider a sequence of binary random variables $(Y_1,\ldots,Y_n)$ from a homogeneous Markov chain with state space $\{0, 1\}$ and transition probability matrix

$$\begin{pmatrix} 1 - \pi_0 & \pi_0 \\ 1 - \pi_1 & \pi_1 \end{pmatrix}$$

where π_0, π_1 are parameters in $(0,1)$. If $Y_1 \sim Bin(1, \pi)$, the parameter is $\theta = (\pi, \pi_0, \pi_1)$ which belongs to $\Theta = (0,1)^3$, if no constraint is present. Notice that, in this case, the Y_j variables are neither independent nor identically distributed. Given a sequence $(y_1,\ldots,y_n)$ of observed values, the likelihood is

$$L(\pi, \pi_0, \pi_1) = \pi^{y_1} (1 - \pi)^{1-y_1} \prod_{j=2}^{n} \pi_{y_{j-1}}^{y_j} (1 - \pi_{y_{j-1}})^{1-y_j}$$

□

Example 2.2.5 It is very commonly the case that there is incomplete knowledge about the value taken by the quantity of interest, and this can occur in a variety of forms. One of the most common situations of incomplete knowledge occurs when it is known that the variable of interest has exceeded a certain value, but its exact value is not known In this case, we say that **censoring** has occurred; the sample containing censored data is said to be a **censored sample.** Censoring itself can appear in number of forms, but we shall restrict ourselves to a rather simple case, illustrating it with a problem from a medical context, where in fact it occurs very often.

The term **survival** denotes the span of time between two given events, an initial and a final event, say. Some typical examples are the span of time between a heart transplant and its rejection, or the interval of time between remission of a patient with cancer of a certain type and the patient's first relapse.

It is, however, quite common that the final event has not yet been observed at the moment when the data are analysed. Suppose, for instance, that n patients had undergone kidney transplant at dates $(d_1, .. , d_n)$, and the quantity of interest is the probability distribution of the random variable U which denotes the survival time of the transplanted kidney, that is the time interval between transplant and rejection.

Time considerations prevent us from waiting until all patients reject the kidney; moreover, some patients may die before rejection for reasons totally unrelated to the transplant. Therefore, when the data are analysed, for some patients rejection will already have occurred and the corresponding value u_i of U will be known, while for other patients' rejection will not yet have taken place and it will only be known that u_i belongs to $(v_i, +\infty)$, where v_i is the time interval between date d_i and the time of analysis; the latter group of data is then censored

If we denote by $(y_1, \ldots, y_n)$ the observed survival times (censored or not), it is clear that they cannot be used directly for inference as if they were a sample from U, since censoring of some data would lead to inferential bias. It is then necessary to write down the appropriate likelihood for the problem under consideration.

For the sake of simplicity, consider the 'case of a continuous random variables, and denote by $f(\ ;\theta)$ the density function of the survival time U and by $g(\cdot)$ the density function of the censoring time V, where the notation reflects the fact that only the distribution of U contains parameters of interest and that the distribution of V is not of direct interest. Then the available data are sampled from the random variables

$$Y = \min(U, V), \qquad Z = \begin{cases} 1 & \text{if } U < V, \\ 0 & \text{if } U > V, \end{cases}$$

where the variable Y represents the observed (actual or censored) survival time, and the indicator variable Z marks the value of Y as the actual survival time ($Z = 1$) or a censored datum ($Z = 0$)

To write down the likelihood, we need to obtain first the joint distribution of the pair (Y, Z), since this is the observable quantity, not (U, V). Assuming independence of U and V, i e assuming that the choice of the date for performing the censoring is not influenced by the survival times, we obtain

$$\begin{aligned}
\Pr\{Y \in (y, y + dy), Z = 1\} &= \Pr\{U \in (y, y + dy), V > y + dy\} \\
&= f(y; \theta)\{1 - G(y)\}dy + o(dy), \\
\Pr\{Y \in (y, y + dy), Z = 0\} &= \Pr\{V \in (y, y + dy), U > y + dy\} \\
&= g(y)\{1 - F(y; \theta)\}dy + o(dy),
\end{aligned}$$

where F and G are the distribution functions corresponding to f and g, respectively. Denoting by $((y_1, z_1), \ldots, (y_n, z_n))$ the available data, the likelihood for θ is

$$L(\theta) = c \prod_{i=1}^{n} f(y_i; \theta)^{z_i} \{1 - F(y_i; \theta)\}^{1 - z_i} \qquad (2.3)$$

where we have absorbed in c all terms containing $G(y_i)$ and $g(y_i)$

Since we have assumed that (i) U and V are independent, (ii) the density function of V does not depend on θ, then the censoring times contain no information on θ, either directly or indirectly. In this case, we say that the censoring is **non-informative**.

Clearly, for the practical use of (2.3), it is necessary that a parametric family of density functions is specified for f, but it is not required to specify g, provided the censoring is non-informative.

The above practical example has been taken from the medical context, but this is certainly not the only field where censored data arise, neither it is the only area where survival data are relevant. In particular, survival data often appear in industrial applications, when analysing the reliability of products or components. In this context, the survival time is typically the length of time an item functions before failing.

A comprehensive account on survival data is given by Kalbfleisch and Prentice (1980). □

2.2.2 The likelihood principle

The likelihood function connects the pre-experimental information, expressed by the model choice, with the experimental information contained in y. Therefore, in a sense, it contains all we know about the specific inferential problem (apart for any information about the value of θ which, for whatever reason, has not been accommodated into the model, such as personal opinions or the results of related studies) This remark motivates the introduction of the following concept

Definition 2.2.6 (The likelihood principle) *For a statistical model* $\{f(\ ;\theta)\ .\ \theta \in \Theta\}$, *two points* y, $z \in \mathcal{Y}$ *such that* $L(\theta; y) \propto L(\theta; z)$ *must lead to the same inferential conclusions.*

This statement is the weak version of the likelihood principle. There exists a stronger version, which says that the conclusions coincide even when the two observations refer to distinct statistical models and distinct sample spaces.

Definition 2.2.7 (The strong likelihood principle) *Given an observation* y *from statistical model* $\{f(\cdot\ ; \theta) : \theta \in \Theta\}$ *and an observation* z *from statistical model* $\{g(\cdot\ ; \theta) : \theta \in \Theta\}$ *such that* $L_f(\theta; y) \propto L_g(\theta; z)$, *they must lead to the same inferential conclusions.*

Example 2.2.8 Consider two experimental settings from which repeated trials with a binary outcome are performed, each trial having the same probability θ of success, independent of the outcome of other trials. In the first setting, the overall number n of trials is fixed in advance, and the outcome of the experiment is given by the number of successes. In the second setting, the number of required successes is chosen in advance, and the outcome is given by the number of failures observed before stopping the sequence of trials. The probability distributions associated with the two experimental settings are binomial and negative binomial, respectively In detail, denoting by y the number of successes and by z the number of failures observed in any pair of experiments (one from each setting), the respective probability functions are

$$f(y; \theta) = \binom{n}{y} \theta^y (1 - \theta)^z \qquad (y = 0, \ldots, n),$$
$$g(z; \theta) = \binom{y + z - 1}{z - 1} \theta^y (1 - \theta)^z \qquad (z = 0, 1, \ ..)$$

from which we obtain that both likelihood functions are given by

$$L(\theta) = c\,\theta^y(1-\theta)^z$$

if the observed y and z coincide in the two experiments. Therefore, according to the strong likelihood principle, the two experiments must lead to the same inferences.   □

The mere fact of stating Definitions 2.2.6 and 2.2.7 must not imply that everything we will say in this book will comply with them. The theory of statistics has evolved as a compromise among different requirements, attempting to combine logical correctness with practical needs. In particular, it will turn out that a good deal of what we shall say will comply with the weak likelihood principle, but a smaller part will follow the strong likelihood principle.

For an extensive discussion in support of the likelihood principle, see Edwards (1972).

2.3 Sufficient statistics

2.3.1 Statistics

In a very crude description of statistical theory, one could say that its purpose is to select the most appropriate 'operations' to perform on the data. Since a variety of these 'operations' will be considered, it is appropriate to introduce a specific term.

Definition 2.3.1 *A function $T(\cdot)$ from $\mathcal{Y}$ to $\mathbb{R}^r$, for some positive integer r, such that $T(y)$ does not depend on θ, is called a* statistic, *and the value $t = T(y)$ corresponding to the observed value y is called* sample value *of the statistic.*

The condition that $T(y)$ does not depend on θ is needed to make sure that the statistic is actually computable, when the data are given. Some examples of statistics for a sample $(y_1, \ldots, y_n)$ whose elements belong to $\mathbb{R}$ are the following functions: $T_1 = \sum y_i$, $T_2 = \sum \exp(y_i)$, $T_3 = (\sum y_i, \sum y_i^2)$; the first statistic takes values in $\mathbb{R}$, the second in $\mathbb{R}^+$, and the third in $\mathbb{R}^2$.

Sometimes we shall consider the inverse images of the values t of a statistic T, i.e. sets of the type

$$A_t = \{y : y \in \mathcal{Y}, T(y) = t\},$$

which form a partition of the sample space. We shall refer to this partition as the partition induced by $T(y)$. For instance, if

$T = \sum y_i$, the sets $\{A_t\}$ are hyperplanes parallel to each other; a single set A_t is given by all points $y = (y_1, \ldots, y_n)^\top \in \mathbb{R}^n$ satisfying the equation

$$y_1 + \cdots + y_n = t.$$

Example 2.3.2 Among the type of statistics we shall consider, some are used particularly often and they are given specific names. For a sample $(y_1, \ldots, y_n)^\top$, the rth **sample moment** is the statistic from $\mathcal{Y}$ to $\mathbb{R}$ given by

$$m_r = \frac{1}{n} \sum_{i=1}^{n} y_i^r \qquad (r = 1, 2, \ldots).$$

In particular, m_1 is the **arithmetic mean**, commonly called the **sample mean**, and usually denoted by $\bar{y}$. Moreover, we define the **sample variance** to be

$$s_*^2 = \frac{1}{n} \sum_{i=1}^{n} (y_i - \bar{y})^2 = m_2 - m_1^2,$$

which takes values in $\mathbb{R}^+ \cup \{0\}$, All these statistics can be written as expected values of the sample distribution function, introduced in section A.7 4, when it is regarded as a probability distribution. □

Example 2.3.3 The order statistics of y are given by a permutation $(y_{(1)}, \ldots, y_{(n)})$ of y arranged in non-decreasing order. In the case of continuous variables, this permutation is unique with probability 1 In the latter case, the rth component of the sorted vector is called the rth order statistic. See section A.7.1 for further details □

2.3.2 Sufficient statistics

In Example 2.2.2, it was not necessary to know all the individual elements of $(y_1, \ldots, y_n)$ to write down the likelihood function, given $(\sum y_i, \sum y_i^2)$. Therefore, the likelihood function is uniquely identified, among all possible likelihood functions for the chosen statistical model, once the above two values are given. The question now is whether or not such a favourable situation can be extended in general, or at least to some classes of models, and in that case to which classes. It is clearly convenient to be able to reduce the dimensionality of the entity which we work with.

We shall now examine the nature and properties of those statistics which are able to summarize 'all the information' present in the likelihood function.

Definition 2.3.4 *For statistical model (2.1), a statistic $T(y)$ is said to be* sufficient *for θ if it takes the same values at two points in the sample space only if these two points have equivalent likelihoods, i.e. if for all $y, z \in \mathcal{Y}$*

$$T(y) = T(z) \quad \Longrightarrow \quad L(\theta, y) \propto L(\theta, z) \quad \text{for all } \theta \in \Theta.$$

At first sight, it might seem surprising that a non-bijective transformation of the data contains 'all the information' existing in the original data, but one must bear in mind that the sufficiency property is directly related to the choice of the model: if the model changed, a given statistic might no longer be sufficient. This remark stresses the importance of the appropriate selection of the model. The question of model selection will be addressed in a subsequent chapter; for the moment, let us take the model as given.

For any model, there always exists a sufficient statistic, namely y, but this is a trivial sufficient statistic, and in practice it is disregarded.

Example 2.3.5 Suppose that θ can take only two values, 0 and 1 say (but any two numbers would do), with two corresponding probability distributions, as indicated in the following table.

θ	$\Pr\{Y = 0\}$	$\Pr\{Y = 1\}$	$\Pr\{Y = 2\}$
0	8/12	1/12	3/12
1	4/12	2/12	6/12

These two distributions are such that associated with $y = 1$ and $y = 2$ there are equivalent likelihoods, since the last column is three times the column on its left. Therefore $T(y) = I_{\{0\}}(y)$, i e the function equal to 1 if $y = 0$, and equal to 0 if $y \neq 0$, is a sufficient statistic for θ □

Example 2.3.6 Consider a parametric class whose generic element is the density function of an s r s from a random variable with density function $g(\ ; \theta)$, which we do not specify in detail. Then the likelihood for a sample $y = (y_1, \ldots, y_n)$ can be written as

$$L(\theta) = c \prod_{i=1}^{n} g(y_i; \theta) = c \prod_{i=1}^{n} g(y_{(i)}; \theta)$$

where, in the last equality, the terms are multiplied after they have been sorted according to the order statistics. Therefore two samples with the same order statistics (i e equal up to a suitable permutation of the elements) have the same likelihood. It follows that, for *any* density function $g(\cdot\,;\theta)$, the order statistics are sufficient. ☐

If $T(\cdot)$ is a sufficient statistic, then $L(\theta)$ depends on y only through $T(y)$, which means that there exists a function g such that

$$L(\theta) \propto g(T(y); \theta).$$

Since in addition $L(\theta) \propto f(y; \theta)$, it follows that $f(y; \theta)/g(T(y); \theta)$ does not depend on θ; denote this ratio by $h(y)$. Therefore, if T is a sufficient statistic, then the following relationship must hold

$$f(y; \theta) = h(y)\, g(T(y); \theta) \qquad (2.4)$$

for some functions g and h On the other hand, if (2.4) holds, $L(\theta)$ is clearly a function of y only through $T(y)$, so $T(y)$ is sufficient. To summarize, we have obtained the following result.

Theorem 2.3.7 (Neyman's factorization) *For model (2.1), the statistic $T(\cdot)$ is sufficient for θ if and only if $f(y; \theta)$ can be written in the form (2.4).*

Remark 2.3.8 From (2.4), it follows that the density function of T at t is

$$
\begin{aligned}
f_T(t; \theta) &= \int_{\{y:T(y)=t\}} f(y; \theta)\, d\nu(y) \\
&= g(t; \theta) \int_{\{y:T(y)=t\}} h(y)\, d\nu(y) \\
&= g(t; \theta)\, h^*(t),
\end{aligned}
$$

so $g(t; \theta)$ is the density function of T for the part depending on θ. Therefore another way of interpreting the definition of a sufficient statistic is the following: if we only knew the sample value $t = T(y)$ and wrote the likelihood $L_T(\theta; t)$ for the statistical model associated with the distribution of T, then such a likelihood would be equivalent to $L(\theta)$. ☐

Example 2.3.9 To illustrate the previous remark, consider an s r s $y = (y_1, \ldots, y_n)$ from a $Bin(1, \theta)$ random variable, having likelihood function

$$L(\theta) = c \prod_{i=1}^{n} \theta^{y_i} (1-\theta)^{1-y_i} = c\, \theta^{T(y)} (1-\theta)^{n-T(y)}$$

with $T(y) = \sum y_i$ as a sufficient statistic. On the other hand, the distribution of T is $Bin(n, \theta)$ and, if only the overall number of successes was recorded instead of the full sequence of individual outcomes, this would lead to the parametric class of density functions

$$\binom{n}{t} \theta^t (1 - \theta)^{n-t}$$

where t is the number of successes, hence an equivalent likelihood. Therefore, in the case of Bernoulli trials, it will not be necessary to distinguish between the two different ways of recording the outcome, since this is irrelevant for the likelihood. □

Theorem 2.3.10 *For the statistical model (2.1), the statistic $T(\cdot)$ is sufficient for θ if and only if the distribution of Y conditional on the observed value of T does not depend on θ.*

Proof. Suppose that $T(\ )$ is sufficient, so (2.4) holds, by Neyman's factorization theorem. Compute the conditional density function of Y at a point y in $\mathcal{Y}$, given that $T = t$. If $T(y) \neq t$, then this density function is 0; otherwise, on using the density function of T obtained in Remark 2 3.8, we have

$$
\begin{aligned}
f(y|T = t) &= \frac{f(y; \theta)}{f_T(t; \theta)} \\
&= \frac{h(y)\, g(T(y); \theta)}{h^*(t)\, g(t; \theta)} \\
&= h(y)/h^*(t)
\end{aligned}
$$

which does not depend on θ.

Suppose now that $f(y|T = t)$ does not depend on θ, and denote this function by $h_*(y, t)$. Then

$$
\begin{aligned}
f(y; \theta) &= f(y|T = t) f_T(t; \theta) \\
&= h_*(y, t)\, h^*(t)\, g(t; \theta)
\end{aligned}
$$

which is a factorization of type (2.4), implying that $t = T(y)$ is sufficient. □

Remark 2.3.11 One consequence of the above result is that one can regard the outcome y of the experiment as if it was obtained in two stages: the first stage selects the value t, i.e. it selects the set $A_t = \{y : T(y) = t\}$, from the distribution $g(\cdot; \theta)$ which *depends* on θ; the second stage chooses an element $y \in A_t$ according to a distribution h_* which *does not* depend on θ. □

Remark 2.3.12 For a sufficient statistic $T(y)$, consider the induced partition and denote by A_t its generic element. Then, for two points y, $z \in \mathcal{Y}$, we have

$$y, z \in A_t \implies T(y) = T(z) = t \implies L(\theta; y) \propto L(\theta; z)$$

i.e. points belonging to the same partition set have the same associated likelihood. This remark shows that what is relevant for sufficiency is not the numerical value $t = T(y)$, but the set A_t associated with the sample statistic In fact, once A_t is known, it is possible to specify the likelihood associated with y within the set of all possible likelihoods, i.e. $\{L(\theta; y) : y \in \mathcal{Y}\}$ One implication of this fact is the following: if we consider a one-to-one function $U(\cdot)$ of $T(y)$, then $U(T(y))$ identifies the same partition of the sample space, and $U(T(y))$ is again a sufficient statistic.   □

Example 2.3.13 For the likelihood function of Example 2.2 2,

$$(t_1, t_2) = \left(\sum_i y_i, \sum_i y_i^2 \right)$$

is a sufficient statistic. On the other hand, the pair given by the sample mean and the sample variance

$$(\bar{y}, s_*^2) = \left(\frac{t_1}{n}, \frac{t_2 - t_1^2/n}{n} \right) = \left(\frac{\sum y_i}{n}, \frac{\sum(y_i - \bar{y})^2}{n} \right)$$

is a one-to-one function of (t_1, t_2), with inverse transformation

$$t_1 = n\bar{y}, \quad t_2 = ns_*^2 + n\bar{y}^2.$$

Hence $(\bar{y}, s_*^2)$ is also a sufficient statistic   □

Example 2.3.14 Consider an s r s $y = (y_1, \ldots, y_n)$ from a random variable $U(\theta, 2\theta)$, where $\theta > 0$. Arguing as in Example 2.2.3, write the density function for a single observation as

$$\frac{1}{\theta} I_{(\theta, 2\theta)}(t) \qquad (t \in \mathbb{R}),$$

and the likelihood is

$$L(\theta) = c \prod_{i=1}^n \frac{1}{\theta} I_{(\theta, 2\theta)}(y_i)$$

$$= \frac{c}{\theta^n} \prod_{i=1}^n I_{(1,2)}(y_i/\theta)$$

$$= \frac{c}{\theta^n} \prod_{i=1}^{n} I_{(\frac{1}{2},1)}(\theta/y_i)$$

$$= \frac{c}{\theta^n} I_{(y_{(n)}/2,y_{(1)})}(\theta)$$

where the last equality was obtained by noticing that

$$\frac{1}{2} < \frac{\theta}{y_i} \quad \text{for all } i \iff \frac{\max(y_1,\ldots,y_n)}{2} < \theta,$$

$$\frac{\theta}{y_i} < 1 \quad \text{for all } i \iff \theta < \min(y_1,\ldots,y_n).$$

Therefore, the pair $(y_{(1)}, y_{(n)})$ is a sufficient statistic for θ. $\qquad\square$

Example 2.3.15 Very often the sample $(y_1,\ldots,y_n)$ *is not* an s r s insofar its components are not sampled from identically distributed random variables, although the assumption of independence can be maintained. The fact that there are different probability distributions associated with the observations means that there are several underlying statistical sub-populations, in general one for each observation One can think of such a sample as the outcome of a **two-stage sampling** mechanism: in the first stage, a set of sub-populations is selected (according to a random or non-random selection rule); in the second stage the actual sample elements are drawn within each selected population. This sort of situation arises extremely frequently, in a large number of variants, some of which will be discussed in the examples throughout this text.

As an introductory example of this kind, suppose that the variable of interest y is the number of flaws in pieces of cloth produced by a textile machine. If it is known that the length x of the pieces available for inspection assumes N possible values, then x can be used to stratify the population (the pieces of cloth) into N sub-populations, one for each possible value of N Associated with each value of x there is a different probability distribution of the number of flaws

A simple probabilistic argument leads to a Poisson process formulation, where x plays the role of length of 'time', and y_i is then a value drawn from a $Poisson(\theta x_i)$ random variable $(i = 1,\ldots,n)$ with θ an unknown positive parameter. The associated likelihood is

$$L(\theta) = c \prod_{i=1}^{n} \left(h(x_i) \frac{e^{-x_i\theta}(x_i\theta)^{y_i}}{y_i!} \right)$$

where $h(x_i)$ denotes the probability that a piece of length x_i is included in the sample, assuming that this inclusion occurs randomly. Both in the case that $h(\cdot)$ is selected by the experimenter, and can then be regarded as known, as well as if $h(\ )$ is unknown (provided that it is independent of the quantity of interest θ), the term $\prod_i h(x_i)$ can be absorbed into the constant c, giving

$$L(\theta) = c \prod_{i=1}^{n} \left(\frac{e^{-x_i\theta} (x_i\theta)^{y_i}}{y_i!} \right)$$

which is the same likelihood we would have if the x_i values had not been randomly chosen but selected by the experimenter Therefore, in all three cases (non-random selection, random selection with known/unknown sampling distribution), the log-likelihood is

$$\ell(\theta) = c - \theta \sum_i x_i + \ln\theta \sum_i y_i$$

which admits $\sum y_i$ as a sufficient statistic for θ

Although $\sum x_i$ enters into $L(\theta)$, it is regarded as a constant value from the point of view of the likelihood function, not as a part of the sufficient statistic, because the values of the stratifying variable x carry no information on the parameter of interest θ, and the response variable of the experiment is y only. The above argument shows the irrelevance of specifying the sampling rule when the focus of interest is on the parameter relating the x_i to the y_i, namely θ in this case. Since this line of argument holds generally, these aspects will no longer be discussed, and $h(\cdot)$ will not be explicitly introduced in subsequent similar examples having independent and not identically distributed observations.

A deeper understanding of this point will be provided by the discussion of the conditionality principle in section 3.3.7   □

Remark 2.3.16 The logical flow of this text is different from that which is more commonly adopted, although equivalent. Usually, Theorem 2.3 10 is given as the definition of sufficient statistic, and our definition is obtained as a deduction. The present style of presentation is intended to stress the role of the likelihood function.

□

2.3.3 Minimal sufficient statistics

Since we are interested in reduction of data dimensionality, a prominent role will be played by those statistics which achieve maximal reduction.

Definition 2.3.17 *For statistical model (2.1), a statistic $T(y)$ is said to be* minimal sufficient *for θ if it is sufficient for θ and if it takes distinct values only at points in the sample space with non-equivalent likelihood functions, i.e. if for all $y, z \in \mathcal{Y}$*

$$T(y) = T(z) \quad \Longleftrightarrow \quad L(\theta, y) \propto L(\theta, z) \text{ for all } \theta \in \Theta.$$

One could ask whether such a statistic exists. To convince ourselves that it does, consider the partition, called the **likelihood partition**, whose elements are the sets formed by the points in the sample space leading to equivalent likelihoods. Any function T assuming constant value on one element of the likelihood partition and distinct values on distinct elements of this partition is *by construction* a minimal sufficient statistic.

From this argument, it follows that the minimal sufficient statistic is essentially unique, i.e. all minimal sufficient statistics are functions of each other, since they share the induced partition, namely the likelihood partition; for this reason, we shall commonly talk of *the* minimal sufficient statistic. Another implication is that the minimal sufficient statistic is a function of any sufficient statistic.

There is still another question: how to decide in practice whether a certain sufficient statistic is minimal. For the minimality of $T(\cdot)$, it must be true that any two points $y,\ z \in \mathcal{Y}$ have equivalent likelihoods if and only if $T(y) = T(z)$. In other words, the ratio

$$L(\theta; y)/L(\theta; z)$$

must be free of θ if and only if $T(y) = T(z)$.

Example 2.3.18 Continuing Example 2.2.2, for which we have seen that $T = (t_1, t_2) = (\sum y_i, \sum y_i^2)$ is a sufficient statistic, choose two points $z, w \in \mathcal{Y}$, and denote by $(t_1^{(z)}, t_2^{(z)})$ and $(t_1^{(w)}, t_2^{(w)})$, respectively, the corresponding values of the sufficient statistic. Then the ratio

$$\frac{L(\theta; z)}{L(\theta; w)} = c \exp\left(-\frac{1}{2\sigma^2}\{(t_2^{(z)} - t_2^{(w)}) - 2\mu(t_1^{(z)} - t_1^{(w)})\}\right)$$

does not depend on θ if and only if the term inside braces vanishes, i.e. if $(t_1^{(z)}, t_2^{(z)}) = (t_1^{(w)}, t_2^{(w)})$. This means that points in $\mathcal{Y}$ have equivalent likelihoods if and only if the components of T coincide, term by term. Therefore, T is minimal sufficient.   □

Example 2.3.19 Consider an s r s $y = (y_1, \ldots, y_n)^\top$ from a Cauchy random variable with scale parameter 1 and location parameter θ. The corresponding likelihood function is given by

$$L(\theta) = c \prod_{i=1}^{n} \frac{1}{1 + (y_i - \theta)^2},$$

and we already know that the order statistics form a sufficient statistic for θ, from Example 2.3.6. Write the ratio between the above likelihood and the same function for another point z of the sample space, i.e.

$$\frac{L(\theta; y)}{L(\theta; z)} = c \frac{\prod_i \{1 + (z_i - \theta)^2\}}{\prod_i \{1 + (y_i - \theta)^2\}},$$

which is the ratio of two polynomials of degree $2n$ (ignoring the c term), and examine the conditions for which the ratio is independent of θ. In general, the ratio of two polynomials is independent of the variable if and only if the polynomials differ by a scale factor. Since, in the present case, the constant term of both polynomials is 1, the above ratio is independent of θ if and only if the polynomials coincide, i.e they have the same coefficients. In turn, this requires that the elements of $(z_1, \ldots, z_n)$ and $(y_1, \ldots, y_n)$ coincide up to a permutation, which implies that the two samples have the same order statistics Then the order statistics are minimal sufficient for θ.   □

For a more advanced discussion of sufficiency and related concepts, see Zacks (1971, Chapter 2) and Barndorff-Nielsen (1978)

2.4 Exponential families

In the last section, we found that in some cases the minimal sufficient statistic has lower dimensionality than the size n of the sample, while in other cases this is not true We now wish to examine in detail an important class of cases leading to reduction of dimensionality.

2.4.1 Introduction

Definition 2.4.1 *The parametric family (2.1) is said to be an exponential family if its elements can be written in the form*

$$f(y; \theta) = q(y) \exp \left(\sum_{i=1}^{r} \psi_i(\theta) \, t_i(y) - \tau(\theta) \right) \qquad (2.5)$$

where $t_1(y), \ldots, t_r(y), q(y)$ *are functions of y not depending on* θ, *while* $\psi_1(\theta), \ldots, \psi_r(\theta), \tau(\theta)$ *are functions of* θ *not depending on y.* Correspondingly, if a sample y is drawn from a random variable with distribution (2.5), the associated likelihood function is said to be of exponential form.

Since the term $\exp(\cdot)$ in (2 5) is always positive, $f(y; \theta)$ can vanish only where $q(y)$ does so; this fact implies that all elements of an exponential family share the same support. Therefore, the parametric classes of Examples 2.1.3, 2.2.3 and 2.3.14 are ruled out

Example 2.4.2 Some of the parametric classes examined in previous examples can be represented in exponential form, by appropriately choosing the components of (2.5).

(a) If y is sampled from a $Bin(n, \theta)$ random variable as in Example 2 1 1, then we have $r = 1$ and

$$t(y) = y, \qquad\qquad q(y) = \binom{n}{y},$$

$$\psi(\theta) = \ln \left(\frac{\theta}{1 - \theta} \right), \quad \tau(\theta) = -n \ln(1 - \theta)$$

(b) In Example 2 2 2, we have $r = 2$ and

$$t_1(y) = \sum y_i, \quad t_2(y) = \sum y_i^2, \quad q(y) = (2\pi)^{-n/2},$$

$$\psi_1(\theta) = \frac{\mu}{\sigma^2}, \quad \psi_2(\theta) = -\frac{1}{2\sigma^2}, \quad \tau(\theta) = \frac{n\mu^2}{2\sigma^2} + \frac{n}{2} \ln \sigma^2$$

(c) For the Poisson distributions of Example 2.3.15, $r = 1$ and

$$t(y) = \sum y_i, \quad q(y) = \prod x_i^{y_i} / \prod y_i!,$$

$$\psi(\theta) = \ln \theta, \quad \tau(\theta) = \theta \sum x_i.$$

$\square$

At first sight, searching for a representation of type (2.5) seems a mere nuisance, but we shall see later that there are major advantages, since it is then possible to draw relevant conclusions for a given family without explicit calculations for each specific case

To avoid fruitless complications, suppose without loss of generality that the functions $\{1, \psi_1(\theta), \ldots, \psi_r(\theta)\}$ are linearly independent in Θ, i e there exists no linear combination of them which is identically 0 for all $\theta \in \Theta$ (except the trivial linear combination). This property implies that none of them can be written as a linear combination of the others If linear independence of the $\psi(\theta)$ functions did not hold, for instance if

$$\psi_r(\theta) = a_0 + a_1\psi_1(\theta) + \cdots + a_{r-1}\psi_{r-1}(\theta)$$

where $a_0, a_1, \ldots, a_{r-1}$ are constants, then we could replace the dependent term by a linear combination of the others, decreasing r by 1. When no reduction of dimensionality is possible, we say that the exponential family is written in **reduced form**, and call r the **order** of the family.

Theorem 2.4.3 *If the exponential family (2.5) is in reduced form, the statistic $(t_1(y), \ldots, t_r(y))$ is minimal sufficient for θ.*

Proof. Sufficiency of $T = (t_1, \ldots, t_r)$ is obvious, since (2.5) is of type (2.4) To show minimality, consider two points $w, z \in \mathcal{Y}$ whose corresponding values of T are equal to $(t_1^{(w)}, \ldots, t_r^{(w)})$ and $(t_1^{(z)}, \ldots, t_r^{(z)})$, respectively. Then the difference between the two log-likelihoods is

$$\ell(\theta; z) - \ell(\theta; w)$$
$$= [t_0^{(z)} - t_0^{(w)}] + \psi_1(\theta)[t_1^{(z)} - t_1^{(w)}] + \cdots + \psi_r(\theta)[t_r^{(z)} - t_r^{(w)}]$$

where $t_0^{(\,)} = \ln(c q(\cdot))$ The right-hand side is identically 0 when all terms $[t_j^{(z)} - t_j^{(w)}]$ are 0 $(j = 0, 1, \ldots, r)$, because of the linear independence of the functions $\{1, \psi_1(\theta), \ldots, \psi_r(\theta)\}$. Then $L(\theta; z) \propto L(\theta; w)$ implies $[t_j^{(z)} - t_j^{(w)}] = 0$ for $j = 1, \ldots, r$, and this fact means that $(t_1, \ldots, t_r)$ is minimal sufficient. $\qquad\square$

This theorem enables us to establish minimal sufficiency of a statistic as soon as the parametric class is written in exponential form, without explicit use of Definition 2.3.17, whose direct application leads to a rather cumbersome argument, as we saw in Examples 2.3.18 and 2.3 19.

Example 2.4.4 In several examples we have dealt with Bernoulli trials with constant probability of success. In many practical cases, however, the probability of success varies from trial to trial, hence

$Y_i \sim Bin(1, \pi_i)$ say $(i = 1, \ldots, n)$. Notice the similarity with Example 2 3 15, apart for the different type of probability distribution involved.

Assume, without further discussion for the moment, that

$$\pi_i = \Pr\{Y_i = 1\} = \frac{\exp(\alpha + \beta x_i)}{1 + \exp(\alpha + \beta x_i)} \qquad (i = 1, \ldots, n)$$

where $(x_1, \ldots, x_n)$ are known constants and α, β are unknown parameters On inverting the above relationship, we may also write

$$\text{logit}(\pi_i) = \alpha + \beta x_i$$

where the function **logit** is defined as

$$\text{logit}(p) = \ln \frac{p}{1-p} \qquad (0 < p < 1).$$

Then each point (x_i, π_i) lies on the curve

$$\pi(x) = \frac{e^{\alpha + \beta x}}{1 + e^{\alpha + \beta x}} \qquad (-\infty < x < \infty),$$

which is called the **logistic function**. Figure 2.2 shows the plot of the logistic curve for a few choices of the pair (α, β).

If the values $(y_1, \ldots, y_n)$ have been observed, the likelihood function is

$$L(\alpha, \beta) = c \prod_{i=1}^{n} \left(\frac{\exp(\alpha + \beta x_i)}{1 + \exp(\alpha + \beta x_i)} \right)^{y_i} \left(\frac{1}{1 + \exp(\alpha + \beta x_i)} \right)^{1-y_i}$$

$$= c \exp\left(\alpha \sum_i y_i + \beta \sum_i x_i y_i - \sum_i \ln\{1 + \exp(\alpha + \beta x_i)\} \right)$$

which has exponential form of order 2. Then we can state immediately that $(\sum y_i, \sum x_i y_i)$ is the minimal sufficient statistic for the parameter (α, β)

The above relationship between the π_i and the x_i is said to represent a **logistic regression** model, which is often adopted when one believes that the probability of success of a certain event is influenced by the value taken by another variable x For instance, in a study of the toxicity of a certain substance, the random variable Y_i could be 0 or 1 depending on whether an animal treated with that substance dies within 24 hours of treatment or survives for at least 24 hours, and x_i could represent the poison dosage for a specific experimental unit. $\qquad \square$

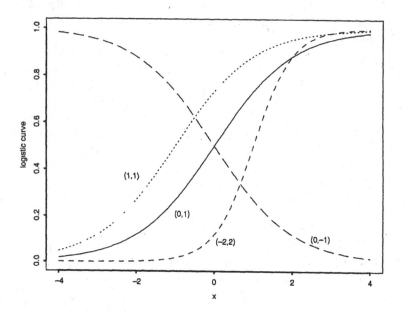

Figure 2 2 *The logistic curve for some choices of* (α, β)

Another important property of exponential families is reproducibility with respect to replications of the experiment, provided the replicates are independent Suppose n observations $(y_1, \ldots, y_n)$ are taken from model (2.5), all independent of each other. Then the likelihood for the whole set of observations is

$$L(\theta) = c \prod_{j=1}^{n} f(y_j; \theta)$$

$$= c \exp \left(\psi_1(\theta) \sum_j t_1(y_j) + \cdot \ \cdot + \psi_r(\theta) \sum_j t_r(y_j) - n\tau(\theta) \right)$$

which is still of type (2.5), and its minimal sufficient statistic is

$$\left(\sum_j t_1(y_j), \ . \ ., \sum_j t_r(y_j) \right),$$

again of dimension r, irrespective of the number n of replicates. Example 2 4 2(b) could be considered an instance of this situation. Therefore, in the case of independent and identically distributed random variables, the phrase 'exponential family' applies equally to a single observation and to a set of observations, since they reproduce the likelihood structure of a single datum.

Considering the strong connection between exponential families and sufficient statistics, it is reasonable to ask whether the existence of a non-trivial sufficient statistic (with dimension independent of n) implies that the parametric class is an exponential family. Without further hypotheses, the answer is negative, as shown by Examples 2 2.3 and 2 3 14. Adding appropriate assumptions on the distributions of the variables, in particular that they are independent and identically distributed and that their support does not depend on θ, it can be shown that the existence of a sufficient statistic with dimension smaller that the dimension of Y implies that the distribution of Y is of type (2 5).

Further properties of the exponential families which we state without proof are the following:

- The function $\tau(\theta)$ of (2.5) allows differentiation of arbitrary order, with respect to θ, if the functions $\psi_j(\theta)$ are differentiable.

- In expressions of type

$$\frac{d^s}{d\theta^s} \int_{\mathcal{Y}} g(y)\, f(y; \theta)\, d\nu(y)$$

  where f is given by (2 5), one can interchange the integration and differentiation operators, provided the integral exists.

2.4.2 Regular exponential families

The mathematical theory of exponential families is very extensive and quite complex, preventing a proper development in a text of this kind. We shall restrict ourselves to sketching some concepts, in an especially simplified form, suitable for the level of this book.

We shall say that an exponential family of type (2.5) is **regular** if the following conditions are satisfied.

- The parameter space Θ coincides with the set where (2.5) is meaningful; in other words, Θ coincides with

$$\left\{ \theta : \int_{\mathcal{Y}} q(y) \exp\{\sum \psi_i(\theta) t_i(y)\}\, d\nu(y) < \infty \right\},$$

and this set is an open interval in $\mathbb{R}^k$.

- The dimension of Θ and of the minimal sufficient statistic coincide, i e $k = r$.

- The transformation from θ to $\psi = (\psi_1(\theta), \ . \ ., \psi_r(\theta))$ is invertible

- The functions $\psi_1(\theta), \ . \ . \ , \psi_r(\theta)$ are infinitely differentiable with respect to the components of θ

The examples of exponential families presented so far have dealt with regular families, provided Θ included all admissible values of θ.

Example 2.4.5 Consider now a case where the sample y is not an s r s since the components are not independent, although they have identical distribution Let $(Y_1, \ . \ ., Y_n)$ be random variables satisfying the stochastic difference equations

$$Y_t = \rho Y_{t-1} + \varepsilon_t \qquad (t = 2, \ . \ ., n; \ -1 < \rho < 1)$$

where $Y_1 \sim N(0, \sigma^2/(1-\rho^2))$ and $\varepsilon_2, \ldots, \varepsilon_n$ are $N(0, \sigma^2)$ independent and identically distributed random variables which are also independent of Y_1. The somewhat surprising value chosen for the variance of Y_1 has the virtue that the distribution of Y_2 still has the same variance $\sigma^2/(1 - \rho^2)$, and the same property holds for all subsequent variables. It can then be checked algebraically that

$$Y = (Y_1, \ . \ ., Y_n)^\top \sim N_n\left(0, \frac{\sigma^2}{1 - \rho^2} \, \Omega\right),$$

where

$$\Omega = \begin{pmatrix} 1 & \rho & \rho^2 & . \ . & \rho^{n-1} \\ \rho & 1 & \rho & . \ . & \rho^{n-2} \\ \rho^2 & \rho & 1 & . \ . & \rho^{n-3} \\ \vdots & . & \vdots & \ddots & \vdots \\ \rho^{n-1} & \rho^{n-2} & \rho^{n-3} & . \ . & 1 \end{pmatrix}$$

is the correlation matrix. We then say that Y is an **autoregressive process** of order 1, with **autocorrelation** parameter ρ. It is possible to introduce autoregressive processes of order 2, 3, ..., satisfying difference equations of the type

$$\begin{aligned} Y_t &= \alpha_1 Y_{t-1} + \alpha_2 Y_{t-2} + \varepsilon_t, \\ Y_t &= \alpha_1 Y_{t-1} + \alpha_2 Y_{t-2} + \alpha_3 Y_{t-3} + \varepsilon_t, \end{aligned}$$

$$\vdots$$

which form a fundamental class of statistical models for the analysis of time series A systematic treatment of statistical methods for the analysis of time series is given by Priestley (1981)

Suppose now that a sample $y = (y_1, \ldots, y_n)^\top$ from an autoregressive process of order 1 has been observed. The log-likelihood is

$$\ell(\rho, \sigma^2) = c - \tfrac{1}{2} \ln |\Omega| - \tfrac{1}{2} n \ln \left(\frac{\sigma^2}{1 - \rho^2} \right) - \frac{1 - \rho^2}{2\sigma^2} y^\top \Omega^{-1} y$$

which requires inversion of Ω to be computable. In the present case, however, one can avoid explicit inversion of Ω, by noticing that $Y_t \sim N(\rho Y_{t-1}, \sigma^2)$, conditionally on $Y_1, \ldots, Y_{t-1}$. Starting from the distribution of Y_1 and multiplying repeatedly by the successive conditional density functions of Y_t given the value of Y_{t-1}, we obtain that the joint density function of Y at y is

$$\frac{1}{\sqrt{2\pi\sigma^2/(1 - \rho^2)}} \exp\left(-\frac{1 - \rho^2}{2\sigma^2} y_1^2 \right)$$

$$\times \prod_{t=2}^{n} \frac{1}{\sqrt{2\pi\,\sigma^2}} \exp\left(-\frac{(y_t - \rho y_{t-1})^2}{2\sigma^2} \right).$$

After some algebra, it turns out that the log-likelihood, up to a factor 2, is

$$2\ell(\rho, \sigma^2) = c + \ln(1 - \rho^2) - n \ln \sigma^2 - \sigma^{-2}(d_{00} - 2\rho d_{01} + \rho^2 d_{11})$$

where

$$d_{rs} = \sum_{t=s+1}^{n-r} y_t y_{t+r-s}$$

The above likelihood is of exponential form but not regular, since the minimal sufficient statistic (d_{00}, d_{01}, d_{11}) has dimension 3, and the parameter (ρ, σ^2) has dimension 2.

In real data applications, the process frequently has a mean value μ not necessarily equal to 0. In this case, the Y_t are assumed to satisfy the stochastic difference equations

$$Y_t - \mu = \rho(Y_{t-1} - \mu) + \varepsilon_t \qquad (t = 2, \ldots, n)$$

and $Y_1 \sim N(\mu, \sigma^2/(1 - \rho^2))$. When μ is known, its subtraction from the Y_t reproduces the case considered previously. When μ is unknown, an extension of the statistical model is effectively introduced, now having a three-dimensional parameter (μ, ρ, σ^2). It is left to the reader to check that the minimal sufficient statistic is now five-dimensional: (d_{00}, d_{01}, d_{11}) as above, plus (d_0, d_1) where

$$d_r = \sum_{t=r+1}^{n-r} y_t .$$

□

We now develop some interesting properties for regular exponential families of order 1. In this case, $k = r = 1$ and the density is

$$f(y; \theta) = q(y) \exp\{\psi(\theta) t(y) - \tau(\theta)\}. \qquad (2.6)$$

Since the integration and differentiation operators can be interchanged, we can write

$$0 = \frac{\partial}{\partial \theta} \int_{\mathcal{Y}} f(y; \theta) \, d\nu(y) = \int_{\mathcal{Y}} \frac{\partial}{\partial \theta} f(y; \theta) \, d\nu(y)$$

and, substituting expression (2.6), obtain

$$\int_{\mathcal{Y}} f(y; \theta)\{t(y) \psi'(\theta) - \tau'(\theta)\} \, d\nu(y) = \mathrm{E}[\, t(Y) \psi'(\theta) - \tau'(\theta)\,] = 0.$$

Then we conclude that

$$\mathrm{E}[t(Y)] = \frac{\tau'(\theta)}{\psi'(\theta)}, \qquad (2.7)$$

and the denominator does not vanish since $\psi(\theta)$ is a monotonic function. Apart from its intrinsic interest, (2 7) will be useful in Chapters 3 and 6.

The above argument can be repeated for the second derivative of the integral in (2.6), leading to

$$\int_{\mathcal{Y}} [\{t(y) \psi'(\theta) - \tau'(\theta)\}^2 + \{t(y) \psi''(\theta) - \tau''(\theta)\}] \, f(y; \theta) \, d\nu(y) = 0$$

The first part of the integrand gives the second moment of the random variable $\{t(Y) \psi'(\theta) - \tau'(\theta)\}$; this random variable has 0 mean value, because of (2.7), so the second moment is equal to the variance. Therefore we can write

$$\begin{aligned}
\mathrm{var}[\, t(Y) \psi'(\theta) - \tau'(\theta)\,] &= -\mathrm{E}[\, t(Y) \psi''(\theta) - \tau''(\theta)\,] \\
&= -\frac{\tau'(\theta)}{\psi'(\theta)} \psi''(\theta) + \tau''(\theta) \\
&= \frac{\psi'(\theta) \tau''(\theta) - \psi''(\theta) \tau'(\theta)}{\psi'(\theta)}
\end{aligned}$$

and finally obtain

$$\text{var}[t(Y)] = \frac{\psi'(\theta)\,\tau''(\theta) - \psi''(\theta)\,\tau'(\theta)}{\psi'(\theta)^3}. \qquad (2.8)$$

By applying the same argument to higher-order derivatives of the integral in (2.6), one can derive expressions for the higher moments of $t(Y)$

Clearly, (2.7) and (2.8) reduce to much simpler expressions if $\psi(\theta) = \theta$, and the simplification is even more dramatic for the higher moments. Setting $\psi(\theta) = \theta$ means that a specific parametrization, called canonical, is selected for (2 6), and ψ is then called the **canonical parameter** of the exponential family.

Example 2.4.6 If Y is a $Bin(n, \theta)$ random variable, the canonical parameter is

$$\psi = \text{logit}(\theta) = \ln\left(\frac{\theta}{1-\theta}\right)$$

and the density function of Y is correspondingly written as

$$f^*(y; \psi) = \binom{n}{y} \exp\left(y\,\psi - n\ln(1 + e^\psi)\right),$$

where $-\infty < \psi < \infty$ □

Relationships of type (2.7) and (2.8) can be extended to higher-order regular exponential families, but we shall not deal with them here. Even the concept of canonical parametrization carries over to higher-order families, and the canonical parameter can be shown to be $(\psi_1(\theta), \ldots, \psi_r(\theta))$

For an extended treatment of exponential families, see the books by Barndorff-Nielsen (1978) and Brown (1986).

Exercises

2.1 Prove the statement, given shortly after Definition 2.2.1, that the likelihood function is an equivalence class of functions.

2.2 If an s r s of size $n = 2$ is taken from a $U(0, \theta)$ random variable, what is the sample space? What are the elements A_t of the likelihood partition?

2.3 Consider an s r s of size n from a random variable with density function at t given by

$$g(t; \theta) = \begin{cases} e^{-(t-\theta)} & \text{if } t > \theta, \\ 0 & \text{otherwise,} \end{cases}$$

where $\theta \in \mathbb{R}$.

(a) What is the sample space?

(b) Write down the corresponding likelihood function.

2.4 For the parametric class (2.1), let θ_0 be a fixed element in Θ. Prove that, if the relationship

$$\frac{f(y; \theta)}{f(y; \theta_0)} = g(u(y), \theta)$$

holds for functions g and u not depending on θ, then $u(y)$ is a sufficient statistic for θ.

2.5 A finite population has N_A individuals of type A and N_B individuals of type B. The overall number $N_A + N_B$ is known to be 100, but the size of the two groups is unknown. If a sample of 5 individuals is taken without replacement, write down the likelihood for N_A in the following three situations:

(a) the observed sequence is A, B, A, B, A;

(b) the observed sequence is A, A, B, B, A;

(c) the precise sequence of the outcomes is not given, but it is known that three elements are As, two are Bs.

Compare the likelihood functions and comment on your findings

2.6 Denote by θ the probability that a child is female, and assume independence among successive births. In a sample of n families with 3 children each, denote by y_k the number of families having k daughters (with $y_0 + \cdots + y_3 = n$).

(a) Write down the corresponding likelihood for θ.

(b) Is this an exponential family?

(c) What is the minimal sufficient statistic?

2.7 Consider a discrete random variable with probability function

$$f(t; \theta) = \frac{\theta^t}{c_\theta \, t} \qquad (t = 1, 2, \ldots)$$

where c_θ is a suitable normalizing constant.

(a) Determine c_θ.

(b) Explain why this distribution is called the **logarithmic series distribution**.

(c) Write this distribution in exponential form

2.8 Does the likelihood of Example 2.2.4 admit of a non-trivial sufficient statistic? Does the likelihood have an exponential family structure? Answer the same questions for the modified problem obtained by sampling y_1 from the stationary distribution of the Markov chain.

2.9 If an observation y is sampled from an $N(\theta, \theta)$ random variable where $\theta > 0$, show that y itself is a sufficient statistic for θ, but y^2 is minimal sufficient. [Therefore, a one-dimensional sufficient statistic is not necessarily minimal.]

2.10 For an s r s $y = (y_1, \ldots, y_n)$ from an $N(\theta, \theta^2)$ random variable with $\theta > 0$, write the corresponding likelihood in exponential form and determine its order. Is this a regular exponential family?

2.11 For the statistical model of Example 2.3.15, obtain the mean value and the variance of the sufficient statistic. Which is the canonical parameter?

2.12 Check the statement at the end of Example 2.4.5 about the expression for the minimal sufficient statistic of an autoregressive process of order 1, with an unknown mean value

2.13 In some applications, the observed variable represents an angle; for instance, it could be the wind direction recorded at a given location over a number of days. For data of this kind, called **directional data**, it is necessary to employ density functions whose support is $(0, 2\pi)$, and in addition they must be periodic functions with period 2π to make sure that the density values at 0 and 2π coincide, since the origin is usually arbitrary. One of the most commonly employed distributions in this context is the **von Mises distribution**, whose density function is

$$f(t; \kappa, \alpha) = \frac{1}{2\pi I_0(\kappa)} \exp\{\kappa \cos(t - \alpha)\} \qquad (0 \le t < 2\pi)$$

where $0 \le \alpha < 2\pi$, $\kappa \ge 0$, and

$$I_0(z) = \frac{1}{\pi} \int_0^\pi \cosh(z \cos \omega)\, d\omega = \sum_{j=0}^\infty \frac{(\frac{1}{2}z)^{2j}}{(j!)^2}$$

is the modified Bessel function of order 0.

(a) Show that α is the mode of the distribution and κ is an index of variability.

(b) Express the density function in exponential form and determine its order.

2 14 Let $(y_1, \ldots, y_n)$ be an s r s from an $N_k(\mu, \Omega)$ random variable where $\mu \in \mathbb{R}^k$ and Ω is a positive definite matrix; this means that the pair (μ, Ω) can vary across the entire admissible parameter space Discuss what this parameter space is. Write down the corresponding likelihood function in exponential form, and determine its order and the minimal sufficient statistic. [In this case, any of the sample elements y_i is an observation from a multivariate random variable; the part of statistics dealing with methods for multivariate observations is called multivariate statistics.]

CHAPTER 3

Maximum likelihood estimation

3.1 Introduction

3.1.1 Estimates and estimators

Our first practical problem is to choose a value of θ close to the true parameter value θ_*. Such a value, called an **estimate** of θ, should be the choice which, in some sense, best explains why the phenomenon under study has produced the observed data y rather than some other point in the sample space. Sometimes, the term **point estimate** is equivalently used.

More specifically, we aim to define and evaluate general criteria for constructing estimates. These criteria should ideally be usable in all estimation problems and with all statistical models

A criterion for constructing estimates associates an element of Θ with every element $y \in \mathcal{Y}$. Therefore, it defines a function from $\mathcal{Y}$ into $\Theta \subseteq \mathbb{R}^k$, i.e. it is a statistic, which in this case is called an **estimator**. There exist several general criteria for generating estimates; we shall deal with only the most important ones

One of the oldest criteria for constructing estimators is the **method of moments** widely used by Karl Pearson and his school at the beginning of the twentieth century. Consider a one-dimensional random variable Y belonging to a k-dimensional parametric family such that there exist k functions connecting the parameter $\theta = (\theta_1, \ldots, \theta_k)$ with the first k moments of Y. More precisely, assume there exist functions $g_1, \ldots, g_k$ such that

$$g_1(\theta_1, \ldots, \theta_k) = \mu_1'$$
$$g_2(\theta_1, \ldots, \theta_k) = \mu_2'$$
$$\vdots$$
$$g_k(\theta_1, \ldots, \theta_k) = \mu_k'$$

where

$$\mu_r' = \mathrm{E}[Y^r] \qquad (r = 1, \ldots, k)$$

is the rth moment of Y. If an s r s from Y is available, the method of moments works by replacing μ'_r with the corresponding sample moment m_r $(r = 1, .. , k)$ and solving the above equations with respect to $\theta_1, . , \theta_k$ to obtain their estimates.

The method of moments has the merit of being very simple to define and usually simple to implement; often, the equations to be solved are algebraic equations leading to explicit solutions There are also important disadvantages, however: (i) the method can be applied only to samples with all components having the same distribution, otherwise the relationship between μ'_r and m_r disappears; (ii) in general, it is less efficient than the method of maximum likelihood, with respect to criteria which we shall describe later in this chapter, (iii) if k is high, the method can lead to unstable functions, inasmuch as a small perturbation of the sample values can be greatly magnified by the kth power transformation, with appreciable effect on the final estimates. For these reasons, we shall direct our attention elsewhere.

3.1.2 Maximum likelihood estimates

Definition 3.1.1 *Given a likelihood function* $L(\theta)$ *for* $\theta \in \Theta$, *à maximum likelihood estimate of* θ *is an element* $\hat{\theta} \in \Theta$ *which attains the maximum value of* $L(\theta)$ *in* Θ, *i.e. such that*

$$L(\hat{\theta}) = \max_{\theta \in \Theta} L(\theta). \tag{3.1}$$

Because of the close connection between the concepts of 'estimate' and 'estimator', we shall often use the abbreviation MLE to denote either the maximum likelihood estimate or the estimator. Which of the two is intended should be clear from the context.

The rationale for introducing the maximum likelihood criterion is that '(other things being equal) we choose the system which gives the highest chance to the facts we have observed' (Ramsey, 1931, p. 209)

The MLE idea was described, albeit under a different name, as a general estimation criterion by Ronald A Fisher in 1912 (when he was a third-year undergraduate!), and developed in Fisher (1922; 1925). However, the concept of the maximum likelihood estimate has been traced back to Daniel Bernoulli and Johann Heinrich Lambert in the eighteenth century. An account of the history of the method is given by Edwards (1974).

There are several features of the above definition which are worth emphasizing

(a) It is not necessary for Θ to be a numeric set, i.e. we need not be dealing with a parametric model, although we shall restrict ourself to this case.

(b) The MLE may not exist.

(c) The MLE need not be unique.

(d) The likelihood function has to be maximized in the space Θ specified by the statistical model, not over the set of the mathematically admissible values of θ.

(e) Often $\hat{\theta}$ cannot be written explicitly as a function of the sample values, i.e. in general the MLE estimator has no closed-form expression, although it may do so in relatively simple problems.

(f) In the above-mentioned case, the MLE has to be obtained numerically, for the observed value y. In real applications, this aspect is very relevant, and often gives rise to interesting questions of numerical methods, or computational statistics, as it is now called. In this book, these issues will be considered only briefly

(g) If $T(y)$ is a sufficient statistic, then $\hat{\theta}$ is a function of $T(y)$. In fact, using (2.4), one can write

$$L(\theta) = L(\theta; y) = c(y)g(T(y); \theta)$$

where the term $c(y)$ does not depend on θ; then, maximising $L(\theta)$ is equivalent to maximizing $g(T(y); \theta)$, which depends on y only via $T(y)$.

In spite of remarks (b) and (c) above, it is nevertheless true that, in most cases, the maximum of the likelihood exists and is unique; moreover, the parameter space Θ is often the entire admissible space for θ. Since the MLE is essentially 'unique', it is common to use the phrase *the* maximum likelihood estimate.

Since the following examples aim to illustrate the above remarks, they inevitably tend to stress somewhat unusual aspects of MLE. More regular examples are given in section 3.1.4.

Example 3.1.2 The likelihood obtained from an s r s of n ele-

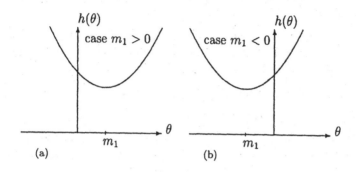

Figure 3 1 *Sum of squared deviations between observations and the pa-rameter value: (a) $m_1 > 0$, (b) $m_1 < 0$*

ments $(y_1, \ldots, y_n)$ from a random variable $N(\theta, 1)$ is

$$L(\theta) = \exp\left(-\tfrac{1}{2}\sum_{i=1}^{n}(y_i - \theta)^2\right).$$

The usual term $c(y)$ on the right-hand side has been omitted here, since it does not affect the maximization problem In the following, we shall do this without further mention

(a) Suppose first that θ varies across the whole real line. Since $\exp(\cdot)$ is a strictly increasing function, the maximum of $L(\theta)$ can be obtained by minimizing the quadratic function

$$h(\theta) = \sum_{i=1}^{n}(y_i - \theta)^2 = n\left(m_2 - 2\theta m_1 + \theta^2\right)$$

where m_1, m_2 are the first two sample moments. We can immediately check that $h(\theta)$ takes its minimum value when θ is equal to

$$\hat{\theta} = m_1 = \frac{\sum_{i=1}^{n} y_i}{n},$$

which is in fact the MLE, and which is unique.

(b) Assume $\theta \in [0, \infty)$. If $m_1 \geq 0$, nothing is changed. When $m_1 < 0$, the global minimum of the parabola, m_1, is not an admissible value, and the constrained minimum is at the origin, as illustrated in Figure 3 1.

To summarize, we have that

$$\hat{\theta} = \begin{cases} m_1 & \text{if } m_1 \geq 0, \\ 0 & \text{if } m_1 < 0. \end{cases}$$

(c) Consider now the case $\theta \in (0, \infty)$. The argument of the previous case holds, except that now 0 is not an admissible value for θ. On the other hand, if $m_1 < 0$, for any given positive value θ' of θ, there exists another positive value θ'' such that $0 < \theta'' < \theta'$ and $L(\theta'') > L(\theta')$. Therefore, we conclude

$$\hat{\theta} = \begin{cases} m_1 & \text{if } m_1 > 0, \\ \text{does not exist} & \text{if } m_1 \leq 0. \end{cases}$$

However, in this case, since $\theta = 0$ is a value for which the likelihood is mathematically meaningful, it is customary to put $\hat{\theta} = 0$; in the following we shall do so without further comment. This conventional adjustment would become unfeasible if the likelihood increased towards a value for which the likelihood itself would not exist (for instance $\theta = \infty$). In such a case, nothing can be done about the non-existence of the MLE.

□

Example 3.1.3 Consider an s r s $(y_1, \ldots, y_n)$ from a random variable with Laplace distribution, i.e. with density function at t given by

$$\exp\{-|t - \theta|\}/2 \qquad (-\infty < t < \infty)$$

where θ is a parameter varying over the whole real line. Since the likelihood function is

$$L(\theta) = \exp\left(-\sum_{i=1}^{n} |y_i - \theta|\right) = \exp\left(-\sum_{i=1}^{n} |y_{(i)} - \theta|\right),$$

where $y_{(i)}$ denotes the ith order statistic $(i = 1, \ldots, n)$, the maximum is achieved by minimizing

$$h(\theta) = \sum_{i=1}^{n} |y_{(i)} - \theta|.$$

First of all, notice that $h(\theta)$ is a convex function since it is the sum of n convex functions. To examine $h(\theta)$, we need to consider separately the case of n even and n odd.

(a) For n odd, say $n = 2k - 1$, where k is a positive integer, we have that

$$h(y_{(k)}) = \sum_{j=1}^{k-1}(y_{(k)} - y_{(j)}) + \sum_{j=k+1}^{2k-1}(y_{(j)} - y_{(k)}).$$

If we choose a positive number ε such that $y_{(k)} + \varepsilon < y_{(k+1)}$, then

$$\begin{aligned}
h(y_{(k)} + \varepsilon) &= \sum_{j=1}^{k}(y_{(k)} + \varepsilon - y_{(j)}) + \sum_{j=k+1}^{2k-1}(y_{(j)} - y_{(k)} - \varepsilon) \\
&= h(y_{(k)}) + 2\varepsilon,
\end{aligned}$$

showing that $h(\theta)$ increases to the right of $y_{(k)}$. By a similar argument, it can be shown that $h(\theta)$ increases to the left of $y_{(k)}$. We therefore conclude that $y_{(k)}$ is a global minimum, taking into account that $h(\theta)$ is a convex function.

(b) For n even, say $n = 2k$, where k is again a positive integer, if we choose two values θ' and $\theta'' = \theta' + \varepsilon$, both in the interval $(y_{(k)}, y_{(k+1)})$, then

$$\begin{aligned}
h(\theta') &= \sum_{j=1}^{k}(\theta' - y_{(j)}) + \sum_{j=k+1}^{2k}(y_{(j)} - \theta'), \\
h(\theta'') &= \sum_{j=1}^{k}(\theta' + \varepsilon - y_{(j)}) + \sum_{j=k+1}^{2k}(y_{(j)} - \theta' - \varepsilon) \\
&= h(\theta'),
\end{aligned}$$

hence $h(\theta)$ is constant in $(y_{(k)}, y_{(k+1)})$. Now choose ε such that $\theta'' = \theta' + \varepsilon \in (y_{(k+1)}, y_{(k+2)})$. We have that

$$\begin{aligned}
h(\theta'') &= \sum_{j=1}^{k+1}(\theta' + \varepsilon - y_{(j)}) + \sum_{j=k+2}^{2k}(y_{(j)} - \theta' - \varepsilon) \\
&= \sum_{j=1}^{k+1}(\theta' - y_{(j)}) + (k+1)\varepsilon + \\
&\qquad \sum_{j=k+2}^{2k}(y_{(j)} - \theta') - (k-1)\varepsilon \\
&= h(\theta') + 2\varepsilon - 2(y_{(k+1)} - \theta') \\
&> h(\theta'),
\end{aligned}$$

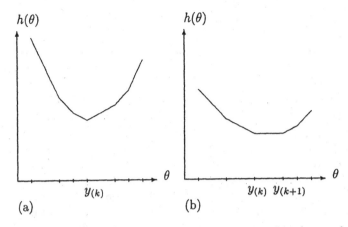

Figure 3.2 *Sum of absolute deviations between sample values and the parameter: (a) n odd; (b) n even*

and a similar result holds if θ'' lies in $(y_{(k-1)}, y_{(k)})$. Therefore $h(\theta)$ increases outside the interval $(y_{(k)}, y_{(k+1)})$, and convexity of h assures that the value taken by the function in this interval is the global minimum. Figure 3.2 summarizes our findings.

We thus have the following conclusions: (a) if n is odd, the MLE exists, is unique and is given by $y_{(k)}$; (b) if n is even, any point in the interval $(y_{(k)}, y_{(k+1)})$ is an MLE In other words, $\hat{\theta}$ is the sample median, as defined in section A.7.1.   □

Example 3.1.4 In Example 2 2.3 we deduced that the likelihood of an s r s from a $U(0, \theta)$ random variable is

$$L(\theta) = \theta^{-n} I_{[y_{(n)}, \infty)}(\theta),$$

where $y_{(n)}$ is the largest sample element. Since θ^{-n} is a strictly decreasing function, the MLE of θ is $y_{(n)}$ itself, if θ is free to move over the positive real half-life. If θ is constrained to a subset of $\mathbb{R}^+$, some adjustment is required, similarly to Example 3.1.2.   □

Example 3.1.5 Consider again Example 2.3.14, in which

$$L(\theta) = \theta^{-n} I_{(y_{(n)}/2, y_{(1)})}(\theta)$$

is positive only in the interval $(y_{(n)}/2, y_{(1)})$, and has $\hat{\theta} = y_{(n)}/2$. In this case, $\hat{\theta}$ has lower dimensionality than the minimal sufficient statistic; therefore $\hat{\theta}$ is a function of the minimal sufficient statistic but itself is *not* a sufficient statistic.   □

Example 3.1.6 To estimate the size of a population there exist strategies called capture–recapture techniques, of which the simplest version is as follows. To estimate the number N of fishes living in a lake, a certain number M of them are caught and, after being marked, are put back in the lake. Some time later, a second catch takes place; denote by n the number of fishes captured, of which a certain number, m say, turn out to be marked

Assume that the probability of being captured is the same for all fishes, and argue conditionally on the outcome of the first catch, when M subjects were marked. Then the probability distribution of the number m of marked fishes in the second catch is of hypergeometric type (A.19), and the corresponding log-likelihood is

$$\ell(N) = c + \ln\left(\frac{(N-M)!\,(N-n)!}{(N-M-n+m)!\,N!}\right)$$

where N is an integer, and $N \geq M + n - m$ It can be shown that this function is increasing if $N < M\,n/m$ and decreasing if $N > M\,n/m$, implying that the maximum is achieved by one of the integers on either side of $M\,n/m$. This algebraic statement has a simple probabilistic interpretation: if the second sample is regarded as of binomial type, which is approximately correct if N is much greater than n, then

$$L(N) \approx \binom{n}{m} p^m (1-p)^{n-m}$$

where $p = M/N$. This approximation leads to the estimate $\tilde{N} = M\,n/m$, but in general this is not an integer

It is worth mentioning that this is another case not of simple random type, since sampling is without replacement.

For a thorough presentation of the various capture–recapture schemes, see Seber (1973).   □

3.1.3 Equivariance

Theorem 3.1.7 *Let $\psi(\cdot)$ be a one-to-one function from the set Θ onto the set Ψ. Then the MLE of $\psi(\theta)$ is $\psi(\hat{\theta})$, if $\hat{\theta}$ denotes the MLE of θ.*

Proof. Denote by $L(\theta)$ the likelihood function on the space Θ, and by $L_\Psi(\cdot)$ the likelihood function on the space Ψ. By the very definition of likelihood, it follows that

$$L_\Psi(\psi_0) = L(\psi^{-1}(\psi_0)),$$

for any element $\psi_0 \in \Psi$. Therefore

$$L_\Psi(\psi(\hat\theta)) = L(\hat\theta)$$

which is the maximum value $\square$

In other words, if the parameter is transformed, the corresponding estimate is changed by the same transformation. Such a property, called **equivariance**, is relevant both from the conceptual point of view, because it avoids inconsistencies when moving from one parametrization to another, and also for practical purposes, since it can save time on computation, as demonstrated by the next example.

Example 3.1.8 Consider again Example 3.1.2 (any of the three cases would do) and derive the estimate of $\psi = \Pr\{Y_1 < 0\} = \Phi(-\theta)$. To estimate ψ directly from the definition is inconvenient, but the equivariance property ensures that $\hat\psi = \Phi(-\hat\theta)$, if $\hat\theta$ exists (otherwise neither $\hat\theta$ nor $\hat\psi$ exist). $\square$

Remark 3.1.9 What happens when ψ is not one-to-one? Consider, for instance, estimation of $\psi = \theta(1-\theta) = \mathrm{var}[Y]/n$, where $Y \sim Bin(n,\theta)$. It would be tempting to say that $\hat\psi = \hat\theta(1-\hat\theta)$, but this conclusion does not follow from the equivariance property because ψ is not in a one-to-one relationship with θ. In fact, it does not even help to start from the definition since for any given value of ψ there exist *two* associated values of θ, and these two values assign different probabilities to the observed value y.

The question can be settled by extending the definition of the MLE, i.e. defining $\hat\psi = \psi(\hat\theta)$ when $\psi(\cdot)$ is not one-to-one, or by introducing the concept of **induced likelihood** defined by

$$L_\Psi(\psi_0) = \max_{\{\theta:\psi(\theta)=\psi_0\}} L(\theta),$$

whose maximum value is at $\hat\psi = \psi(\hat\theta)$ (Zehna, 1966).

Scholz (1980) has proposed an alternative and more general definition of the MLE, capable of handling the non-invertible transformation case without introducing additional concepts such as induced likelihood. Moreover, this definition is also suitable for estimation in nonparametric models. $\square$

3.1.4 Likelihood equations

Since the logarithm is a strictly increasing function, the MLE can be obtained by maximizing the log-likelihood $\ell(\theta)$ If we suppose $L(\theta)$ to be differentiable and assume that Θ is an open subset of $\mathbb{R}^k$, then the MLE must satisfy

$$\frac{d}{d\theta}\ell(\theta) = 0, \qquad (3\ 2)$$

referred to as the **likelihood equation(s)**.

Obviously, a solution of (3.2) is not necessarily the global maximum of the likelihood function In some cases, it can be established that $\ell(\theta)$ is a concave function, implying that the solution of (3.2) is indeed the MLE.

In certain other cases, it can be proved that the log-likelihood is locally concave at every stationary point, that is the matrix

$$\left.\frac{d^2}{d\theta\,d\theta^\top}\,\ell(\theta)\right|_{\theta=\hat{\theta}}$$

is negative definite at every point $\hat{\theta}$ satisfying (3.2). For the one-dimensional parameter case, this condition is sufficient to ensure that the solution of (3.2) is unique and is the MLE

In the multi-dimensional case, local concavity of a function is not a sufficient condition for the existence of a unique maximum, although finding counter-examples requires a lot of imagination. Some additional condition must then be imposed; one such requirement is that the function approaches C when the argument approaches the boundary of the domain, where C is a constant real number or $-\infty$. This additional requirement is very mild if the function of interest is a sufficiently regular log-likelihood, since the condition is satisfied in virtually all practical cases, with $C = -\infty$. See Mäkeläinen, Schmidt and Styan (1981) for details.

If neither global concavity nor local concavity of $\ell(\theta)$ can be established, then it is advisable to explore more closely the behaviour of the log-likelihood, searching for all solutions of (3.2). Plotting the likelihood function itself can be very useful, not only for maximization purposes, but also for other reasons mentioned in sections 4.3 and 4.5. Unfortunately, such plotting is feasible directly only for one- or two-dimensional parameters; in higher dimensions, methods for reducing the dimensionality must be adopted. One simple device for this purpose is to examine only 'slices' of the likelihood, plotting the likelihood as a function of two parameters for fixed values (in particular, the MLEs) of the other $k-2$ parameters. A more refined approach is to use the profile likelihood described in section 4.3.4.

Example 3.1.10 In Example 2.2.2, the log-likelihood of an s r s $(y_1, \ldots, y_n)$ from an $N(\mu, \sigma^2)$ random variable was obtained, namely

$$
\begin{aligned}
\ell(\theta) &= -\tfrac{1}{2} n \ln \sigma^2 - \sum_{i=1}^n (y_i - \mu)^2 / (2\sigma^2) \\
&= -\tfrac{1}{2} n \ln \sigma^2 - n(m_2 - 2m_1\mu + \mu^2)/(2\sigma^2)
\end{aligned}
$$

where $\theta = (\mu, \sigma^2)$, and m_1 and m_2 denote the first and second sample moments, respectively. Let us take as our parameter space Θ the entire admissible space $\mathbb{R} \times \mathbb{R}^+$. The likelihood equations are

$$
\frac{\partial \ell}{\partial \mu} = \frac{n}{\sigma^2}(m_1 - \mu) = 0,
$$

$$
\frac{\partial \ell}{\partial \sigma^2} = -\frac{n}{2\sigma^2} + \frac{n(m_2 - 2m_1\mu + \mu^2)}{2\sigma^4} = 0.
$$

The first equation leads to the unique solution

$$
\hat{\mu}_1 = m_1
$$

which, substituted in the second equation, gives

$$
\hat{\sigma}^2 = m_2 - m_1^2.
$$

Since we can immediately check that $\ell(\mu, \sigma^2) \to -\infty$ as θ approaches the boundary of the parameter space, and

$$
\frac{d^2}{d\theta\, d\theta^{\mathsf{T}}} \ell(\hat{\theta}) \Big|_{\theta = \hat{\theta}} = -n \begin{pmatrix} 1/\hat{\sigma}^2 & 0 \\ 0 & 1/\hat{\sigma}^4 \end{pmatrix}
$$

is negative definite, then $(\hat{\mu}, \hat{\sigma}^2)$ represents the MLE.   □

Example 3.1.11 For the autoregressive process of order 1 in Example 2.4.5, the likelihood equations are

$$
\frac{\partial^2 \ell}{\partial \sigma^2} = -T\sigma^{-2} + \sigma^{-4}(d_{00} - 2\rho d_{01} + \rho^2 d_{11}) = 0,
$$

$$
\frac{\partial^2 \ell}{\partial \rho} = -2\rho/(1 - \rho^2) - \sigma^{-2}(-2d_{01} + 2\rho d_{11}) = 0.
$$

From the first relationship, we obtain

$$
\hat{\sigma}^2 = (d_{00} - 2\hat{\rho} d_{01} + \hat{\rho}^2 d_{11})/T,
$$

which, substituted in the second, leads after some algebra to the cubic equation for $\hat{\rho}$

$$\frac{T-1}{T}d_{11}\hat{\rho}^3 - \frac{T-2}{T}d_{01}\hat{\rho}^2 - (d_{11} + d_{00}/T)\hat{\rho} + d_{01} = 0.$$

Denoting by $C(\hat{\rho})$ the left-hand side of the above expression, we have

$$C(1) = -T^{-1}(d_{00} + d_{11} - 2d_{01})$$

$$= -T^{-1}\sum_{t=2}^{T}(y_t - y_{t-1})^2 < 0 \quad \text{(a.s)},$$

$$C(-1) = T^{-1}(d_{00} + d_{11} + 2d_{01})$$

$$= T^{-1}\sum_{t=2}^{T}(y_t + y_{t-1})^2 > 0 \quad \text{(a.s.)}.$$

Since the coefficient of $\hat{\rho}^3$ is positive a.s., we conclude that $C(\hat{\rho}) = 0$ has three real roots, one for each of the open intervals $(-\infty, -1)$, $(-1, 1)$, $(1, \infty)$; the solution lying in the second interval is the MLE. It would also be possible to obtain the closed-form solution of the equation, using well-known formulae for cubic equations. □

3.1 5 Computational aspects

It has already been mentioned that in many cases it is impossible to obtain a closed-form expression for the MLE as a function of the observations, and one then has to resort to numerical methods

This feature of the MLE was a major hindrance to its widespread use until, with the development of fast digital computers, the burden of computation ceased to be a major issue in practical statistical work.

Numerical methods are employed in a number of ways, either for direct maximization of the log-likelihood or for solving the likelihood equations. When analytic differentiation of $\ell(\theta)$ is feasible, the latter route is preferable, since the solution of nonlinear equations is a simpler numerical problem than direct function maximization.

Clearly, a full account of numerical methods for nonlinear equations is outside the scope of this book, but it is worth mentioning at least the Newton–Raphson algorithm. Suppose that, for a differentiable real-valued function $g(\cdot)$, we have to solve the equation

$$g(x) = 0$$

with respect to the scalar variable x, starting from a tentative solution x_0. The Newton–Raphson method provides an approximate
solution x_1 via the expression

$$x_1 = x_0 - \frac{g(x_0)}{g'(x_0)}. \tag{3.3}$$

This rule follows from the approximation, based on the first-order
Taylor expansion,

$$g(x_1) \approx g(x_0) + (x_1 - x_0)g'(x_0)$$

which gives (3 3) when set equal to 0 and solved for x_1.

By applying (3.3) again with x_0 replaced by x_1, a new approximation x_2 is produced Repeated application of this scheme forms
a sequence of values $x_0, x_1, . \quad , x_s, \ldots$ which under suitable conditions converges to the exact solution. One factor which may be
decisive in achieving convergence is the choice of the initial value
x_0.

In practice, the above sequence of approximate solutions must
be stopped after a finite number of terms, generally at the value s
such that $|x_s - x_{s-1}|$ is less than a pre-assigned quantity, or when
$g(x_s)$ is sufficiently close to 0, or when a combination of these two
criteria is satisfied.

If $g(x)$ in (3.3) denotes a vector of k differentiable functions of
a k-dimensional argument, the Newton–Raphson algorithm takes
the form

$$x_{s+1} = x_s - \left(\frac{d}{d x^\top} g(x_s) \right)^{-1} g(x_s) \qquad (s = 0, 1, ..) \tag{3.4}$$

where the subscript s denotes the sth approximation to the solution
rather than the sth component of x.

In statistical applications, the Newton–Raphson method is used
mainly for solving likelihood equations, or other equations deduced
from them. In particular, replacing $g(x)$ by $d\,\ell(\theta)/(d\,\theta)$ in (3.4), we
obtain the following expression for successive approximations of
the MLE:

$$\hat{\theta}_{s+1} = \hat{\theta}_s - \left(\frac{d^2}{d\theta\, d\theta^\top} \ell(\hat{\theta}_s) \right)^{-1} \frac{d}{d\theta} \ell(\hat{\theta}_s) \qquad (s = 0, 1, \ldots) \tag{3.5}$$

To use this equation, a preliminary estimate $\hat{\theta}_0$ is necessary, usually chosen by some alternative technique such as the method of
moments.

A systematic presentation of the body of computational techniques in use in statistics, a field often referred to as **computational statistics**, is presented by Thisted (1988).

Example 3.1.12 Let $(y_1, \ldots, y_n)$ be an s r s from a gamma random variable with mean value μ and index ω ($\mu > 0, \omega > 0$) i.e. with density function at y given by

$$\frac{e^{-\omega y/\mu} y^{\omega-1}}{\Gamma(\omega)} \left(\frac{\omega}{\mu}\right)^{\omega}$$

if $y > 0$ and $(\omega, \mu) \in \mathbb{R}^+ \times \mathbb{R}^+$. This parametrization is different from that of section A.2.3, but the equivariance property ensures that the reparametrization causes no inconsistencies. The log-likelihood is

$$\ell(\omega, \mu) = -\frac{\omega}{\mu} + (\omega - 1) \sum_i \ln y_i + n\omega(\ln \omega - \ln \mu) - n\ln\Gamma(\omega)$$

and it is easy to check that $\ell(\omega, \mu) \to -\infty$ when (ω, μ) approaches the boundary of $\mathbb{R}^+ \times \mathbb{R}^+$, taking into account Stirling's formula (A.10) The likelihood equations are

$$\frac{\partial \ell}{\partial \mu} = \frac{\omega \sum_i y_i}{\mu^2} - \frac{n\omega}{\mu} = 0,$$

$$\frac{\partial \ell}{\partial \omega} = -\frac{\sum_i y_i}{\mu} + \sum \ln y_i + n(\ln\omega + 1 - \ln\mu) - n\psi(\omega) = 0,$$

where

$$\psi(x) = \frac{d}{dx}\ln\Gamma(x)$$

is the digamma function whose behaviour is qualitatively similar to the logarithmic function (see Abramowitz and Stegun 1965, Chapter 6).

Instead of using a two-dimensional Newton–Raphson search, the problem can be reduced to one dimension, with considerable savings in computation. For the first equation, there exists the unique solution

$$\hat{\mu} = \frac{\sum_{i=1}^{n} y_i}{n}$$

which, substituted in $\partial \ell/\partial \omega$, leads to the equation for ω

$$g(\omega) = \sum \ln y_i + n\ln(\omega/\hat{\mu}) - n\psi(\omega) = 0,$$

which must be solved numerically. The direct proof of the existence and uniqueness of the solution for the above equation is not

Table 3.1 *A few iterations of the Newton–Raphson algorithm for maximizing the likelihood of a sample from a gamma random variable*

s	$\hat{\omega}_s$	$g(\hat{\omega}_s)$	$g'(\hat{\omega}_s)$	$\ell(\hat{\omega}_s, \hat{\mu})$
0	1.00000	1.37853	−6.44934	5 76289
1	1 21375	0.26105	−4.23466	7.59222
2	1.27539	0 01362	−3.80417	8.08001
3	1.27897	0.00004	−3.78118	8.10788
4	1.27898	0.00000	−3 78111	8.10796
5	1 27898	0.00000	−3.78111	8.10796

immediate, but the conclusion follows from a general property of the exponential family which will be described in section 3.3.6.

To demonstrate the use of the Newton–Raphson algorithm for solving $g(\omega) = 0$, consider the following s r s with $n = 10$,

$$0\ 07\quad 0.27\quad 0.15\quad 0.36\quad 1.25\quad 0.26\quad 0.14\quad 1.51\quad 0.25\quad 0.24,$$

and choose the initial value $\hat{\omega}_0 = 1$. The derivative

$$g'(\omega) = \frac{n}{\omega} - n\psi'(\omega)$$

involves the trigamma function $\psi'(\cdot)$ which can be computed using approximate formulae given, for instance, by Abramowitz and Stegun (1965, Chapter 6); the digamma function can be computed similarly.

Table 3.1 reports summary results from using the algorithm for the above data. It will be noticed that a few iterations are sufficient to reach convergence of $\hat{\omega}_s$ and of the likelihood, with $g(\hat{\omega}_s)$ effectively equal to 0. □

Example 3.1.13 Let y_i be a value sampled from an exponential random variable Y_i with scale parameter ρ_i, such that

$$\ln \rho_i = \alpha + \beta x_i \qquad (i = 1, \ldots, n)$$

where the x_i are known constants not all equal, and α and β are real parameters to be estimated. Without loss of generality, assume that $\sum x_i = 0$, since otherwise we could set

$$\ln \rho_i = \alpha^* + \beta(x_i - \overline{x})$$

where $\overline{x} = \sum x_i / n$, $\alpha^* = \alpha + \beta \overline{x}$, and the condition would be met by the new constants $x_i - \bar{x}$. Under independence of the observations, the log-likelihood is

$$\ell(\alpha, \beta) = \sum_{i=1}^{n} \{\alpha + \beta x_i - y_i \exp(\alpha + \beta x_i)\}$$

$$= n\alpha - e^{\alpha} \sum y_i \exp(\beta x_i)$$

which does not lead to non-trivial sufficient statistics. For likelihood maximization, consider

$$\frac{\partial \ell}{\partial \alpha} = n - e^{\alpha} \sum y_i e^{\beta x_i},$$

$$\frac{\partial \ell}{\partial \beta} = -e^{\alpha} \sum x_i y_i e^{\beta x_i}.$$

Equating the first expression to 0, we obtain

$$\hat{\alpha}(\beta) = \ln\left(\frac{n}{\sum y_i e^{\hat{\beta} x_i}}\right),$$

and the second leads to the equality

$$g(\hat{\beta}) = \sum_i x_i y_i e^{\hat{\beta} x_i} = 0$$

whose solution must then be substituted into $\hat{\alpha}(\beta)$. The equation $g(\hat{\beta}) = 0$ does not allow explicit solution, but we can state that the solution exists and is unique because:

1. the left-hand side, $g(\hat{\beta})$, is the sum of n increasing functions;

2. as $\hat{\beta} \to \infty$, all terms $x_i y_i \exp(\hat{\beta} x_i)$ with $x_i > 0$ tend to ∞, while those with $x_i < 0$ vanish, hence $g(\hat{\beta}) \to \infty$;

3. if $\hat{\beta} \to -\infty$, then $g(\hat{\beta}) \to -\infty$, for similar reasons.

The Hessian matrix of the log-likelihood is

$$\begin{pmatrix} -e^{\alpha} \sum y_i e^{\beta x_i} & -e^{\alpha} \sum x_i y_i e^{\beta x_i} \\ -e^{\alpha} \sum x_i y_i e^{\beta x_i} & -e^{\alpha} \sum x_i^2 y_i e^{\beta x_i} \end{pmatrix}$$

whose diagonal elements are negative, and the determinant is

$$e^{2\alpha} \left\{ \sum y_i e^{\beta x_i} \sum x_i^2 y_i e^{\beta x_i} - \left(\sum x_i y_i e^{\beta x_i}\right)^2 \right\}$$

which is positive at least at the solution of the likelihood equations, since the last term vanishes. Therefore the likelihood equations have a unique solution, and this corresponds to a maximum.

Table 3 2 *Some fictitious data used to illustrate the Newton–Raphson algorithm*

x	y	x	y
−0 49	1.26	−1.06	1.76
0 50	0.21	−0.27	1.28
0.67	0.32	0.50	1.09
−0.27	0.18	0 61	0.01
−0 43	0 45	0.24	0.70

Table 3.3 *A few Newton–Raphson iterations for some data from exponential random variables, whose parameters depend upon a second variable*

s	$\hat{\beta}_s$	$g(\hat{\beta}_s)$	$g'(\hat{\beta}_s)$	$\hat{\alpha}_s$	$\ell(\hat{\alpha}_s, \hat{\beta}_s)$
0	1.00000	0 26407	1.88050	0 43604	−5.63958
1	0.85957	−0 00436	1.94638	0.43888	−5 61117
2	0.86181	−0 00000	1.94514	0.43888	−5.61116
3	0.86181	−0 00000	1.94514	0.43888	−5.61116
4	0.86181	0 00000	1.94514	0.43888	−5.61116

Also in this case, the actual solution of the likelihood equation requires numerical methods To demonstrate, the fictitious data of Table 3.2 have been inserted in the above formulae. Table 3.3 shows some summary results of a few Newton–Raphson iterations for these data

In this case, too, a few iterations are sufficient to reach convergence of the algorithm The estimates $(\hat{\alpha}, \hat{\beta})$ turn out to be reasonably close to the actual values of the parameters used for generating the data, namely $(\frac{1}{2}, 1)$, considering that the small number of available data prevents accurate estimation. □

The fact that we have described the Newton–Raphson method here does not mean that it is the only existing technique for solving likelihood equations. In particular, when computation of $\ell''(\cdot)$ is difficult or lengthy, it may be preferable to use some other method which can dispense with it

Even considering the extraordinary development of computers, as well as of numerical methods and algorithms, it remains true that a numerical solution is a *specific* one, meaning that changing even a single datum in the problem makes it necessary to go through the whole maximization procedure again. Moreover, the lack of an explicit solution does not allow certain algebraic manipulations useful in exploring formal properties of the estimators. Therefore, whenever feasible, an explicit solution of the likelihood equations must be searched for.

3.2 Fisher information

So far, certain elegant aspects of the MLE have been presented, such as equivariance However, nothing has been said about the accuracy of this method for estimating the unknown parameter θ_*. After all, the accuracy of the estimate should be of primary concern to us.

Clearly, for any given sample and estimation problem, the exact value of the estimation error $\hat{\theta} - \theta_*$ cannot be known, except in degenerate cases. All we can do is to produce an index which gives an assessment of the plausible amount of estimation error. The development of this point will involve a number of concepts, some of which are of independent interest.

3.2.1 Observed Fisher information

If θ is one-dimensional and the log-likelihood is sufficiently regular, we can expand it as

$$
\begin{aligned}
\ell(\theta) &= \ell(\hat{\theta}) + \ell'(\hat{\theta})(\theta - \hat{\theta}) + \tfrac{1}{2}\ell''(\hat{\theta})(\theta - \hat{\theta})^2 + \cdots \\
&= \ell(\hat{\theta}) + \tfrac{1}{2}\ell''(\hat{\theta})(\theta - \hat{\theta})^2 + \cdots
\end{aligned}
\tag{3.6}
$$

where the terms $\ell(\hat{\theta})$ and $\ell''(\hat{\theta})$ are constants with respect to θ. Therefore the behaviour of $\ell(\theta)$ near $\hat{\theta}$ is largely determined by $\ell''(\hat{\theta})$, which is a measure of the local curvature of the log-likelihood.

The non-negative quantity $\mathcal{I}(\hat{\theta}) = -\ell''(\hat{\theta})$ is called the **observed (Fisher) information** It can be regarded as an index of the steepness of the log-likelihood moving away from $\hat{\theta}$, and as an indicator of the strength of preference for the MLE point with respect to other points of the parameter space.

If θ is multi-dimensional, the corresponding Taylor series expansion of the log-likelihood leads to a matrix expression for the observed information, namely

$$\mathcal{I}(\hat{\theta}) = - \left. \frac{d^2}{d\theta\, d\theta^\mathsf{T}} \ell(\theta) \right|_{\theta=\hat{\theta}}$$

which is positive definite, except in degenerate cases.

3.2.2 Repeated sampling principle

The idea of the curvature of the log-likelihood is useful for comparing different likelihoods and MLEs, but it does not answer, by itself, the question: is maximum likelihood a 'good' strategy for tackling estimation problems?

A first answer in support of the likelihood, and correspondingly of the MLE, as a general framework for inference, might be of logical and philosophical type (see Edwards (1972)).

An alternative way of answering the above question is to obtain some mathematical properties of the MLE concerning the estimation error $\hat{\theta} - \theta_*$. To follow this line, let us return to the fact that y is regarded as a value sampled from a random variable Y, so that $\hat{\theta}$ itself is a value sampled from some random variable, still denoted (for historical reasons) $\hat{\theta}$ Of this random variable it makes sense to consider the expected value, the variance, and so on. This view represents the **repeated sampling principle**.

Suppose, for instance, that the random variable $\hat{\theta} - \theta_*$ has a probability distribution highly concentrated around the value 0. Then, a priori with respect to the sampling, it can be expected that the subsequent estimate $\hat{\theta}$ will be in some sense close to θ_*, although this fact cannot be ensured for each specific sample taken.

A high proportion of the statistical literature is concerned with the evaluation of the properties of various statistical methods, including MLE, from the viewpoint of the repeated sampling principle.

The main goal of the rest of this chapter is in fact the study of the distribution of the random variable $\hat{\theta} - \theta_*$ A simple, and in a sense most obvious, requirement is that this random variable is located around 0. One way of translating this qualitative requirement into a mathematical property is to impose

$$\mathrm{E}\left[\hat{\theta}; \theta\right] = \theta \qquad \text{for all } \theta \in \Theta$$

provided that the expected value exists. If this condition is satisfied, the estimator is said to be **(mean) unbiased**; otherwise,

the difference $\mathrm{E}\left[\hat{\theta}\right] - \theta$ is called the **bias**. A slightly less stringent
condition is that the bias vanishes as n increases without limit, i e.

$$\lim_{n \to \infty} \mathrm{E}\left[\hat{\theta}, \theta\right] = \theta \qquad \text{for all } \theta \in \Theta,$$

where n is the sample size. If this weaker requirement is satisfied,
the estimator is said to be **asymptotically unbiased.**

As mentioned earlier, the term 'estimate' is often used for 'estimator', so the phrase 'unbiased estimate' is commonly employed,
and similarly for other properties

Example 3.2.1 For the situation of Example 3.1 10, it follows
that $\hat{\mu}$ is an unbiased estimate of μ and

$$\begin{aligned}
\mathrm{E}\left[\hat{\sigma}^2\right] &= \frac{1}{n}\mathrm{E}\left[\sum_{i=1}^{n}(Y_i - m_1)^2\right] \\
&= (1 - 1/n)\sigma^2
\end{aligned}$$

taking into account the results of section A 5.7. Therefore, $\hat{\sigma}^2$ is
a biased estimated of σ^2, but the bias goes to 0 at rate $1/n$ as
$n \to \infty$. Moreover, the bias can be removed by considering the
corrected sample variance

$$s^2 = \frac{n}{n-1}\hat{\sigma}^2 = \frac{\sum_i (y_i - m_1)^2}{n-1}. \tag{3.7}$$

$\square$

An exact computation, such as the one just developed, is feasible only in simple cases, but it becomes prohibitive when $\hat{\theta}$ is a
complicated function of the data (especially when an explicit representation of $\hat{\theta}$ as a function of the data does not exist!). Our
targets are

1. to obtain general results about the properties of the MLE, or
   results at least of some degree of generality, in order to avoid
   studying its behaviour for each specific estimation problem;

2. to evaluate other aspects of the random variable $\hat{\theta} - \theta_*$, not
   just its mean value.

Before pursuing such an analysis, some additional concepts outside
the context of the MLE must be introduced; these are the subject
of the rest of this section.

3.2.3 Regular estimation problems

An estimation problem is said to be **regular** when the following conditions are satisfied by the statistical model (2.1) under consideration

(a) The statistical model is identifiable, in accordance with the definition in section 2.1.3.

(b) The parameter space Θ is an open interval in $\mathbb{R}^k$.

(c) All (probability) density functions specified by the model have the same support.

(d) For the function f in (2.1), differentiation with respect to θ and integration with respect to y can be interchanged twice, specifically

$$\text{(i)} \qquad \int_y \frac{\partial}{\partial \theta} f(y; \theta)\, d\nu(y) = \frac{\partial}{\partial \theta} \int_y f(y; \theta)\, d\nu(y),$$

$$\text{(ii)} \qquad \int_y \frac{\partial^2}{\partial \theta^2} f(y; \theta)\, d\nu(y) = \frac{\partial^2}{\partial \theta^2} \int_y f(y; \theta)\, d\nu(y).$$

If $k > 1$, then (ii) becomes

$$\text{(ii')} \qquad \int_y \frac{d^2}{d\theta\, d\theta^{\mathsf{T}}} f(y; \theta)\, d\nu(y) = \frac{d^2}{d\theta\, d\theta^{\mathsf{T}}} \int_y f(y; \theta)\, d\nu(y)$$

signifying that integration of a matrix-valued function is performed component-wise.

All these assumptions are met in a large number of practical situations. Condition (a) requires that different values of θ lead to distinct probability distributions, for the reasons explained in section 2.1.3. With highly complex models, especially when $\dim(Y_i) > 1$, it may happen that identifiability problems occur; it is then necessary to parametrize the model differently, or to restrict Θ to an appropriate subset. Except for unusual situations with peculiar shapes of Θ, condition (b) is very often met, possibly by changing the coordinate system for Θ; the more relevant case ruled out is when θ is on the boundary of the parameter space. Conditions (c) and (d) simply express a form of mathematical regularity, and are satisfied in most practical cases.

3.2.4 Expected Fisher information

The quantity called the **(Fisher) score function**,

$$u(\theta) = u(\theta; y) = \frac{d}{d\theta} \ell(\theta; y) \qquad (3.8)$$

plays a very important role in statistical theory. On using assumption (d)(i) above, it follows that, for $k = 1$,

$$
\begin{aligned}
\mathrm{E}[\,u(\theta);\theta\,] &= \mathrm{E}\left[\frac{\partial}{\partial\theta} \ln f(Y;\theta); \theta\right] \\
&= \int_{\mathcal{Y}} \left(\frac{1}{f(y;\theta)} \frac{\partial}{\partial\theta} f(y;\theta)\right) f(y;\theta)\, d\nu(y) \\
&= \frac{\partial}{\partial\theta} \int_{\mathcal{Y}} f(y;\theta)\, d\nu(y) = \frac{\partial}{\partial\theta} 1 = 0;
\end{aligned}
$$

for general k, $u(\theta)$ is a vector with 0 mean vector. The quantity

$$\mathrm{var}[\,u(\theta; Y);\theta\,] = \mathrm{E}\left[u(\theta; Y)^2;\theta\right]$$

is called the **expected (Fisher) information**, and is denoted by $I(\theta)$. For generic k, $I(\theta)$ is a symmetric $(k \times k)$ matrix

$$I(\theta) = \mathrm{E}\left[u(\theta)\, u(\theta)^{\mathsf{T}}; \theta\right], \qquad (3.9)$$

which is positive semidefinite since

$$a^{\mathsf{T}} I(\theta) a = \mathrm{E}\left[\|u(\theta)^{\mathsf{T}} a\|^2\right] \geq 0$$

for any $a \in \mathbb{R}^k$.

Using assumption (d)(ii), we obtain

$$
\begin{aligned}
I(\theta) &= \mathrm{E}\left[u(\theta)^2; \theta\right] \\
&= \int_{\mathcal{Y}} \left(\frac{1}{f(y;\theta)} \frac{\partial}{\partial\theta} f(y;\theta)\right)^2 f(y;\theta)\, d\nu(y) \\
&= \int_{\mathcal{Y}} \left(\frac{1}{f(y;\theta)} \frac{\partial^2}{\partial\theta^2} f(y;\theta) - \frac{\partial^2}{\partial\theta^2} \ln f(y;\theta)\right) f(y;\theta)\, d\nu(y) \\
&= -\mathrm{E}\left[\frac{\partial^2}{\partial\theta^2} \ln f(Y;\theta)\right] \\
&= -\mathrm{E}\left[\frac{\partial}{\partial\theta} u(\theta; Y); \theta\right]
\end{aligned}
$$

(for $k = 1$) which provides an alternative representation of $I(\theta)$. Notice the formal similarity with the expression for the observed information. For generic k, we have

$$\begin{aligned} I(\theta) &= -\mathrm{E}\left[\frac{d^2}{d\theta\, d\theta^\mathsf{T}} \ln f(Y;\theta)\right] \\ &= -\mathrm{E}\left[\frac{d}{d\theta^\mathsf{T}} u(\theta;Y)\right]. \end{aligned}$$

An important property of the expected information is additivity, when the components are independent. In fact, denoting by $I_1(\theta)$ and $I_2(\theta)$ the information associated with independent random variables Y_1 and Y_2, the information for the pair (Y_1, Y_2) is

$$I(\theta) = I_1(\theta) + I_2(\theta)$$

by the additivity property for the variance of independent random variables. Repeating the argument for the case of n observations from independent and identically distributed random variables, it follows that

$$I(\theta) = n\, i(\theta)$$

where $i(\theta)$ is the information for a single observation

Another useful property of the information is the simple rule for its transformation after reparametrization. If we reparametrize from θ to $\psi(\theta)$, where $\psi(\cdot)$ is a monotonic differentiable function, the expected information is modified according to a simple rule, namely

$$I_\Psi(\psi) = \{\psi'(\theta)\}^{-2}\, I(\theta)\Big|_{\theta=\theta(\psi)} = \left(\frac{d\,\theta(\psi)}{d\,\psi}\right)^2 I(\theta)\Big|_{\theta=\theta(\psi)} \qquad (3\,10)$$

if θ is scalar.

For generic k, suppose that $\psi(\theta)$ represents an invertible transformation from $\mathbb{R}^k$ to $\mathbb{R}^k$, such that every transformation from θ to the generic component ψ_j of ψ is differentiable $(j = 1,\ ..,k)$. Then (3 10) takes the general form

$$I_\Psi(\psi) = \Delta^\mathsf{T} I(\theta)\Delta \qquad (3.11)$$

where $\Delta = (\delta_{ij}) = (d\theta_i/d\psi_j)$. The proof is left to the reader; see Exercise 3.10.

3.2.5 Cramér–Rao inequality

Theorem 3.2.2 *In a regular estimation problem with $k = 1$, let*

$T(y)$ be an estimator such that the mean value

$$a(\theta) = E[T(Y); \theta]$$

and its derivative

$$a'(\theta) = \frac{d}{d\theta} \int_{\mathcal{Y}} T(y) f(y; \theta) \, d\nu(y) = \int_{\mathcal{Y}} T(y) \frac{\partial}{\partial \theta} f(y; \theta) \, d\nu(y)$$

$$(3.12)$$

exist. Then

$$\text{var}[T(Y); \theta] \geq \{a'(\theta)\}^2 / I(\theta) \qquad (3\ 13)$$

if $0 < I(\theta) < \infty$.

Inequality (3.13) is called the Cramér–Rao inequality or lower bound.

Proof. From

$$
\begin{aligned}
a'(\theta) &= \int_{\mathcal{Y}} T(y) \frac{\partial}{\partial \theta} f(y; \theta) \, d\nu(y) \\
&= E\left[T(Y) \frac{\partial}{\partial \theta} \ln f(Y; \theta) \right] \\
&= \text{cov}[T(Y), u(\theta; Y)]
\end{aligned}
$$

and the Schwarz inequality, it follows that

$$\text{cov}[T(Y), u(\theta; Y)]^2 \leq \text{var}[T(Y)] \, \text{var}[u(\theta; Y)]$$

which is equivalent to

$$\frac{a'(\theta)^2}{\text{var}[u(\theta; Y)]} \leq \text{var}[T(Y)].$$

$\square$

In particular, if $T(y)$ is an unbiased estimator for θ, then $a'(\theta) = 1$ for all θ and

$$\text{var}[T(Y)] \geq 1/I(\theta)$$

which explains the name 'expected information' for $I(\theta)$. It represents an index of the maximal (mean) precision, at least for unbiased estimators. Notice however that such an estimator with maximal precision may not exist, or it may not be unique.

If $k > 1$ and $T(y)$ is an unbiased estimate, the statement and the proof of the Cramér–Rao inequality are quite similar. Assuming that $I(\theta)$ is a positive definite matrix, then it can be shown that

$$\text{var}[T(Y)] \geq I(\theta)^{-1} \qquad (3.14)$$

meaning that the matrix difference of the two sides is positive semidefinite. In particular, (3.14) implies that

$$\text{var}\left[c^{\mathsf{T}}T(Y)\right] \geq c^{\mathsf{T}}I(\theta)^{-1}c,$$

for any linear combination $c^{\mathsf{T}}T(y)$ of the components of $T(y)$.

In the case of an s r s (3.14) becomes

$$\text{var}[T(Y)] \geq \{n\, i(\theta)\}^{-1}$$

taking into account the results of section 3.2.4.

Example 3.2.3 Continuing Example 3.1.2, we have

$$
\begin{aligned}
\ell(\theta) &= -\tfrac{1}{2}\sum_{i=1}^{n}(y_i - \theta)^2, \\
u(\theta) &= \ell'(\theta) = \sum_{i=1}^{n}(y_i - \theta), \\
u'(\theta) &= \ell''(\theta) = -n, \\
I(\theta) &= n.
\end{aligned}
$$

Therefore, for an estimator $T(y)$ such that

$$E[T(Y);\theta] = \theta,$$

there exists the lower bound

$$\text{var}[T(Y);\theta] \geq 1/n.$$

It has already been shown that the MLE is the arithmetic mean for which we know

$$E\left[\hat{\theta};\theta\right] = \theta, \quad \text{var}\left[\hat{\theta};\theta\right] = 1/n.$$

Since $\hat{\theta}$ is unbiased and attains the Cramér–Rao lower bound, then it cannot be improved, i.e there exists no unbiased estimate with lower variance. Rather, we could wonder whether or not there exist unbiased estimators with the same variance $1/n$. The answer is essentially negative, but its explanation involves concepts outside the scope of this book; for a discussion of this point, see, for instance, Lehmann (1983, section 2.1)

If both parameters are unknown, as in Examples 2.2.2 and 3.1 10, the expected information is

$$I(\mu, \sigma^2) = n\begin{pmatrix} 1/\sigma^2 & 0 \\ 0 & 1/(2\sigma^4) \end{pmatrix}.$$

□

Example 3.2.4 For the setting of Example 3 1 13, we want to obtain $I(\alpha, \beta)$. Since the Hessian matrix has already been computed, it suffices to compute its expected value, changing sign. Taking into account

$$\exp(\alpha + \beta x_i)\,E[\,Y_i\,] = 1,$$

we obtain

$$I(\alpha, \beta) = \begin{pmatrix} n & 0 \\ 0 & \sum x_i^2 \end{pmatrix}.$$

If the model is extended from two parameters to k parameters, namely

$$\ln \rho_i = x_i^\top \beta \qquad (i = 1, \dots, n),$$

where x_i is a k-dimensional vector of known constants and β a k-dimensional parameter, then

$$
\begin{aligned}
\ell(\beta) &= \sum_{i=1}^{n} x_i^\top \beta - \sum_{i=1}^{n} y_i \exp(x_i^\top \beta), \\
\frac{d\ell}{d\beta} &= \sum_{i=1}^{n} x_i - \sum_{i=1}^{n} y_i \exp(x_i^\top \beta)\, x_i, \\
\frac{d^2\ell}{d\beta\, d\beta^\top} &= -\sum_{i=1}^{n} y_i \exp(x_i^\top \beta)\, x_i x_i^\top
\end{aligned}
$$

and

$$I(\beta) = \sum_{i=1}^{n} x_i x_i^\top = X^\top X$$

where X is the $n \times k$ matrix whose ith row is $x_i^\top$. $\square$

3.2.6 Efficiency

In regular estimation problems, the Cramér–Rao inequality provides a reference point, useful in determining whether an estimator cannot be improved upon, at least among the unbiased ones If an estimator does not achieve the Cramér–Rao lower bound, the quantity

$$\{\mathrm{var}[\,T(Y)\,]\, I(\theta)\}^{-1}$$

indicates how far it is from the optimum, which explains why it is called the **efficiency** of $T(y)$, provided $E[\,T(y)\,] = \theta$. This definition makes sense only when the Cramér–Rao lower bound is actually achievable.

Clearly, to compare two estimators by considering just their variances makes sense only if they are both unbiased, or at least the bias is negligible; otherwise, an absurd estimator which is identically equal to 0 would always appear to be superior. Therefore, a more appropriate criterion is to consider the **mean squared error**

$$E\left[(T(Y) - \theta)^2 \right] = \{E[T(Y)] - \theta\}^2 + var[T(Y)],$$

which takes into account both the variance and the bias of an estimator.

In many cases, however, the square of the bias of any reasonable estimator has a much smaller order of magnitude than its variance, so only the latter quantity is relevant. Typically, when present, the squared bias is of order $O(n^{-2})$ while $var[T(Y)] = O(n^{-1})$; therefore, the latter is the predominant term. For specific cases, see Examples 3.2.1 and 3.2 3.

Example 3.2.5 None of the above discussion makes sense outside regular estimation problems. For the setting of Example 3.1.4, we have

$$\Pr\left\{ \hat{\theta}/\theta \le t \right\} = t^n \qquad (0 < t < 1),$$

using the results of section A.7.2; hence

$$E\left[\hat{\theta}\right] = \frac{n}{n+1}\theta, \quad var\left[\hat{\theta}\right] = \frac{n}{(n+1)^2\,(n+2)}\theta^2.$$

The corrected estimator

$$T(y) = \frac{n+1}{n}\hat{\theta}$$

is unbiased and has variance $O(n^{-2})$, i.e. a smaller order of magnitude than expected from the Cramér–Rao inequality. In fact, conditions (c) and (d) of section 3.2.3 fail here.   □

Remark 3.2.6 The reader could object that we have implicitly adopted the criterion of ordering estimators according to their mean squared error, and this ordering would possibly change if another criterion was chosen.

In particular, instead of the square, one could consider some other power of the deviation between estimate and estimand, i.e.

$$E[\,|T(Y) - \theta|^p\,]$$

for some positive p. This more general optimality criterion includes the mean squared error when $p = 2$, but other values of p can produce interesting methods.

In particular, if $1 \leq p < 2$, the estimator minimizing the above expression can be shown to enjoy good **robustness** properties, meaning that it is less sensitive to inadequacy of the selected statistical model. Unfortunately, the resulting estimators have no explicit representation even in simple cases. Only the case $p = 1$ lends itself, in a limited form, to any analytic results. Some brief additional aspects on this issue will be mentioned in section 5.5.
□

3.3 Properties of the MLE

Although this may seem a paradox,
all exact science is dominated by the idea of approximation

(B Russell)

3.3.1 The role of asymptotic theory

The aim of this section is to study the accuracy of the MLE from the repeated sampling principle viewpoint.

While most of our discussion so far has been conducted in very general terms, apart from some regularity conditions, somewhat severe restrictions will be introduced in pursuing the present analysis. Only cases with simple mathematical structure will be considered here, both for the sake of clarity of exposition, and because the argument for the simpler case is the basis for more advanced problems. Some of the results we will describe could be achieved under weaker mathematical assumptions, but this would require more complex arguments, obscuring the thread of the reasoning.

All general results on the repeated sampling properties of the MLE are of asymptotic type, i.e. assuming that the sample size tends to ∞ Obviously, for any given problem, the number of available observations is what it is; it does not tend to ∞. To suppose that the sample size tends to ∞ is a mathematical 'trick' which has to be adopted because of the extreme difficulty of obtaining *exact* results, valid for finite sample size It is then necessary to consider the problem assuming $n \rightarrow \infty$, hoping that the *asymptotic* result provides a satisfactory *approximation* when used for finite n.

This approach leaves open the question of checking the adequacy of such approximations. In fact, a relevant portion of the statistical literature is concerned with the problem of establishing the quality of these approximations, and the corrections to be made, when appropriate. Luckily, in a large number of cases, these asymptotic approximations turn out to be very effective even for moderate sample size, in some cases even error-free.

The meaning of 'moderate sample size' is difficult to state precisely, because it depends partly on the required level of approximation. A rule of thumb, applicable in a fair number of practical situations, is to require $n > 30$. However, it must be borne in mind that sometimes asymptotic results are usable only for very large sample sizes; in other cases, a sample size substantially smaller than 30 is perfectly adequate.

3.3.2 Consistent estimates

An intuitively essential requirement for an estimation procedure to be acceptable is that, as the number of observations tends to ∞, the estimate should get better and better, until eventually the true parameter and the estimate should coincide. This requirement is formally expressed as follows.

Definition 3.3.1 *For the observable random variable Y, denote by $\mathcal{F}_n$ a statistical model of type (2.1) formed by a set of probability distributions with dimension some multiple of n. An estimator $T_n(\cdot)$ of the true parameter θ_* is said to be (weakly) consistent if $T_n(Y) \xrightarrow{p} \theta_*$, i e if*

$$\lim_{n \to \infty} \Pr\left\{|T_n(Y) - \theta_*| > \varepsilon; \theta_*\right\} = 0 \quad \text{for all } \varepsilon > 0$$

for any $\theta_* \in \Theta$; the corresponding sample value is called a consistent estimate. If $T_n(Y) \to \theta_*$ a.s., then $T_n(y)$ is said to be strongly consistent.

Example 3.3.2 In Example 3.1.2(a), Y is an n-dimensional random variable with probability distribution obtained by the n-fold product of the $N(\theta, 1)$ density function. The MLE $\hat{\theta}$ is the sample arithmetic mean, and $\hat{\theta} \to \theta_*$ a.s. by the strong law of large numbers, for any value of θ_* provided Θ includes θ_*; see Theorem A.8.2(i). Therefore, $\hat{\theta}$ is strongly consistent. $\qquad \square$

In general, when the estimate is a continuous function of sample moments of the s r s y taken from the random variable Y with independent and identically distributed components, convergence of the estimate to the corresponding function of the parent distribution moments (provided they exist) is ensured by well-known theorems on limits of random variables; see section A.8.2 for a small selection of them, in particular result (e).

In certain other cases, the strong law of large numbers cannot be employed, but it may happen that the expected value and variance of $T_n(Y)$ can be computed. If it turns out that

$$E[T_n(Y)] \to \theta, \quad \text{var}[T_n(Y)] \to 0,$$

then (weak) consistency follows from Theorem A.8.2(b).

However, there are cases where none of the above lines of argument is feasible, or at least they cannot be followed very easily, and it is then useful to have general results available to ensure consistency of the estimates.

We shall now consider the question of strong consistency of the MLE in a particularly simple but important situation: suppose that conditions (a) to (d) of section 3.2.3 are satisfied and, in addition, that $Y = (Y_1, \ldots, Y_n)^\top$ is a random variable with independent and identically distributed components. The log-likelihood is

$$\ell_n(\theta) = \sum_{i=1}^{n} \ln g(Y_i; \theta)$$

where $g(\cdot)$ denotes the density function of a single component, and the subscript n has been added as a reminder that in fact we are dealing with a sequence of likelihood functions. As n tends to ∞, the quantity

$$\frac{1}{n} \left(\ell_n(\theta) - \ell_n(\theta_*) \right) = \frac{1}{n} \sum_{i=1}^{n} \ln \frac{g(Y_i; \theta)}{g(Y_i; \theta_*)} \tag{3.15}$$

converges a.s. to

$$E\left[\ln \frac{g(Y_1; \theta)}{g(Y_1; \theta_*)}; \theta_* \right]$$

from the strong law of large numbers. If $\theta \neq \theta_*$, then it follows that

$$E\left[\ln \frac{g(Y_1; \theta)}{g(Y_1; \theta_*)}; \theta_* \right] < \ln E\left[\frac{g(Y_1; \theta)}{g(Y_1; \theta_*)}; \theta_* \right] = 0 \tag{3.16}$$

using the Jensen inequality. This implies, for all $\theta \neq \theta_*$,

$$\ell_n(\theta_*) - \ell_n(\theta) \to \infty$$

a.s. when $n \to \infty$.

The last relationship means that, for large n, the likelihood at θ_* is expected to be much higher that at any point different from

θ_*. This fact suggests that $\hat{\theta} \to \theta_*$, but the formal proof requires a careful explanation

Examine the very restrictive case of Θ formed by a finite set of points, i.e. $\Theta = \{\theta_*, \theta_1, \ldots, \theta_m\}$ say, and, for some choice of $\varepsilon > 0$, define

$$A_j = \{\ell_n(\theta_*) - \ell_n(\theta_j) > \varepsilon \text{ for all } n > n_0\}, \quad (j = 1, \ldots, m),$$

whose probability can be made greater than $1 - \delta$ by choosing n_0 sufficiently large. Then

$$\begin{aligned}
\Pr\left\{\bigcap_{j=1}^m A_j\right\} &= 1 - \Pr\left\{\bigcup_{j=1}^m \bar{A}_j\right\} \\
&\geq 1 - \sum_j \Pr\{\bar{A}_j\} \\
&\geq 1 - m\,\delta,
\end{aligned}$$

implying that

$$\Pr\{\ell_n(\theta_*) - \ell_n(\theta_j) > \varepsilon \text{ for all } \theta_j \neq \theta_*, \text{ for all } n > n_0\} \geq 1 - m\,\delta,$$

where δ is arbitrary; this conclusion is equivalent to strong consistency.

Formal complications arise when Θ is an interval, which is clearly the more relevant case The proof of strong consistency in this case has been given by Wald (1949); a relatively simpler argument is presented by Zacks (1971, p. 233). These proofs are technically more complicated than the previous one for finite Θ, but the line of reasoning is similar; in particular, a.s. consistency of (3.15) and inequality (3.16) provide the key ingredients of the argument. Additional assumptions required for the proof are mild; the more relevant of these is that $g(y; \theta)$ must be continuous with respect to θ, for all y.

Remark 3.3.3 In our restricted proof of consistency of the MLE, as well as in the full version, the space Θ is fixed with respect to n. In some situations, it may happen instead that the number of unknown parameters k increases with n; the MLE then generally fails to be consistent. In practice, when Θ is of finite dimension but the sample size is not much larger than the number of parameters, it is very hazardous to rely on consistency of the MLE, since the translation of the formal condition '$n \to \infty$ and k fixed' into practical terms is 'n very much larger that k'. $\qquad\square$

3.3.3 Asymptotic distribution of the MLE

Consistency of the MLE is a fundamental property, but it is also important to establish the order of magnitude (in probability) of the deviation $(\hat{\theta} - \theta_*)$, and to obtain the asymptotic distribution of this random variable, after suitable normalization. To develop such an analysis, for the sole case $k = 1$, assume the following.

(a) The estimation problem is regular

(b) The components of $Y = (Y_1, \ldots, Y_n)^\top$ are independent and identically distributed random variables with marginal density function $g(y; \theta)$.

(c) The expected information $i(\theta)$ for an individual observation is finite and positive.

(d) The MLE $\hat{\theta}$ is consistent.

(e) There exists a function $M(y; \theta)$ such that

$$\left| \frac{\partial^3}{\partial \theta^3} \ln g(y; \theta) \right| < M(y; \theta),$$

and $M(y; \theta)$ is integrable, with bounded integral, with respect to g, i e there exists a number M_0 such that

$$\mathrm{E}[\, M(Y_1; \theta); \theta\,] < M_0 < \infty.$$

Under these assumptions $\ell'(\hat{\theta}) = 0$ and, expanding $\ell'(\hat{\theta})$ as a Taylor series about θ_*, it follows that

$$
\begin{aligned}
0 &= \ell'(\hat{\theta}) \\
&= \ell'(\theta_*) + \ell''(\theta_*)(\hat{\theta} - \theta_*) + \tfrac{1}{2}\ell'''(\tilde{\theta})(\hat{\theta} - \theta_*)^2,
\end{aligned}
\qquad (3.17)
$$

where $\tilde{\theta}$ lies between $\hat{\theta}$ and θ_*, so that

$$\sqrt{n}(\hat{\theta} - \theta_*) = \frac{-\dfrac{1}{\sqrt{n}}\ell'(\theta_*)}{\dfrac{1}{n}\ell''(\theta_*) + \dfrac{\ell'''(\tilde{\theta})}{2n}(\hat{\theta} - \theta_*)}. \qquad (3.18)$$

From the central limit theorem, the numerator

$$-\frac{1}{\sqrt{n}}\ell'(\theta_*) = -\frac{1}{\sqrt{n}} \sum_{i=1}^{n} \frac{\partial}{\partial \theta} \ln g(Y_i; \theta) \Big|_{\theta = \theta_*}$$

converges in distribution to $N(0, i(\theta_*))$ as $n \to \infty$. The denominator of (3.18) is such that $\ell''(\theta_*)/n \to -i(\theta_*)$ a.s. by the strong law of large numbers. The quantity $(\hat{\theta} - \theta_*) = o_p(1)$ (in the notation defined in section A.8.3) is multiplied by

$$\frac{1}{n}|\ell'''(\tilde{\theta})| \le \frac{1}{n}\sum_{i=1}^{n}\left|\frac{\partial^3}{\partial\theta^3}\ln g(Y_i;\theta)\right|\Bigg|_{\theta=\tilde{\theta}} < \frac{1}{n}\sum_i M(Y_i;\theta)$$

which is $O_p(1)$, so the product is $o_p(1)$, using Theorem A.8.6. The final conclusion is

$$\sqrt{n}(\hat{\theta} - \theta_*) \xrightarrow{d} N(0, i(\theta_*)^{-1}), \qquad (3.19)$$

which says that $\hat{\theta} = \theta_* + O_p(n^{-1/2})$, from Theorem A.8.8, and it gives the asymptotic distribution of the deviation, after suitable normalization.

The practical implication of the above results is that, for n finite but large, the distribution of $\hat{\theta}$ can be approximated as

$$\hat{\theta} \rightsquigarrow N\left(\theta_*, \frac{1}{n\,i(\theta_*)}\right).$$

If the parameter is k-dimensional, it is again possible to prove the multi-dimensional version of (3.19), under suitable generalization of the previous regularity conditions. For detailed mathematical aspects, see Zacks (1971, p. 247) and Serfling (1980, Chapter 4)

3.3.4 Efficiency of the MLE

The definition of efficiency, given in section 3 3.6, is not useful if one cannot compute the mean and variance of an estimator, so some adaptation is necessary in the asymptotic theory context

An estimator $T(Y)$ such that

$$\sqrt{n}(T(Y) - \theta) \xrightarrow{d} N(0, v(\theta)) \qquad (3.20)$$

for some positive quantity $v(\theta)$ is said to be **best asymptotically normal** if $v(\theta)$ is equal to $i(\theta)^{-1}$.

The MLE satisfies the above requirement, but so do many other estimators. So it remains to be explained why the MLE is given such a prominent role. Besides the arguments already discussed, there are additional reasons, related in fact to efficiency, which justify the primary role assigned to the MLE.

The following discussion will leave out substantial mathematical aspects. For a wide class of estimation problems, it can be proved that any best asymptotically normal estimator $T(Y)$ is such that

$$E[T(Y)] = \theta + \frac{b(\theta)}{n} + O(n^{-2}),$$

$$\text{var}[T(Y)] = \frac{1}{n\,i(\theta)} + O(n^{-2}),$$

for some function $b(\theta)$ which determines the asymptotic bias. Therefore, the mean squared error is dominated by the term $1/(n\,i(\theta))$.

Suppose now we compute a higher-order approximation of the mean squared error, obtaining an expression such as

$$E\left[(T(Y) - \theta)^2\right] = \frac{1}{n\,i(\theta)} + \frac{a_2(\theta)}{n^2} + o(n^{-2}).$$

The leading term on the right-hand side is common to many competing estimators, so any comparison has to be based on the next term, effectively $a_2(\theta)$. The estimator having uniformly minimum $a_2(\theta)$ is to be preferred as the one with higher **second-order efficiency**.

Analyses of this kind have in fact been conducted in great detail, and the MLE has been found to have higher second-order efficiency, provided that the bias has been removed either completely (as in Example 3 2.1) or almost completely using for instance

$$\hat{\hat{\theta}} = \hat{\theta} - \frac{b(\hat{\theta})}{n}$$

to obtain a new estimator with bias $O(n^{-2})$. For detailed discussions of these problems, see Rao (1961), Efron (1975) and Amari (1985).

We can thus conclude by saying that the MLE enjoys both wide practical applicability and good formal properties. This explains the wide interest it receives both in practical and in theoretical work. Predictably, however, there do exist dissenting views; see, for instance, Berkson (1980).

Remark 3.3.4 In order to stress certain subtle mathematical aspects of asymptotic theory, rather than for their real importance, we mention there exist estimators with asymptotic variance $v(\theta)$ in (3.20) smaller than $i(\theta)^{-1}$, even in regular estimation problems.

This phenomenon, called **super-efficiency**, can, however, occur only for a countable number of points of the parameter space; moreover, it causes a completely unsatisfactory behaviour of the estimator over a neighborhood of each of those points. Since the improvement is limited to a 'small' set in the parameter space and the problems occur over a 'much larger' set, such an estimator is of no interest.

However, we still need to appreciate the apparently paradoxical existence of estimators which seem to violate the Cramér-Rao lower bound. The explanation lies in the fact that (3.20) and the subsequent definition of efficiency refer to the variance of the asymptotic distribution, while the Cramér-Rao lower bound applies to the variance of the estimator, hence to the limit of this variance. In the overwhelming majority of cases, these two limit quantities coincide, but the super-efficiency phenomenon shows the existence of exceptions. For an extended discussion of this point, see Lehmann (1983, section 6.1).   □

3.3.5 Change of parametrization

The MLE property of best asymptotic normality carries over to transformations of the parameter. More specifically, if θ ($\theta \in \mathbb{R}^k$) is reparametrized into $\psi = \psi(\theta)$, where $\psi(\cdot)$ is a one-to-one transformation such that each component $\psi_j(\cdot)$ is a differentiable function, then

$$\sqrt{n}(\hat{\psi} - \psi_*) \overset{d}{\longrightarrow} N_k(0, (\Delta^\top i(\theta_*)\Delta)^{-1}), \qquad (3.21)$$

where Δ is the matrix of the partial derivatives, as defined shortly after (3.11), and $\hat{\psi}$ is best asymptotically normal among the estimators of ψ.

Although both $\hat{\psi}$ and $\hat{\theta}$ are asymptotically normal, the speed of convergence to the asymptotic distribution does not need to be the same. Therefore, an appropriate choice of parametrization can substantially improve this speed of convergence The importance of this phenomenon will be apparent in Chapter 4.

Example 3.3.5 Once more, consider an s r s of size n from an $N(\mu, \sigma^2)$ random variable. Using the expression for $I(\mu, \sigma^2)$ obtained in Example 3.2.3, it follows that

$$\sqrt{n}(\hat{\sigma}^2 - \sigma^2) \overset{d}{\longrightarrow} N(0, 2\sigma^4)$$

as $n \to \infty$, leading to the approximation for fixed n

$$\hat{\sigma}^2 \rightsquigarrow N(\sigma^2, 2\sigma^4/n).$$

Since the exact distribution of $\hat{\sigma}^2$ is known to be $\sigma^2 \chi_{n-1}^2/(n-1)$

from the results of section A.5.7, it is possible to appreciate the closeness of the approximation to the exact distribution Figure 3.3(a) shows a graph of the two curves for a specific choice of n and σ^2.

Consider in addition the reparametrization $\psi = \psi(\sigma^2) = \sqrt{\sigma^2}$, whose MLE is $\psi(\hat{\sigma}^2) = \sqrt{\hat{\sigma}^2}$, with asymptotic distribution such that

$$\sqrt{n}(\hat{\psi} - \psi) \xrightarrow{d} N(0, \tfrac{1}{2}\psi^2)$$

using (3.21) or Corollary A.8.10; hence, for fixed n, the approximate distribution is

$$\hat{\psi} \rightsquigarrow N\left(\psi, \frac{\psi^2}{2n}\right).$$

Again, the exact distribution of $\hat{\psi}$ is available, by transforming the exact distribution of $\hat{\sigma}^2$. Figure 3.3(b) provides a graphical comparison between these two densities.

It is apparent that the approximation to the exact distribution provided by asymptotic theory works appreciably better when ψ is used as parameter, and the improvement is more relevant in the tails of the distribution, where the closeness of approximation is more relevant for the purposes of the methods presented in Chapter 4.

The present example has no practical relevance, since the exact distribution of the estimator is known, but it illustrates that an appropriate choice of parametrization can be very advantageous in those cases where the exact distribution of $\hat{\theta}$ is unknown. The problem is to find the appropriate parametrization. □

3.3.6 Maximum likelihood and exponential families

Consider the special case of a sample y from a random variable Y belonging to a regular exponential family of order 1. Then

$$L(\theta) = \exp\{\psi(\theta)\,t(y) - \tau(\theta)\}$$

and the likelihood equation is

$$\psi'(\theta)\,t(y) = \tau'(\theta).$$

Then $\hat{\theta}$, if it exists, is a value satisfying the equality

$$\mathrm{E}\left[t(Y); \hat{\theta}\right] = t(y), \qquad (3.22)$$

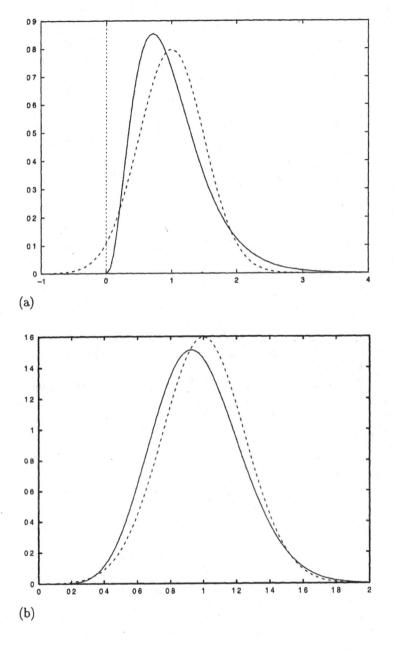

Figure 3 3 *Exact (continuous line) and approximate (dashed line) density of the MLE using two alternative parametrizations: (a) exact and approximate density of $\hat{\sigma}^2$ when $\sigma^2 = 1$, $n = 8$; (b) exact and approximate density of $\hat{\psi}$ when $\psi = 1$, $n = 8$*

taking into account (2.7). In other words, $\hat{\theta}$ is the value of θ equating the mean value of the sufficient statistic to the sample value of this statistic.

As for the existence of a solution to (3.22), it can be shown that the equation has a solution if and only if $t(y)$ is an interior point of the convex hull of the support of $t(Y)$. This condition is relevant only for discrete random variables, since otherwise it is satisfied with probability 1.

Differentiating the log-likelihood twice, we obtain

$$\ell''(\theta) = \psi''(\theta)t(y) - \tau''(\theta),$$

and the expected information is

$$I(\theta) = \mathrm{E}[-\ell''(\theta;Y);\theta] = -\psi'' \mathrm{E}[t(Y);\theta] + \tau''(\theta).$$

Using (3.22), notice that

$$-\ell''(\hat{\theta}) = I(\theta)\Big|_{\theta=\hat{\theta}} \qquad (3.23)$$

which implies $\ell''(\hat{\theta}) \le 0$, and in non-degenerate cases $\ell''(\hat{\theta}) < 0$ Since every solution to (3.22) is a maximum point of the likelihood, it follows that the solution is unique.

These results can be extended to regular exponential families of order k. This explains in particular the statement about the uniqueness of the solution in Example 3.1.12.

Example 3.3.6 The sufficient statistic for the logistic regression model of Example 2.4.4 is the pair $(\sum y_i, \sum x_i y_i)$ with expected value $(\sum \pi_i, \sum x_i \pi_i)$, where

$$\pi_i = \frac{\exp(\alpha + \beta x_i)}{1 + \exp(\alpha + \beta x_i)} \qquad (i = 1, \dots, n).$$

Then the likelihood equations are

$$\sum_i y_i = \sum_i \frac{\exp(\alpha + \beta x_i)}{1 + \exp(\alpha + \beta x_i)},$$

$$\sum_i x_i y_i = \sum_i x_i \frac{\exp(\alpha + \beta x_i)}{1 + \exp(\alpha + \beta x_i)},$$

with exactly one solution for α and β, due to the above-mentioned results, unless $t(y)$ lies on the boundary set of the convex hull of the support (see Exercise 3.18). □

3.3.7 Conditionality principle

Suppose that a minimal sufficient statistic s can be split into two components, $s = (t, a)$ say, where a is a statistic with distribution not depending on θ; a is then called an **ancillary statistic**. By itself, a cannot provide any information about θ; on the contrary, its intrinsic variability is a nuisance to inference. This remark motivates the **conditionality principle**, which stipulates that inference should be based not on the distribution of S, but on the distribution of T conditionally on the observed value a of the random variable A, i.e. one must use the likelihood

$$L_a(\theta) = c(t)\, f_{T|A}(t|a; \theta), \tag{3.24}$$

where the final term is the distribution of T conditional on $A = a$.

Actually, the likelihood $L_a(\theta)$ is no different from the usual likelihood, since $L(\theta)$ is obtained multiplying (3.24) by the density function of A, $f_A(a)$ say, but this quantity does not depend on θ; hence $L_a(\theta) \propto L(\theta)$. Therefore, until one operates within the likelihood principle, it makes no difference whether we use $L_a(\theta)$ or $L(\theta)$; in particular, the MLE is the same. What changes is the evaluation of $\hat\theta$, or any other estimator (and, more generally, any inferential method), from the point of view of the repeated sampling principle. The following examples are intended to illustrate these points.

Example 3.3.7 (Cox, 1958) A fair coin is tossed, and the value $a = 0$ is attached to the outcome 'heads', $a = 1$ to 'tails'. From an $N(\theta, 1)$ random variable, 10^{2a} independent observations are then taken, with sample mean $\bar{y}$. The probability distribution of the pair $(\bar{Y}, A)$ is

$$\frac{1}{2} \times \frac{1}{\sqrt{2\pi}\, 10^{-a}} \exp\left\{ -\frac{1}{2}\left(\frac{\bar{y} - \theta}{10^{-a}}\right)^2 \right\}.$$

Clearly, $(\bar{y}, a)$ is a minimal sufficient statistic for θ, and a is ancillary. The MLE of θ is $\bar{y}$, with

$$\mathrm{var}\left[\bar{Y}|a\right] = 10^{-2a}, \quad \mathrm{var}\left[\bar{Y}\right] = \frac{1}{2} \times \frac{1}{100} + \frac{1}{2} \times 1 = \frac{1}{2}\left(\frac{101}{100}\right).$$

The first of these expressions provides the variance of the observation actually performed, while the second gives the variance to be expected a priori, before tossing the coin.

This apparently weird example presents schematically a situation where the experimenter cannot control a priori the amount of information provided by the experiment. In this case, $\mathrm{var}\left[\bar{Y}\right]$ is the appropriate quantity for comparing the precision of this experimental design with some other design, while $\mathrm{var}\left[\bar{Y}|a\right]$ measures a posteriori the precision actually achieved.   □

Example 3.3.8 For the cases considered in Examples 2.3.14 and 3.1.5, the minimal sufficient statistic is $(y_{(1)}, y_{(n)})$, with density function at (x, y) given by

$$f(x, y) = \frac{n(n-1)}{\theta^n}(y - x)^{n-2} \qquad (\theta < x < y < 2\theta)$$

using (A.32) An equivalent representation of this statistic is (t, a), where $t = y_{(n)}$, $a = y_{(n)}/y_{(1)}$. The corresponding density function of (T, A), at the point (t, a), is given by

$$g(t, a) = \frac{n(n-1)}{\theta^n} \frac{t^{n-1}(b - 1)^{n-2}}{a^n} \qquad (1 < a < 2, a\theta < t < 2\theta).$$

Then the marginal density function of A and the conditional density function of T, given $A = a$, are

$$\begin{aligned} g_A(a) &= (n - 1)(a - 1)^{n-2}(2^n - a^n)/a^2 & (1 < a < 2), \\ g_{T|a}(t) &= \frac{n\, t^{n-1}}{\theta^n\, (2^n - a^n)} & (a\theta < t < 2\theta), \end{aligned}$$

respectively; hence a is an ancillary statistic The likelihood associated with (3.24) is

$$L_a(\theta) = \frac{c(a, t)}{\theta^n} \qquad (t/2 < \theta < t/a).$$

The MLE of θ obtained by maximizing $L_a(\theta)$ is $\hat{\theta} = t/2$, exactly the same result as in Example 3.1.5 using the unconditional likelihood, but the variance of the estimator is different. In fact, after some algebra, one obtains

$$\mathrm{var}\left[\hat{\theta}\right] = \frac{n\theta^2}{4(n+1)^2(n+2)},$$

$$\mathrm{var}\left[\hat{\theta}|a\right] = \mathrm{var}\left[\hat{\theta}\right] \frac{(2^{n+1} - a^{n+1})^2 - (n+1)^2(2 - a)^2 (2a)^n}{(2^n - a^n)^2}.$$

The dependence of $\mathrm{var}\left[\hat{\theta}|a\right]$ on a reflects the fact that a is close to 2 when $y_{(1)}$ and $y_{(n)}$ are close to the boundaries of the interval $(\theta, 2\theta)$, and that if a is close to 1 the individual values of $y_{(1)}$

and $y_{(n)}$ are more vaguely located. In particular, it can be shown that $\mathrm{var}\left[\hat{\theta}|a\right] \to 0$ when $a \to 2$, as is reasonable, since $a = 2$ implies $\hat{\theta} = \theta$. Instead, $\mathrm{var}\left[\hat{\theta}\right]$ does not take into consideration the information contained in a.   □

Use of $L_a(\theta)$ instead of $L(\theta)$ allows a more accurate evaluation of the 'quality' of the estimator, since the conditioning operation has the effect of evaluating the variability of the estimator within a subset of the sample space in some sense similar to the observed sample, namely a subset of points sharing the same value of the ancillary statistic.

While there is a general consensus about the conditionality principle, there are some difficulties in its implementation. Sometimes the ancillary statistic on which to condition is not unique, and clearly the results are not the same for the two choices. Although examples of this kind are rare, the fact that they can occur is irritating from the theoretical viewpoint. Far more often, however, there is no known ancillary statistic to use. Methods for constructing *approximate* ancillary statistics have thus been searched for; see, for instance, Barndorff-Nielsen and Cox (1994, Chapter 7) for a systematic development.

3.3.8 The observed information again

Once the MLE has been found, an evaluation of its 'quality' is required, to be able to form some idea about the plausible deviation between the estimate and the true parameter. Since the asymptotic variance of $\hat{\theta}$ is $I(\theta_*)^{-1}$, one possibility is to use $I(\hat{\theta})^{-1}$. At least for the case discussed in section 3.3.3, $i(\hat{\theta})$ is a consistent estimate of $i(\theta_*)$, by Theorem A.8.2(e), if $i(\theta)$ is a continuous function of θ; hence $I(\hat{\theta})$ is a reasonable estimate of $I(\theta_*)$.

Alternatively, the observed information

$$\mathcal{I}(\hat{\theta}) = -\ell''(\hat{\theta}) = -\frac{d^2}{d\theta^2}\ell(\theta)\bigg|_{\theta=\hat{\theta}}$$

can be used. At least in the case of an s r s , the observed information is also a plausible estimate of $I(\theta)$, as is easy to see on writing

$$-\frac{1}{n}\frac{d^2}{d\theta^2}\ell(\theta) = \frac{1}{n}\sum_{i=1}^{n}\left(-\frac{d^2}{d\theta^2}\ell_i(\theta)\right)$$

where $\ell_i(\theta)$ is the contribution to the log-likelihood from the ith observation. The term on the right-hand side converges a.s. to the expected value of the generic term, $i(\theta)$, by the strong law of large numbers.

So the question arises: which of $I(\hat{\theta})$ and $\mathcal{I}(\hat{\theta})$ is better? Asymptotically, the problem vanishes, and it does not even exist for regular exponential families, bearing in mind (3.23); but in other cases the two expressions are not equivalent.

It is possible to present reasons for preferring the observed information, which provides a measure of the information actually present in the sample, while $I(\theta)$ provides an sort of average value over the whole sample space; a detailed discussion is given by Efron and Hinkley (1978). Notice the connection with the conditionality principle. The observed information falls within the strong likelihood principle, while the same is not true for $I(\hat{\theta})$.

Example 3.3.9 Suppose that an s r s $(y_1, \ldots, y_n)$ from a exponential distribution with expected value $1/\theta$ contains some censored data with fixed censoring time equal to C, supposed to be known. This kind of censoring is a special case of the general type considered in Example 2.2.5, obtained when the distribution of censoring time $G(\cdot)$ is degenerate at point C. Then, using (2.3), the likelihood is

$$L(\theta) = c \prod_{i=1}^{n} \theta^{z_i} \exp(-\theta\, y_i)$$

leading to

$$\mathcal{I}(\hat{\theta}) = \frac{R}{\hat{\theta}^2}, \qquad I(\hat{\theta}) = \left.\frac{\mathrm{E}[\,R\,]}{\theta^2}\right|_{\theta=\hat{\theta}}$$

where $R = \sum_i z_i$ is the number of sample elements not censored. The only difference between these two quantities is in the numerator: one has R, the actual number of uncensored sample elements, while $I(\hat{\theta})$ has

$$\mathrm{E}[\,R\,]\big|_{\theta=\hat{\theta}} = n\{1 - \exp(-C\hat{\theta})\},$$

an estimate of $\mathrm{E}[\,R\,]$ evaluated over the whole sample space.   □

3.4 Some numerical examples

Example 3.4.1 The data in Table 3.4, reproduced from Hald

Table 3.4 *Distribution of the diameter of the head of 500 rivets (Data from Hald (1952) Statistical Theory with Engineering Applications, Table 6 3, 135, John Wiley, reproduced with permission)*

Diameter (class midpoint) z_i	Frequency f_i
13 07	1
13 12	4
13.17	4
13 22	18
13 27	38
13.32	56
13 37	69
13 42	96
13 47	72
13 52	68
13.57	41
13 62	18
13.67	12
13 72	2
13 77	1

(1952, p. 135), are of very simple structure and of commonly occurring form They refer to a single variable, and values are available in the form of a frequency distribution, rather than individually; this means that a partition of the real axis is assigned and the frequency of observations falling in each interval is given.

Specifically, the data refer to the diameter in millimetres of the head of $n = 500$ rivets, divided into $k = 15$ intervals, each of width $h = 0.05$ mm. The frequencies reported in the table refer to the number of measurements falling in the interval centred on the corresponding value in the first column.

The batch of 500 rivets can be regarded as a simple random sample from a probability distribution. We assume that this underlying distribution is of normal type; in the present case, this choice is made without further discussion apart from noticing that similar kinds of measurements often exhibit good empirical agreement with a normal distribution.

A simple and important method for representing graphically

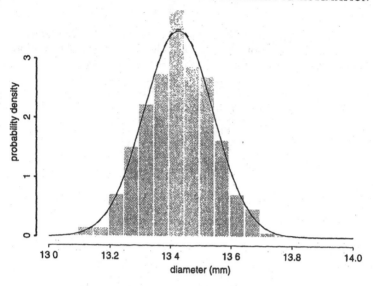

Figure 3.4 *Histogram of the rivets data and fitted normal curves by approximate (continuous line) and exact (dashed line) MLE*

data of tabular form is the **histogram**, obtained by drawing for each class a rectangle with height equal to $f_i/(n\,h)$, which constitutes a simple estimate of the density near the midpoint z_i $(i = 1, \ldots, k)$. The histogram for the data in Table 3.4 is plotted in Figure 3.4, and its shape corroborates the normality assumption. The meaning of the continuous curves will be explained shortly.

To estimate the parameters of the underlying $N(\mu, \sigma^2)$ distribution, one simple device is to impute all the data in a class to its midpoint, leading to the weighted sample mean and variance

$$\tilde{\mu} = \frac{\sum_{i=1}^{k} z_i f_i}{n}, \qquad \tilde{\sigma}^2 = \frac{\sum_{i=1}^{k} (z_i - \tilde{\mu})^2 f_i}{n},$$

whose values for the given data are $\tilde{\mu} = 13.4264$ and $\tilde{\sigma}^2 = 0.013149$. The normal density curve corresponding to these values is superimposed on the histogram in Figure 3 4.

While the above method is extremely simple and adequate in many practical cases (as we shall soon see for the present data set), it does not provide the exact MLE, however. To obtain the latter estimate, we need to consider the probability of observing

the frequency distribution of Table 3.4 as a function of $\theta = (\mu, \sigma^2)$
Since we are modelling the probability of a certain number of successes for a certain set of events, the multinomial distribution is appropriate, and the resulting likelihood is

$$L(\theta) = c \prod_{i=1}^{k} p_i(\theta)^{f_i}, \qquad (3.25)$$

where

$$p_i(\theta) = \Phi\left(\frac{z_i + h/2 - \mu}{\sigma}\right) - \Phi\left(\frac{z_i - h/2 - \mu}{\sigma}\right),$$

for $i = 1, \ldots, k$, gives the probability that an observation falls in the ith class.

Notice that (3.25) provides the expression for the likelihood in all cases where data are grouped, as in the case of Table 3.4; the only adjustment required is in the expression for the p_i as a function of the parameters, if a different parametric family is considered instead of the normal one.

Numerical maximization of (3 25) leads to the MLE $\hat{\mu} = 13.4264$ and $\hat{\sigma}^2 = 0.012941$, identical to the previous estimate in respect of the first component, and nearly so for the second. In fact, the corresponding normal density, represented by the dashed curve in Figure 3.4, is hardly distinguishable from the previous one.

We have obtained the MLE through direct maximization of $\log L(\theta)$, but it is interesting to note that, for these data, the exact MLE for σ^2 is virtually the same as would be obtained using **Sheppard's correction for continuity**,

$$\tilde{\sigma}^2 - \frac{h^2}{12},$$

an old method used for correcting the variance when data are grouped in classes; see, for instance, Cramér (1946, pp. 359–363) for details and further developments. □

Example 3.4.2 The data in Table 3.5 are reproduced from Patil (1962), who attributes them to Karl Pearson in his studies on albinism. For a sample of 60 families, each with 5 children, the table gives the number f_k of families with exactly k albinotic children, for $k = 1, \ldots, 5$.

If we denote by π the probability that a child is albinotic in one of the families described above, then the distribution of the number

Table 3.5 *Number of albinotic children in families with five children (Data from* Biometrika, **49**, *231, reproduced with the permission of the* Biometrika *trustees.)*

number of albinos (k)	1	2	3	4	5
frequency (f_k)	25	23	10	1	1

of albinos in one family is a truncated binomial, of the form

$$\Pr\{Y_1 = k\} = \frac{\binom{m}{k}\pi^k(1-\pi)^{m-k}}{1-(1-\pi)^m}, \qquad k = 1, \ldots, m,$$

where $m = 5$, and the corresponding log-likelihood for the parameter π is

$$\ell(\pi) = c + \sum_{k=1}^{m} f_k\left[k\log\pi + (m-k)\log(1-\pi) - \log\{1-(1-\pi)^m\}\right]$$

$$= c + n\left[\bar{y}\log\pi + (m-\bar{y})\log(1-\pi) - \log\{1-(1-\pi)^m\}\right]$$

where

$$n = \sum_k f_k, \qquad \bar{y} = \sum_k \frac{k f_k}{n},$$

and the latter quantity is the sufficient statistic for $\ell(\pi)$. Equating to 0 the score function

$$\ell'(\pi) = n\left(\frac{\bar{y}}{\pi} - \frac{m-\bar{y}}{1-\pi} - \frac{m(1-\pi)^{m-1}}{1-(1-\pi)^m}\right)$$

it emerges, after some algebra, that the MLE $\hat{\pi}$ satisfies the equation

$$\hat{\pi} = \frac{\bar{y}}{m}\left(1 - (1-\hat{\pi})^m\right)$$

which must be solved numerically. The particular form of this equation allows the use of a very simple algorithm, sometimes called **repeated substitution**, which works by computing the right-hand side of the equation with the current approximation for $\hat{\pi}$, obtaining the next value of $\hat{\pi}$ (as indicated by the left-hand side of the equation), and repeating these steps until convergence. A reasonable initial value for $\hat{\pi}$ is $\bar{y}/m$, as it is easy to check that

$$\mathrm{E}[Y_1] = \frac{m\pi}{1-(1-\pi)^m},$$

and the denominator can temporarily be ignored if π and m are not very small. For the data in Table 3.5, $\bar{y}/m = 0.3667$ and, after about ten iterations of the repeated substitution method, we obtain $\hat{\pi} = 0.3088$. Alternatively, the Newton–Raphson algorithm could be used.

To compute the information in the sample, it is convenient to use the expression

$$
\begin{aligned}
I(\pi) &= \operatorname{var}[\,\ell'(\pi)\,] \\
&= n^2 \operatorname{var}[\,\bar{Y}\,] \left(\frac{1}{\pi} + \frac{1}{1-\pi} \right)^2 \\
&= n \operatorname{var}[\,Y_1\,] \left(\frac{1}{\pi(1-\pi)} \right)^2 ,
\end{aligned}
$$

where $\operatorname{var}[\,Y_1\,]$ must be obtained by direct computation from the probability distribution given above, which leads to

$$
\operatorname{var}[\,Y_1\,] = \frac{(m\pi)^2 + m\pi(1-\pi)}{1-(1-\pi)^m} - \left(\frac{m\pi}{1-(1-\pi)^m} \right)^2 .
$$

Substitution of $\hat{\pi}$ in $I(\pi)$ gives the value 972.59, whose reciprocal 0.0010304 gives an estimate of $\operatorname{var}[\,\hat{\pi}\,]$. □

Example 3.4.3 Population genetics is a field where probability models and statistical methods are widely used, due to the inherent unpredictability of gene combination when mating occurs. A few introductory genetic concepts and some simple applications of probability methods are presented by Feller (1968, pp. 132ff.); for a more systematic account, see, for instance, Sbr, Owen and Edgar (1965).

The following is a very simple, but in a sense typical, example of stochastic models involved in genetics. Individual characteristics are controlled by genes; these are entities capable of assuming one possible form out of a small number of given alternatives, called alleles. For instance, the blood group of an individual is controlled by a gene with three alleles, denoted by A, B, O. Since alleles appear in pairs, each individual belongs to one of six possible types, called genotypes: AA, AO, BB, BO, AB, OO; here genotypes of the form AO and OA are regarded as equivalent. An individual's membership of a genotype is not directly observable, however, since both forms AA and AO appear to belong to a single class (forming a phenotype), in this case the 'blood group A'; this confounding of the two genotypes is related to the dominant nature of the A allele over the recessive O. There are therefore four observable blood groups: A, B, AB, O.

Denote by p, q and r the proportions of the three alleles A, B, O in a given population (so that $p + q + r = 1$). Combination of two genes to form a pair takes place under mutual independence; hence the genotype AA occurs with probability p^2, and similarly for the other genotypes. The situations is summarized in the following table.

genotype	AA	AO	BB	BO	AB	OO
probability	p^2	$2pr$	q^2	$2qr$	$2pq$	r^2
blood group	A	A	B	B	AB	O

It is of interest to estimate the overall proportions p, q, r of the three type of alleles in the population. Given a sample of individuals, we cannot directly observe the frequencies of occurrence of the three alleles, for the reasons explained, and we must resort to a form of indirect count based on the blood types. To be specific, suppose now that a sample of $n = 345$ subjects has been classified according to their blood group, obtaining the following frequency table.

blood group	A	B	AB	O
frequency	150	29	6	160
	(n_A)	(n_B)	(n_{AB})	(n_O)

The data can then be thought of as being generated by a four-cell multinomial distribution with probabilities

$$p^2 + 2pr, \quad q^2 + 2qr, \quad 2pq, \quad r^2.$$

The corresponding log-likelihood function is

$$
\ell(p, q) = n_A \log(p^2 + 2pr) + n_B \log(q^2 + 2qr)
$$
$$
+ n_{AB} \log(2pq) + 2n_O \log(1 - p - q)
$$

where $r = 1 - p - q$, $p \in (0, 1)$, $q \in (0, 1)$, $p + q < 1$. A plot of this function using contour levels is given in Figure 3.5

It appears from the plot that the maximum is not far from $p = 0.3$, $q = 0.1$. For the precise detection of the MLE, we use the Newton–Raphson algorithm, which requires

$$
\frac{\partial \ell}{\partial p} = (n_A + n_{AB})/p - n_A/\bar{p} - 2n_B/\bar{q} - 2n_O/r,
$$

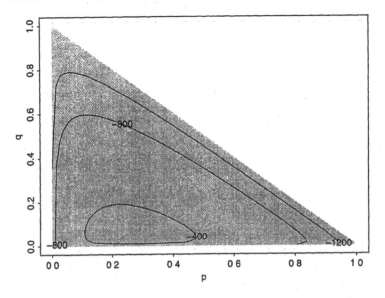

Figure 3.5 *Log-likelihood function for the blood groups example*

$$\frac{\partial \ell}{\partial q} = (n_B + n_{AB})/q - 2n_A/\bar{p} - n_B/\bar{q} - 2n_O/r$$

and

$$-\frac{\partial^2 \ell}{\partial p^2} = (n_A + n_{AB})/p^2 + n_A/\bar{p}^2 + 4n_B/\bar{q}^2 + 2n_O/r^2,$$

$$-\frac{\partial^2 \ell}{\partial q^2} = (n_B + n_{AB})/q^2 + 4n_A/\bar{p}^2 + n_B/\bar{q}^2 + 2n_O/r^2,$$

$$-\frac{\partial^2 \ell}{\partial p \partial q} = 2n_A/\bar{p}^2 + 2n_B/\bar{q}^2 + 2n_O/r^2,$$

where

$$\bar{p} = 2r + p, \qquad \bar{q} = 2r + q.$$

To initialize the Newton–Raphson algorithm we could simply choose $p = 0.3$, $q = 0.1$, as suggested by Figure 3.5. However, it is instructive to devise a simple alternative method which does not require plotting the log-likelihood over its full range, as this is not always possible. A very simple rule is as follows. Estimate r by $\sqrt{160/345}$, and then $p + q = 1 - r \approx 0.32$; since there are about five more A cases than B cases, choose initially $p = 0.32 \times 5/6 =$

Table 3.6 *Newton–Raphson for blood groups example*

Cycle	p	q	$\partial\ell/\partial p$	$\partial\ell/\partial q$
0	0.27000	0.05000	-25.97	24.80
1	0.26057	0.05226	0.67	1.31
2	0.26076	0.05226	$\approx 2 \times 10^{-4}$	$\approx 5 \times 10^{-3}$

0.27, $q = 0.32/6 = 0.05$. The first two iterations are reported in Table 3.6.

The observed information matrix and its inverse, computed at the final $(\hat{p}, \hat{q})$ point, are

$$\mathcal{I}(\hat{p}, \hat{q}) = \begin{pmatrix} 3085.4 & 818.8 \\ 818.8 & 13772.8 \end{pmatrix},$$

$$\mathcal{I}(\hat{p}, \hat{q})^{-1} = 10^{-4} \begin{pmatrix} 3.2930 & -0.1958 \\ -0.1958 & 0.7377 \end{pmatrix}.$$

Notice that, for the same data, a different choice of the initial values can cause the algorithm to fall outside the feasible region; this happens in particular to the uniform distribution $p = q = r = 1/3$.

$\square$

Exercises

3.1 For an s r s $(y_1, \ldots, y_n)$ from a *Gamma*(ω, λ) random variable, estimate the parameters using the method of moments.

3.2 Are the estimators constructed using the method of moments equivariant?

3.3 In the setting of Example 3.1.3, obtain the MLE when $\theta > 0$

3.4 Let $(y_1, \ldots, y_N)$ be an s r s from a Poisson random variable with mean value θ $(\theta > 0)$.

    (a) Obtain the MLE of θ.

    (b) Supposing that only the first n $(n < N)$ elements of the sample are known explicitly, while for the other $N - n$ only their sum, t say, is known, determine the MLE of θ.

    (c) Compare the answers to (a) and (b), and comment.

3 5 For an s r s from an $N(\mu, \sigma^2)$ random variable, obtain the MLE of (μ, σ^2) under the constraint $a < \mu < b$, where a, b are known constants.

3.6 For the setting of Example 3.1.6,

(a) state whether the likelihood belongs to an exponential family;

(b) verify that the expression for $\ell(N)$ is correct integer and $N \geq M + n - m$;

(c) check the statement in the text that $\ell(N)$ is increasing to the left of $M\,n/m$, decreasing to the right.

3.7 For the autoregressive process of Examples 2.4.5 and 3 1 11, obtain the expected information of (ρ, σ^2). [Notice the lack of additivity of the information in this case.]

3.8 Let $(y_1, \ldots, y_n)$ be an s r s such that the generic component is sampled from a random variable Y_1 with geometric distribution of the form

$$\Pr\{Y_1 = y; \theta\} = \begin{cases} \left(\dfrac{\theta}{1+\theta}\right)^y \dfrac{1}{1+\theta} & \text{if } y = 0, 1, \ldots, \\ 0 & \text{otherwise,} \end{cases}$$

for some positive θ, and let $(x_1, \ldots, x_m)$ be an s r s from a Poisson distribution with mean θ, independent of the first sample.

(a) Write down the likelihood equation for θ, given the joint sample of $n + m$ elements.

(b) Show that this equation has one and only one admissible solution, except in the degenerate case.

3.9 The discrete random variable Y_1 has probability distribution

$$\Pr\{Y_1 = r\} = \frac{\theta^r a_r}{f(\theta)} \qquad (r = 0, 1, \ldots),$$

which is called **power series distribution**; here θ is a positive parameter, and $\{a_r\}$ is a sequence of non-negative constants independent of θ.

(a) Is this an exponential family distribution?

(b) State whether or not the binomial, geometric and Poisson distributions are of this type.

(c) Show that f is indefinitely differentiable.

(d) Obtain $E[Y_1]$ and $\text{var}[Y_1]$.

(e) For an s r s $(y_1, \ldots, y_n)$ from Y_1 with $\bar{y} = \sum y_i/n$, show that the MLE of θ satisfies the equation

$$\bar{y} = \frac{\hat{\theta} f'(\hat{\theta})}{f(\hat{\theta})}.$$

(f) Show that the expected information is $n\sigma^2(\theta)/\theta^2$, where $\sigma^2(\theta) = \text{var}[Y_1]$.

3.10 Prove (3.10) and extend it to the multiparameter case (3.11).

3.11 Show that, for an s r s $(y_1, \ldots, y_n)$ from a beta random variable (A.15), the sufficient statistic for (p, q) is (t_1, t_2), where

$$t_1 = \left(\prod_{i=1}^{n} y_i\right)^{1/n}, \quad t_2 = \left(\prod_{i=1}^{n}(1 - y_i)\right)^{1/n}$$

are the geometric means of the y_i and the $(1 - y_i)$, respectively. Write down the likelihood equations for (p, q)

3.12 The s r s $(y_1, \ldots, y_n)$ has been taken from a random variable with density function given by

$$f(y; \theta) = k(\theta)\, g(y)\, I_{[0,\infty)}(y - \theta),$$

where $\theta > 0$ and $g(\cdot)$ is a non-negative function integrable over the real line.

(a) Find the minimal sufficient statistic for θ

(b) Show that $k(\theta)$ is a non-decreasing function

(c) Obtain the MLE of θ.

3 13 Prove (3.21).

3.14 For an s r s from a random variable with density function $f_4(y - \theta)$ obtained by setting $r = 4$ in Exercise A.4, and θ a real number,

(a) obtain the sufficient statistic for θ;

(b) write down the corresponding likelihood equation;

(c) discuss whether or not this equation has a unique solution.

3.15 An s r s from a bivariate normal random variable is formed by n pairs (x_i, y_i) for $i = 1, \ldots, n$. Obtain the MLE of the five parameters involved (expected values, variances and correlation coefficient), and their asymptotic variances.

3.16 From an s r s from an $N(\mu, \sigma^2)$ random variable with both parameters unknown, obtain an unbiased estimator for σ.

3.17 Obtain the four expressions for the curves plotted in Figure 3.3.

3.18 For the logistic regression model of Examples 2.4.4 and 3.3.6, suppose $x = (1, 2, 3, 4, 5)^\top$. Examine numerically and compare the different behaviours of the likelihood equations for (α, β) when $y = (0, 0, 1, 0, 1)^\top$ and $y = (0, 0, 0, 1, 1)^\top$.

   [Hard] Establish general conditions under which the likelihood equations have a solution.

3 19 A certain kind of plant cell may appear in any of four versions. According to a genetic theory, the four versions have the following probabilities of appearance:

$$1/2 + \theta/4, \quad (1 - \theta)/4, \quad (1 - \theta)/4, \quad \theta/4,$$

where θ is a parameter not specified by the genetic theory $(0 < \theta < 1)$. If a sample of n cells had the observed frequencies (n_0, n_1, n_2, n_3) where $n = n_0 + n_1 + n_2 + n_3$,

   (a) write down the minimal sufficient statistic for θ,

   (b) obtain the MLE of θ, and an estimate of $\mathrm{var}\left[\hat{\theta}\right]$.

CHAPTER 4

Hypothesis testing

4.1 General aspects

4.1.1 Statement of the problem

Besides obtaining point estimates of parameters θ of interest, statistical inference deals with other questions. In particular, we are often interested in deciding whether the data are compatible with a given reference value of θ. This problem was mentioned in Chapter 1, where it was tackled by using the function

$$\tilde{L}(\theta) = \frac{L(\theta)}{L(\hat{\theta})} \qquad (4.1)$$

called the **relative likelihood** function. Ignoring the cases when $L(\theta)$ is unbounded, the relative likelihood falls in the interval $[0, 1]$. Since the denominator of (4.1) depends on the data only, it plays the role of $1/c(y)$ in the definition of likelihood (2.2); therefore, (4.1) is just a special version of the likelihood.

Denoting by θ_0 the values of θ specified by the hypothesis, the quantity $\tilde{L}(\theta_0)$ is an index of the agreement between data and hypothesis; if $\tilde{L}(\theta_0) = 1$, there is perfect agreement (assuming that the statistical model is appropriate); on the other hand, if $\tilde{L}(\theta_0) = 0$, the hypothesis appears completely unsatisfactory.

Since the observed value of $\tilde{L}(\theta_0)$ usually lies somewhere between 0 and 1, we have the problem of choosing a general strategy. As in Chapter 1, one way out is to select a **critical value**, r say, such that when $\tilde{L}(\theta_0) \leq r$ the degree of agreement between data and hypothesis is too low, and we must then declare that the conjectured value θ_0 is inadequate, i e we *reject* the hypothesis; otherwise, if $\tilde{L}(\theta_0) > r$, the hypothesis is *accepted*.

However, one could object that the introduction of such a crude accept/reject rule, is inappropriate in practical terms and an oversimplification on general grounds. We take the view that this approach to the problem allows us to analyse the formal properties of

the associated procedures, but their practical usage will not necessarily match the formal decision-theoretic rule, since actual 'decisions' may not necessarily be so dichotomous, and they may also
be influenced by additional considerations, such as the outcome of
similar studies and physical plausibility

Even accepting the present view, there still remain various issues
to be tackled

- How should the critical value be chosen? In the example considered in Chapter 1, the choice $r = 1/5$ appeared intuitively
  acceptable, but can it be justified rationally? Can it be used
  generally?

- Can this kind of strategy be evaluated from the point of view
  of the repeated sampling principle, as with the MLE?

- How can the above criterion be adapted to more complex situations, for instance when the hypothesis does not select a specific
  value θ_0 but, for instance, states that the two components of a
  two-dimensional parameter θ are equal?

The overall purpose of this chapter is attempt to answer the
above questions. To this end, our tour will be a long one, even
somewhat contorted at places. This is partly due to the historical development of the theory of statistical inference, especially
of hypothesis testing, and to the already mentioned necessity of
combining logical correctness and practical requirements. It is also
intended in the following account to stress the primary role played
in this context by the likelihood function, and to illustrate the
connection with the theory of estimation.

4.1.2 Statistical test procedures

Definition 4.1.1 *If the parameter space* Θ *of model (2.1) is partitioned into two subsets* Θ_0 *and* Θ_1 *such that* $\Theta = \Theta_0 \cup \Theta_1$, *a*
statistical test procedure *is a function* $T(y)$ *from the sample
space* $\mathcal{Y}$ *onto the set* $\{\Theta_0, \Theta_1\}$.

In plain words, a test procedure is a rule for choosing one and
only one of two alternatives on the basis of the observed data.
Often these two alternatives are summarized by the notation

$$\begin{cases} H_0 : \theta_* \in \Theta_0, \\ H_1 : \theta_* \in \Theta_1. \end{cases} \tag{4.2}$$

Statement H_0 is called the **null hypothesis**, and statement H_1 is

called the **alternative hypothesis**. Some typical specifications of
(4.2) are

$$\begin{cases} H_0 : \theta_* = 3, \\ H_1 : \theta_* \neq 3, \end{cases} \quad \begin{cases} H_0 : \theta_* \leq 8, \\ H_1 : \theta_* > 8, \end{cases} \quad \begin{cases} H_0 : \theta_{*1} = \theta_{*2}, \\ H_1 : \theta_{*1} \neq \theta_{*2}, \end{cases}$$

where in the third case θ_{*1} and θ_{*2} denote the components of a
two-dimensional vector θ_*.

A test procedure is essentially a partition of the sample space; the
elements of $\mathcal{Y}$ such that $T(y) = \Theta_0$, denoted by $\mathcal{Y}_0$, form the **acceptance region** of the null hypothesis, while the other elements
of $\mathcal{Y}$ form the **rejection region** or **critical region**, denoted by
$\mathcal{Y}_1$. If the sample value y belongs to the acceptance region, the
conclusion is that the null hypothesis is supported by the data and
we *accept* the null hypothesis, otherwise we *reject* the hypothesis.

In most cases, this partition of the sample space is induced by a
real-valued statistic, called the **test statistic**, and the two regions
are associated with an appropriate partition of the real line. For
instance, (4 1) defines a test statistic $T(y) = \tilde{L}(\theta_0; y)$, where the
notation emphasizes the dependence of the log-likelihood on the
sample value y If the critical value $r = 1/5$ is chosen, the rejection
region is

$$\mathcal{Y}_1 = \{y : T(y) \leq 1/5\},$$

and the complementary set $\mathcal{Y} \backslash \mathcal{Y}_1$ is the acceptance region; then the
test procedure essentially equates to the indicator function of the
set $\mathcal{Y}_1$. The same test statistic can be used with different critical
values to define different test procedures

Although there exists a logical distinction between test statistics
and test procedures, they are extremely closely related entities.
For this reason, it is customary to talk about 'tests', blurring the
distinction.

The selection of the null and alternative hypothesis, and thus
the partition of the parameter space, is a problem which depends
on subject-matter considerations, and so can be regarded as given,
at least in the ideal situation that the statistical model has already
been specified in advance. On the other hand, the selection of the
partition of the sample space deals with the problem of appropriate exploitation of available information; therefore, this operation
properly belongs to the realm of statistical theory.

A test procedure cannot be expected to be infallible, however.
When $\theta_* \in \Theta_0$, it can happen that $y \in \mathcal{Y}_1$, leading to selection of
the alternative hypothesis; this results in a **type I error**. Sym-

metrically, when $\theta_* \in \Theta_1$ and $y \in \mathcal{Y}_0$, so that the decision is in favour of the null hypothesis, a **type II error** results. Our target is therefore to choose the partition of the sample space which minimizes the probability of making the two types of error

4.1.3 Tests procedures with given significance level

From a strictly formal point of view, a test procedure $\mathcal{T}(y)$ is characterized by its **power function**

$$\gamma(\theta) = \Pr\{\mathcal{T}(Y) = \Theta_1; \theta\}. \qquad (4.3)$$

Sometimes the equivalent concept of the **operating characteristic function**, defined as $1 - \gamma(\theta)$, is used.

From the power function, various characteristics of the test procedure can be obtained, for instance the **(significance) level**

$$\alpha = \sup_{\theta \in \Theta_0} \gamma(\theta) \qquad (4.4)$$

which is the maximum probability of a type I error.

Example 4.1.2 A value y sampled from a random variable $Y \sim N(\theta, 1)$ is used to test the hypotheses

$$\begin{cases} H_0 : \theta_* \leq 0, \\ H_1 : \theta_* > 0 \end{cases}$$

using a test procedure with acceptance region $\{y : y < \frac{1}{2}\}$. The power function is

$$\begin{aligned} \gamma(\theta) &= \Pr\{Y > \tfrac{1}{2}; \theta\} \\ &= \Pr\{Z > \tfrac{1}{2} - \theta\} \\ &= \Phi(\theta - \tfrac{1}{2}), \end{aligned}$$

where $Z = Y - \theta \sim N(0, 1)$. In this case, the power function is a strictly increasing function of θ; moreover, $\gamma(\theta) \to 1$ when $\theta \to \infty$, and $\gamma(\theta) \to 0$ when $\theta \to -\infty$. The significance level is $\gamma(0) = \Phi(-\tfrac{1}{2}) \approx 0.31$, from (A.8) or using statistical tables. $\square$

Ideally, one would like to have $\gamma(\theta)$ as large as possible when $\theta \in \Theta_1$, and as small as possible when $\theta \in \Theta_0$. These two requirements are clearly conflicting: enlarging the acceptance region decreases the probability of type I error, but inevitably increases the probability of type II error, and the opposite effect occurs when enlarging the rejection region. The extreme situations are obtained

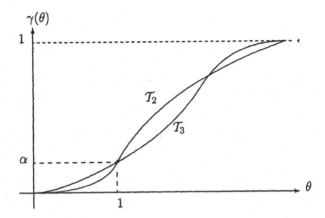

Figure 4.1 *Two intersecting power functions*

by the test procedures $T_0(y) \equiv \Theta_0$ and $T_1(y) \equiv \Theta_1$; each of these procedures is the best possible when θ belongs to a certain subset of Θ, but the worst possible otherwise

One possible way out of this dilemma is to fix the significance level and select the test procedure with highest power function for $\theta \in \Theta_1$ among those with the same significance level. This is the 'classical' approach to the problem, introduced by Jerzy Neyman and Egon S. Pearson.

Actually, even using the Neyman–Pearson approach, the problem does not need to have a unique solution Consider for instance Figure 4.1, showing the power function of two test procedures, T_2 and T_3, to test the hypotheses

$$\begin{cases} H_0 : \theta_* \leq 1, \\ H_1 : \theta_* > 1. \end{cases}$$

Both test procedures have significance level α, but neither is uniformly superior to the other. Luckily, in a number of important cases such an ambiguous situation does not occur, but in many others it does, and an additional criterion must then be introduced to choose among the various test procedures.

To tackle this sort of situation, several optimality criteria have been considered, whose suitability depends both on the practical and on the mathematical structure of the problem. The standard account of the theory is Lehmann (1986). We shall not explore optimality questions in greater detail than hitherto, although in fact the methods to be discussed do satisfy, in most cases, these additional optimality requirements.

An important point to notice is that, in this approach to the problem, the two statistical hypotheses are treated asymmetrically. This is justified by the special role usually played, in some sense, by the null hypothesis; for instance, it may be supported by some scientific theory. In view of this special role of the null hypothesis, we prefer to control the significance level, keeping low the probability of incorrect rejection of this hypothesis This point will be expanded in section 4.4.3 For a critical discussion of this and other aspects of statistical tests, see Cox (1977).

4.2 Three test statistics related to the likelihood

4.2.1 Likelihood ratio

For a random variable Y with density function $f(y; \theta)$, consider first an extremely special case, assuming that Θ has only two elements, θ_0 and θ_1 say, and work with the associated testing problem

$$\begin{cases} H_0 : \theta_* = \theta_0, \\ H_1 : \theta_* = \theta_1. \end{cases} \tag{4.5}$$

Given the observed value y, it is reasonable to base our decision on the ratio of the likelihoods

$$\lambda^* = \lambda^*(y) = \frac{L(\theta_0; y)}{L(\theta_1; y)} \tag{4.6}$$

called the **likelihood ratio**, which serves as a test statistic for our problem If λ^* is high, we shall be inclined towards H_0; if λ^* is low, then our preference will be for H_1. In the Neyman–Pearson approach, the rejection region is

$$\mathcal{Y}_L = \{y : \lambda^*(y) \le \lambda_\alpha\} \tag{4.7}$$

where the critical value λ_α is chosen so that

$$\Pr\{\lambda^*(Y) \le \lambda_\alpha; \theta_0\} = \alpha. \tag{4.8}$$

If Y and, correspondingly, $\lambda^*(Y)$ are discrete random variables, then in general there exists no λ_α which solves equation (4 8) exactly. In this case, λ_α is usually taken to be equal to the value with highest possible significance level among the achievable levels not greater than α.

The test function (4.6) and the test procedure (4.7) have been introduced for the sake of intuitive motivation only. The following statement provides a formal justification for them.

Theorem 4.2.1 (Neyman–Pearson Lemma) *For the hypotheses (4.5) on the parameter θ of the density function $f(y; \theta)$, the test procedure (4.7) has maximum power among all test procedures with significance level α, defined by (4.8).*

Proof. Denote by $\mathcal{Y}_A$ any other rejection region with significance level not greater than α, i e

$$\alpha = \int_{\mathcal{Y}_L} f(y; \theta_0)\, d\nu(y) \geq \int_{\mathcal{Y}_A} f(y; \theta_0)\, d\nu(y),$$

implying

$$\int_{\mathcal{Y}_L \setminus \mathcal{Y}_A} f(y; \theta_0)\, d\nu(y) \geq \int_{\mathcal{Y}_A \setminus \mathcal{Y}_L} f(y; \theta_0)\, d\nu(y)$$

since the integral over $\mathcal{Y}_L \cap \mathcal{Y}_A$ is common. If $y \in \mathcal{Y}_L \setminus \mathcal{Y}_A$, hence $y \in \mathcal{Y}_L$, then $f(y; \theta_1)\lambda_\alpha > f(y; \theta_0)$; on the other hand, if $y \in \mathcal{Y}_A - \mathcal{Y}_L$, we have $f(y; \theta_0) > \lambda_\alpha f(y; \theta_1)$. Therefore

$$\begin{aligned} \lambda_\alpha \int_{\mathcal{Y}_L \setminus \mathcal{Y}_A} f(y; \theta_1)\, d\nu(y) &\geq \int_{\mathcal{Y}_L \setminus \mathcal{Y}_A} f(y; \theta_0)\, d\nu(y) \\ &\geq \lambda_\alpha \int_{\mathcal{Y}_A \setminus \mathcal{Y}_L} f(y; \theta_1)\, d\nu(y), \end{aligned}$$

where the second inequality is strict unless the two regions are essentially equivalent. On dividing by λ_α and adding to both sides of the inequality the integral over $\mathcal{Y}_A \cap \mathcal{Y}_L$, we obtain

$$\int_{\mathcal{Y}_L} f(y; \theta_1)\, d\nu(y) \geq \int_{\mathcal{Y}_A} f(y; \theta_1)\, d\nu(y).$$

$\square$

Remark 4.2.2 As remarked earlier, a required level α may not be exactly achievable if Y is discrete; in that case let α' be the achievable level. The theorem shows that region (4.7) is optimal among all regions with level α'. $\square$

Remark 4.2.3 Although the above results have been presented in a parametric setting, (4.7) in fact provides the optimal test procedure for testing whether a density function (or a probability function) is equal to a certain function f_0 rather than an alternative function f_1 If suffices to rewrite the theorem with different notation.   □

4.2.2 Three test statistics related to the likelihood

Testing problem (4.5) is essentially of theoretical interest, while in practical situations we are more likely to deal with more complex cases; for instance, a frequently occurring problem is

$$\begin{cases} H_0 : \theta_* = \theta_0, \\ H_1 : \theta_* \neq \theta_0. \end{cases} \tag{4.9}$$

For this testing problem, a reasonable extension of criterion (4.6) is given by the likelihood ratio

$$\begin{aligned} \lambda(y) = \lambda &= \frac{L(\theta_0; y)}{\sup_{\theta \neq \theta_0} L(\theta; y)} \\ &= \frac{L(\theta_0; y)}{\sup_{\theta \in \Theta} L(\theta; y)} \\ &= \frac{L(\theta_0; y)}{L(\hat{\theta}; y)}, \end{aligned}$$

where we have assumed the continuity of $L(\theta)$ with respect to θ for all $y \in \mathcal{Y}$. Notice that $\lambda(y) = \tilde{L}(\theta_0)$ is the relative likelihood function (4.1) in this case.

A monotonic transformation of a test statistic together with a corresponding transformation of the critical value does not change the partition of the sample space into acceptance and rejection regions, hence it does not change the test procedure. Therefore, an equivalent test statistic is

$$W(y) = -2 \ln \lambda(y) \tag{4.10}$$

which is still called, with slight abuse of terminology, the likelihood ratio. Obviously, the interpretation of extreme values of $W(y)$ will be reversed compared to $\lambda(y)$, since the transformation is decreasing. Since $\Pr\{0 < \lambda(Y) < 1\} = 1$ in the overwhelming majority of cases, then effectively $W(y) \in (0, \infty)$.

We now wish to derive some other test statistics associated with the likelihood function and closely related to $W(y)$. Assume the same five conditions of section 3.3.3, and expand $\ell(\theta_0)$ as a Taylor series about $\hat{\theta}$, obtaining

$$
\begin{aligned}
W(y) &= -2\{\ell(\theta_0) - \ell(\hat{\theta})\} \\
&= -2\{(\theta_0 - \hat{\theta})\ell'(\hat{\theta}) + \tfrac{1}{2}(\theta_0 - \hat{\theta})^2 \ell''(\tilde{\theta})\}, \quad (4.11)
\end{aligned}
$$

where $\tilde{\theta} \in (\hat{\theta}, \theta_0)$ and $\ell'(\hat{\theta}) = 0$. Since $\hat{\theta}$ is a consistent estimate of θ_0 under H_0, then so is $\tilde{\theta}$, leading to the expansion

$$
\begin{aligned}
W(y) &= -n(\hat{\theta} - \theta_0)^2 \frac{\ell''(\theta_0)}{n} + o_p(1) \\
&= n(\hat{\theta} - \theta_0)^2 \{i(\theta_0) + o_p(1)\} + o_p(1) \quad (4.12) \\
&= n(\hat{\theta} - \theta_0)^2 i(\theta_0) + o_p(1).
\end{aligned}
$$

If $i(\theta)$ is a continuous function, then $i(\theta_0) = i(\hat{\theta}) + o_p(1)$ under H_0 Then we obtain the approximation

$$
W(y) = W_e(y) + o_p(1),
$$

where

$$
W_e(y) = n(\hat{\theta} - \theta_0)^2 i(\hat{\theta}), \quad (4.13)
$$

which is called the **Wald test statistic**. From (3.18), it follows that, under H_0,

$$
\sqrt{n}(\hat{\theta} - \theta_0) = \ell'(\theta_0)/\{\sqrt{n}\, i(\theta_0)\} + o_p(1),
$$

which, substituted into (4.12), gives another approximation to $W(y)$, namely

$$
W(y) = W_u(y) + o_p(1)
$$

where

$$
W_u(y) = \frac{\ell'(\theta_0)^2}{n\, i(\theta_0)}, \quad (4.14)
$$

which is called the **score test statistic**.

There are now three test functions, each one differing from the others by $o_p(1)$ terms. These test functions quantify three aspects of the log-likelihood functions, as illustrated by Figure 4.2:

- $W(y)$ measures the difference on the ordinate axis between the log-likelihood computed at $\hat{\theta}$ and at θ_0;

- $W_e(y)$ measures the deviation of the abscissae of $\hat{\theta}$ and θ_0, suitably normalized;

- $W_u(y)$ measures the slope of the log-likelihood at θ_0, again suitably normalized.

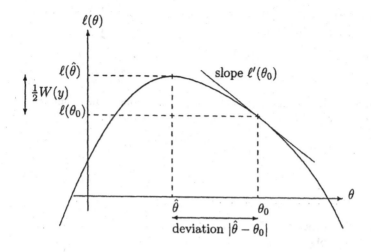

Figure 4 2 *Three test functions connected to the log-likelihood*

Example 4.2.4 Suppose that the s r s $y = (y_1, \ldots, y_n)$ from a $N(\theta, \sigma^2)$ random variable, with σ^2 known, is available for testing the hypotheses

$$\begin{cases} H_0 : \theta_* = \theta_0, \\ H_1 : \theta_* \neq \theta_0, \end{cases}$$

where θ_0 is a given value. Taking into account Example 3.2.3, the three test functions above take the form

$$\begin{aligned}
W(y) &= 2\{\ell(\hat{\theta}) - \ell(\theta_0)\} \\
&= -\sum (y_i - \hat{\theta})^2/\sigma^2 + \sum (y_i - \theta_0)^2/\sigma^2 \\
&= n(\hat{\theta} - \theta_0)^2/\sigma^2, \\
W_e(y) &= (\hat{\theta} - \theta_0)^2 I(\hat{\theta}) \\
&= n(\hat{\theta} - \theta_0)^2/\sigma^2, \\
W_u(y) &= \{\ell'(\theta_0)\}^2/I(\theta_0) \\
&= \{n(\hat{\theta} - \theta_0)/\sigma^2\}^2/(n/\sigma^2) \\
&= n(\hat{\theta} - \theta_0)^2/\sigma^2.
\end{aligned}$$

where $\hat{\theta}$ is the sample mean of the data. In this case the three test functions are exactly coincident, as also are the test procedures, at any given significance level. The critical value λ_α is obtained from

$$\Pr\left\{n(\hat{\theta} - \theta_0)^2/\sigma^2 \geq -2\ln\lambda_\alpha; \theta_0\right\} = \alpha.$$

Since, under H_0,

$$n(\hat{\theta} - \theta_0)^2/\sigma^2 \sim \chi_1^2$$

the rejection region is

$$\{y : (\hat{\theta} - \theta_0)^2 n/\sigma^2 \geq c_\alpha\},$$

where c_α is the $(1 - \alpha)$-level quantile of the χ_1^2 distribution.

Under H_1, $\mathrm{E}\left[\hat{\theta}\right] = \theta_* \neq \theta_0$ and the distribution of the test statistic is a non-central χ_1^2 with non-centrality parameter $n(\theta_* - \theta_0)^2/\sigma^2$. For computing the power function, notice that the acceptance region can be written as

$$\{y : -z_{\alpha/2} < (\hat{\theta} - \theta_0)\sqrt{n}/\sigma < z_{\alpha/2}\},$$

where $\Phi^{-1}(1 - \alpha/2) = z_{\alpha/2} = \sqrt{c_\alpha}$. Then the power function is obtained from

$$\begin{aligned}
1 - \gamma(\theta) &= \Pr\left\{-z_{\alpha/2} < \sqrt{n}\left((\hat{\theta} - \theta) + (\theta - \theta_0)\right)/\sigma < z_{\alpha/2}\right\} \\
&= \Pr\left\{-z_{\alpha/2} < Z + \delta < z_{\alpha/2}\right\}
\end{aligned}$$

where $Z \sim N(0, 1)$ and $\delta = \sqrt{n}(\theta - \theta_0)/\sigma$, and finally

$$\gamma(\theta) = 1 - \Phi(z_{\alpha/2} - \delta) + \Phi(-z_{\alpha/2} - \delta),$$

which is an increasing function of $|\delta|$. Hence $\gamma(\theta)$ increases when either n or $|\theta - \theta_0|$ increases, while it decreases for increasing σ. $\square$

Remark 4.2.5 It is most natural to ask why three different test functions W, W_e and W_u have been developed instead of just one, and whether one of the three is preferable. The answer is not so clear-cut, unfortunately.

First of all, one must bear in mind the historical development of the body of statistical theory and methodology. Some methods have been put forward independently of others, and only later have their connections been discovered. In general terms, a ranking among these three methods cannot be established, based on criteria such as the speed of convergence to the asymptotic null distribution and the power function. In some cases and in some aspects, one method is superior; in some other situations, the same method is overtaken by the others. The Wald test statistic has the disadvantage of not being invariant with respect to reparametrizations, while the other two are invariant The score test statistic can be computed without obtaining the MLE.   □

4.3 Likelihood ratio test

For the rest of this chapter, our interest will focus on $W(y)$, as this is more directly related to the likelihood function, although $W_u(y)$ and $W_e(y)$ will still figure in our discussion sometimes.

4.3.1 Definition of LRT

The likelihood ratio test (LRT) procedure for the general testing problem (4.2) at level α has rejection region

$$R = \{y : \lambda(y) \leq \lambda_\alpha\}, \tag{4.15}$$

where

$$\lambda(y) = \frac{\sup_{\theta \in \Theta_0} L(\theta; y)}{\sup_{\theta \in \Theta} L(\theta; y)} \tag{4.16}$$

and λ_α is chosen so that

$$\sup_{\theta \in \Theta_0} \Pr\{\lambda(Y) \leq \lambda_\alpha; \theta\} \leq \alpha. \tag{4.17}$$

In general, $\lambda(y) \in [0, 1]$, but in most cases $\lambda(y) \in (0, 1]$.

4.3.2 Asymptotic distribution

To obtain the critical value λ_α, or equivalently $-2 \ln \lambda_\alpha$, one should in principle compute the distribution of $W(Y) = -2 \ln \lambda(Y)$, and then its appropriate percentile. This was the path followed in Example 4.2.4; additional examples of this type will be presented in section 4.4.

In many other cases, however, exact computation of the distribution of $W(Y)$ is not feasible, and approximate methods must be employed. Even here, the chief technique for providing an approximate distribution is to compute the asymptotic distribution of the test statistic, and then use it for finite sample sizes.

In the special case considered in section 4.2.2, it follows that, under H_0,

$$W(Y) \xrightarrow{d} \chi_1^2, \tag{4.18}$$

taking into account (4 12), (3.19) and Theorems A.8.2(e) and (f) Since $W_e(Y)$ and $W_u(Y)$ differ from $W(Y)$ by $o_p(1)$ terms, they share the same asymptotic distribution, to the first order of approximation.

The use of an approximate distribution, such as (4.18), implies that the probability of exceeding the critical value is not exactly what was chosen; in other words the **actual level** is not equal to the **nominal level**. A persistent theme in the literature is the evaluation of this deviation, in various testing problems

Since, generally, we are concerned with inappropriate rejection of the null hypothesis, for the reasons explained at the end of section 4 1 3, there is a tendency to tolerate a low actual level (compared to the nominal) more readily than a high actual level, within acceptable variations. A test procedure with actual level lower than the nominal is said to be **conservative**.

To examine the behaviour of test procedure under H_1, recall that in Chapter 3 we obtained, for the case of independent and identically distributed random variables, $\ell(\theta_*) - \ell(\theta) \to \infty$ when $n \to \infty$ for all $\theta \neq \theta_*$. Since the probability that $W(Y)$ belongs to any bounded interval tends to 0, it follows that the power $\gamma(\theta)$ tends to 1 for any $\theta \neq \theta_*$ as $n \to \infty$; a test which has this property is said to be **consistent**.

A closer analysis of the distribution of $W(Y)$ under the alternative hypothesis shows that it can be approximated by a non-central χ_1^2 with non-centrality parameter $n(\theta_* - \theta_0)^2 i(\theta_0)$, which tends to infinity as $n \to \infty$, in agreement with the conclusion of the paragraph above. The same statement also holds for the other two statistics.

If θ in (4.9) is k-dimensional, a multivariate expansion similar to (4.11) leads to

$$W(Y) = n(\hat{\theta} - \theta_0)^{\top} i(\theta_0)(\hat{\theta} - \theta_0) + o_p(1),$$

whose asymptotic distribution is χ_k^2, under the null hypothesis.

In addition, it can be shown that, if Θ_0 is defined as the set of elements $\theta = (\theta_1, \ldots, \theta_k)^{\top}$ which satisfy m $(m \leq k)$ equations of the form

$$\begin{cases} g_1(\theta_1, \ldots, \theta_k) = 0, \\ \quad\vdots \\ g_m(\theta_1, \ldots, \theta_k) = 0, \end{cases} \tag{4.19}$$

where the functions $g_1, \ldots, g_m$ satisfy some mild regularity conditions, then

$$W(Y) \xrightarrow{d} \chi_m^2 \tag{4.20}$$

under the null hypothesis For instance, if the hypothesis is of type (4.9), but θ is k-dimensional, equations (4.19) take the form

$$\theta_{*i} - \theta_{0i} = 0 \qquad (i = 1, \ldots, k)$$

so that there are k degrees of freedom, as seen earlier.

The multiparameter extensions of W_e and W_u are given by

$$W_e = n(\hat{\theta} - \theta_0)^\mathsf{T} i(\hat{\theta}_0)(\hat{\theta} - \theta_0)$$

$$W_u = \left(\frac{d\,\ell(\theta)}{d\theta} \bigg|_{\theta=\theta_0} \right)^\mathsf{T} I(\theta_0)^{-1} \left(\frac{d\,\ell(\theta)}{d\theta} \bigg|_{\theta=\theta_0} \right)$$

respectively. For these test statistics a result similar to (4.20) also holds for hypotheses of type (4.19).

Proofs and technical details on the asymptotic distributions of W, W_e, W_u are given, for example, by Serfling (1980, Chapter 4).

Remark 4.3.1 Statement (4.20) assumes that the regularity conditions of section 3.3.3 hold, and that the null hypothesis is expressed in the form of equalities (4.19) against the alternative that at least one equality does not hold.* The next counterexample shows that the distribution of $W(Y)$ can otherwise be different.

Suppose that an s r s $y = (y_1, \ldots, y_n)$ from a random variable with density function

$$g(t; \theta) = \begin{cases} e^{-(t-\theta)} & \text{if } t > \theta, \\ 0 & \text{otherwise,} \end{cases}$$

is used to test the hypothesis

$$\begin{cases} H_0 : \theta_* = \theta_0, \\ H_1 : \theta_* > \theta_0, \end{cases}$$

where θ_0 is a given value. The setting does not define a regular estimation problem and the alternative is one-sided.

The LRT statistic takes the form

$$\lambda(y) = \prod_{i=1}^{n} \frac{e^{-(y_i - \theta_0)}}{e^{-(y_i - \hat{\theta})}}, = e^{-n(\hat{\theta} - \theta_0)},$$

---

* The regularity conditions on the g_i in (4 19) prevent us from transforming an inequality constraint into an equality

where $\hat{\theta} = y_{(1)}$, and

$$W(y) = -2\ln\lambda(y) = 2n(\hat{\theta} - \theta_0).$$

Under the null hypothesis, $(\hat{\theta} - \theta_0)$ is distributed as the smallest element of an s r s from an exponential random variable with mean value 1; hence $\hat{\theta} - \theta_0$ is distributed as an exponential random variable with mean $1/n$. Then, taking into account the connection between the exponential and the χ_2^2 distribution, we conclude that under the null hypothesis

$$W(Y) \sim \chi_2^2$$

without approximation, not χ_1^2.   □

4.3.3 Observed significance level

In practical statistical work, it is seldom the case that a test statistic is used in the formal application of a test procedure, involving a rigid, uncritical comparison of the sample value of the statistic with the critical value, followed by an automatic decision for or against the null hypothesis. After all, the critical value depends on the significance level, which is often arbitrary, except for the requirement of being 'small'.

A more commonly adopted attitude is as follows. For an observed value of the test statistic, the minimum significance level for which the null hypothesis would be rejected is computed; this value is called the **observed significance level** or simply *p*-value. In the case of the LRT, if the sample value of the test statistic is $W(y)$, then the *p*-value is

$$\alpha_{\text{obs}} = \sup_{\theta \in \Theta_0} \Pr\{W(Y) \geq W(y); \theta\}. \qquad (4.21)$$

The closer α_{obs} is to 0, the smaller is the probability that $W(Y)$ produces, under H_0, a *p*-value equal to or smaller than that observed, while the probability of such event is higher under H_1. In this sense, the *p*-value is an indicator of the 'plausibility' of the null hypothesis.

Usually, if $\alpha_{\text{obs}} < 0.05$, the null hypothesis is abandoned, while $\alpha_{\text{obs}} > 0.10$ is regarded as lack of evidence against the null hypothesis. There is no special virtue attached to these two values, and they must be taken as reference points in a broad sense. As α_{obs} gets closer to either boundary point of the interval $(0.05, 0.10)$, so this is taken as increasing evidence for one or other alternative. Situations with α_{obs} within the interval (0.05, 0 10) are usually the most difficult to handle.

A final judgement takes into account not only the p-value, but also a number of other aspects. The following is a non-exhaustive list. Can a significant loss of goodness of fit to the data be compensated by a substantial simplification in the fitted model? Is there is a strong physical motivation for the assumed statistical model? If not, what about sensitivity of the results to slight modifications of the assumed model? And, finally, do the data contain enough information to reach *some* conclusion?

The first question is related to the distinction between statistical significance and importance There can be sufficient evidence in the data to conclude, say, that $\theta \neq 0$ with p-value less than 0.05, but it may still happen that the numerical value of θ is so small that, for a specific practical problem, it can be ignored with tolerable loss of goodness of fit, more than compensated by model simplification.

From a purely formal point of view, to accept the null hypothesis when $W(y) < -2 \ln \lambda_\alpha$ is equivalent to accepting it when $\alpha_{obs} > \alpha$. Therefore, if the outcome of a statistical analysis is reported by using the observed significance level, then this information can be used to perform the formal test procedure at any desired significance level.

Example 4.3.2 In the setting of Example 4.2 4, suppose $n = 22$, $\hat{\theta} = 1.39$ and $\theta_0 = 1$. Then, all three test statistics described earlier are equal to $22 \times (1.39 - 1)^2 = 3.346$, which falls between the upper 10% and 5% points of the χ_1^2 distribution, namely 2.706 and 3.841. Therefore at level 10% the null hypothesis would be rejected, but at level 5% it would be accepted.

Alternatively, from the relationship between χ_1^2 and $N(0,1)$, the p-value is $2\Phi(-\sqrt{3.346}) \approx 0.067$, using the approximate expression (A.8) or tables to compute $\Phi(\cdot)$.   □

Remark 4.3.3 Consider the connection between the p-value and $\tilde{L}(\theta_0)$ in the simple but important case of (4.9). Here $\tilde{L}(\theta_0) = \lambda(y)$ and, since there exists a monotonic relationship between $\lambda(y)$ and the p-value, then a similar connection holds for $\tilde{L}(\theta_0)$ and the p-value; moreover, they both vary in $(0,1)$. Therefore, there is a similarity between these two quantities; however, there is also a major difference, as the p-value stems from the repeated sampling principle while the use of $\tilde{L}(\theta_0)$ falls within the strong likelihood principle.   □

4.3.4 Nuisance parameters and profile likelihood

Quite often, the hypotheses will address only some of the components of the parameter vector More specifically, if θ is partitioned in two components, $\theta = (\psi, \omega)$ say, then the null hypothesis may only be concerned with ψ, called the **parameter of interest**, while ω is regarded as a **nuisance parameter**. In some cases, the separation into two such components can be achieved after suitable reparametrization.

If $\hat{\omega}_\psi$ denotes the value of ω which maximizes the log-likelihood ℓ for a given value of ψ, we define

$$\ell^*(\psi) = \ell(\psi, \hat{\omega}_\psi)$$

as the **profile log-likelihood function**. The term 'profile' comes about through thinking of the usual log-likelihood function as a hill being observed from a viewpoint with abscissa $\omega = \infty$, so that, for any fixed ψ, only the highest value $\ell(\psi, \hat{\omega}_\psi)$ is seen.

In a number of ways, the profile log-likelihood behaves like a proper log-likelihood, and its relevance in not confined to testing problems, although we have introduced it within this context. Clearly, the maximum of ℓ^* is achieved at $\hat{\psi}$, the usual MLE of ψ; moreover, the likelihood ratio for the hypotheses

$$\begin{cases} H_0 : \psi = \psi_0 \\ H_1 : \psi \neq \psi_0 \end{cases}$$

can be written as

$$\begin{aligned} W &= 2\left(\ell(\hat{\psi}, \hat{\omega}) - \ell(\psi_0, \hat{\omega}_{\psi_0})\right) \\ &= 2\left(\ell^*(\hat{\psi}) - \ell^*(\psi_0)\right), \end{aligned}$$

which under the null hypothesis has a χ_q^2 asymptotic distribution, where $q = \dim(\psi)$.

Many of the problems to be considered in section 4.4 are examples of the above situations, hence no specific example is presented here. Notice instead that the asymptotic distribution of W, i e χ_q^2, is the same as obtained when the component ω is not present; in other words, the profile likelihood behaves like a regular likelihood where the ω component is known. This remark suggests that, when the number $k - q$ of nuisance components is high, the χ_q^2 approximation to the distribution of W is likely to be not so accurate. When the distribution of W cannot be computed exactly, **higher-order asymptotic theory** becomes relevant for obtaining accurate approximations, especially when the dimension of the nuisance parameter is not small. A solid account of this topic is given by Barndorff-Nielsen and Cox (1994)

4.4 Some important applications

In this section, we shall develop some applications of the LRT criterion. Formally, they are just specific instances of the general method, but they do provide some essential tools for practical statistical work.

4.4.1 One-sample two-sided Student's t

The s r s $(y_1, \ldots, y_n)$ from an $N(\mu, \sigma^2)$ random variable can be used to test the hypotheses

$$\begin{cases} H_0 : \mu = \mu_0, \\ H_1 : \mu \neq \mu_0, \end{cases} \tag{4.22}$$

for some given real value μ_0; hence in this case

$$\Theta_0 = \{(\mu_0, \sigma^2) : \sigma^2 \in \mathbb{R}^+\}, \quad \Theta = \{(\mu, \sigma^2) : \mu \in \mathbb{R}, \sigma^2 \in \mathbb{R}^+\}.$$

Here μ represents the parameter of interest, and σ^2 is the nuisance parameter.

The likelihood when $\theta \in \Theta$ has already been considered in Example 3.1.10, and its maximum is at

$$\hat{\mu} = \sum_{i=1}^n y_i/n = \overline{y}, \quad \hat{\sigma}^2 = \sum_{i=1}^n (y_i - \hat{\mu})^2/n,$$

providing the value $L(\hat{\mu}, \hat{\sigma}^2)$ for the denominator of (4.16). To compute its numerator, no maximization with respect to μ is required, as only one value is feasible; the maximum of

$$\sup_{\sigma^2} L(\mu_0, \sigma^2),$$

occurs at

$$\hat{\sigma}_0^2 = \sum_{i=1}^n (y_i - \mu_0)^2/n$$

Thus, the LRT statistic is

$$
\begin{aligned}
\lambda(y) &= \frac{L(\mu_0, \hat{\sigma}_0^2)}{L(\hat{\mu}, \hat{\sigma}^2)} = \left(\frac{n^{-1} \sum (y_i - \mu_0)^2}{n^{-1} \sum (y_i - \overline{y})^2} \right)^{-n/2} \\
&= \left(\frac{\sum (y_i - \overline{y})^2 + n(\overline{y} - \mu_0)^2}{\sum (y_i - \overline{y})^2} \right)^{-n/2} \\
&= \left(1 + \frac{t^2}{n-1} \right)^{-n/2}
\end{aligned}
$$

where

$$
t = \frac{\sqrt{n}(\overline{y} - \mu_0)}{\sqrt{\dfrac{\sum (y_i - \overline{y})^2}{n-1}}} = \frac{\sqrt{n}(\overline{y} - \mu_0)}{s}.
$$

Since λ is a decreasing function of $|t|$, a criterion which rejects the null hypothesis for small values of $\lambda(y)$ is equivalent to one which rejects for large values of $|t|$ or t^2, provided that the various critical values determine the same significance level.

The test statistic t^2 is analogous to $W(y) = W_e(y) = W_u(y)$ in Example 4.2 4, but with σ replaced by its estimate s. This substitution modifies the distribution of the statistic, which is no longer χ_1^2 under the null hypothesis. However, an asymptotic computation gives

$$
\begin{aligned}
-2 \ln \lambda &= n \ln(1 + t^2/(n-1)) \\
&= n(t^2/(n-1) + o_p(1/n)) \\
&= t^2 + o_p(1) \\
&\xrightarrow{d} \chi_1^2
\end{aligned}
$$

by expanding the log function as a stochastic Taylor series, as explained in section A.8.4, and taking into account that $t \xrightarrow{d} N(0,1)$ and Theorem A.8.2(e). This direct computation confirms the general result (4.18).

In fact, for this specific case, approximations in computing the distribution of the statistic can be avoided altogether. By the results of section A.5.9, the null distribution of t is Student's t with $n-1$ degrees of freedom, ie $t \sim t_{n-1}$. Under the alternative hypothesis, the distribution is non-central t with non-centrality parameter $\delta = \sqrt{n}(\mu - \mu_0)/\sigma$.

Therefore the rejection region at level α consists of the two intervals $(-\infty, -t_{\alpha/2})$ and $(t_{\alpha/2}, \infty)$, where $t_{\alpha/2}$ is the upper quantile of level $\alpha/2$ of the t_{n-1} distribution. The observed significance level, for the sample value t of the test statistic, is $2\Pr\{T > |t|\}$, where $T \sim t_{n-1}$.

4.4.2 One-sample one-sided Student's t

In the same setting of section 4.4.1, consider the hypotheses

$$\begin{cases} H_0 : \mu \le \mu_0, \\ H_1 : \mu > \mu_0, \end{cases}$$

so that now

$$\Theta_0 = \{(\mu, \sigma^2) : \mu \in \mathbb{R}, \mu \le \mu_0, \sigma^2 \in \mathbb{R}^+\}.$$

Only the estimates for $\theta \in \Theta_0$ must be recomputed. Arguing as in Example 3.1.2(b), the parameter estimates in Θ_0 are

$$\hat{\mu}_0 = \min\{\mu_0, \overline{y}\}$$

and

$$\hat{\sigma}_0^2 = \sum_i (y_i - \hat{\mu}_0)^2 / n$$

respectively, such that

$$\begin{cases} \hat{\sigma}_0^2 = \hat{\sigma}^2 & \text{if } \hat{\mu}_0 = \overline{y}, \\ \hat{\sigma}_0^2 > \hat{\sigma}^2 & \text{if } \hat{\mu}_0 = \mu_0. \end{cases}$$

Then

$$\lambda(y) = \left(\frac{\hat{\sigma}_0^2}{\hat{\sigma}^2}\right)^{-n/2} = \left(\frac{\sum(y_i - \hat{\mu}_0)^2}{\sum(y_i - \hat{\mu})^2}\right)^{-n/2}$$

such that

$$\begin{cases} \lambda(y) = 1 & \text{if } \hat{\sigma}^2 = \hat{\sigma}_0^2, \\ \lambda(y) < 1 & \text{if } \hat{\sigma}^2 < \hat{\sigma}_0^2 \end{cases}$$

If $\lambda(y) = 1$, the null hypothesis is accepted; if $\lambda(y) < 1$, it is accepted if $|t|$ is sufficiently small, where t is defined as in section 4.4.1. Therefore, the critical region is

$$\begin{aligned} \{\lambda(y) < \lambda_\alpha\} &= \{\lambda(y) < 1\} \cap \{t^2 > t_\alpha^2\} \\ &= \{\overline{y} > \mu_0\} \cap \{|t| > |t_\alpha|\} \\ &= \{t > t_\alpha\} \end{aligned}$$

where t_α is the α-level upper quantile of the t_{n-1} distribution. Since $\Pr\{\lambda(Y) = 1\} = \Pr\{Y \le \mu_0\} = \frac{1}{2}$ under H_0, then the constraint $\alpha \le \frac{1}{2}$ holds, but this is irrelevant in practice.

Remark 4.4.1 This is another example where the alternative hypothesis is one-sided, with respect to the value θ_0 specified by the null hypothesis, so it cannot be expressed in the form (4.19). Therefore, the asymptotic distribution of the LRT statistic is not of χ^2 type, as is clear also from the fact that the LRT statistic is equivalent to the t statistic, which does not converge to χ_1^2 as $n \to \infty$.

□

4.4.3 Two-sample two-sided Student's t

Before describing the technical aspects of the next case to consider, we present its practical motivation. It is often the case, in real-world applications, that one needs to compare the effects of two treatments, where the term 'treatment' may denote a therapy, a training course, or anything else which can, or should, modify a relevant characteristic of the subjects under considerations.

To perform this comparison, a group of subjects, randomly selected from the population, is given one treatment, while another similar group is given the other treatment. The treatment is allowed to develop its supposed effect, and then observations are collected for the statistical analysis.

The present context allows us to explain in greater detail a point mentioned at the end of section 4.1.3. Often, one of the two treatments is a widely used and well-established technique, for instance the usual therapy for a given pathology, or possibly a null treatment (a placebo, in medical applications), while the other treatment is the actual focus of interest. Since generally we do not wish to claim that the new treatment is effective, or more effective than the standard one, unless there is sufficiently strong evidence in this direction, we wish to protect ourselves from making an incorrect statement in favour of the new treatment. This requirement is met by constructing the null hypothesis to represent the equality of the two treatment mean effects, so as to be able to control the probability of type I error, which we can choose as small as appropriate.

We shall deal here with two-sided alternatives. In practice, the one-sided alternative case is also often relevant; see Exercise 4.7.

Suppose that two independent samples are available: $(z_1, \ldots, z_n)$ are sampled from an $N(\mu, \sigma^2)$ random variable and $(x_1, \ldots, x_m)$ from an $N(\eta, \sigma^2)$ random variable. Not only must the elements from one population be sampled independently, but the two underlying random variables are assumed to be independent, expressing the fact that the experiment is carried out in such a way that subjects must not influence each other in any way, either within the same treatment group or across groups.

The hypotheses to be tested are

$$\begin{cases} H_0 : \mu = \eta, \\ H_1 : \mu \neq \eta. \end{cases} \tag{4.23}$$

Under both hypotheses, the two distributions are supposed to have the same unknown variance, In other words, it is assumed that the effect of the treatment is only to shift the mean of the probability distribution.

The log-likelihood function is

$$\ell(\mu, \eta, \sigma^2) = -\frac{n+m}{2} \ln \sigma^2 - \frac{1}{2\sigma^2} \sum_{i=1}^{n} (z_i - \mu)^2 - \frac{1}{2\sigma^2} \sum_{i=1}^{m} (x_i - \eta)^2$$

with associated MLEs

$$\hat{\mu} = \sum z_i / n = \bar{z}, \qquad \hat{\eta} = \sum x_i / m = \bar{x},$$
$$\hat{\sigma}^2 = \frac{\sum (z_i - \bar{z})^2 + \sum (x_i - \bar{x})^2}{n + m}.$$

when θ is unrestricted. In Θ_0, we must maximize

$$\ell(\mu, \mu, \sigma^2) = -\frac{n+m}{2} \ln \sigma^2 - \frac{1}{2\sigma^2} \left(\sum_{i=1}^{n} (z_i - \mu)^2 + \sum_{i=1}^{m} (x_i - \mu)^2 \right)$$

leading to the estimates

$$\hat{\mu}_0 = \frac{\sum z_i + \sum x_i}{n + m} = \frac{n\bar{z} + m\bar{x}}{n + m},$$
$$\hat{\sigma}_0^2 = \frac{\sum (z_i - \hat{\mu}_0)^2 + \sum (x_i - \hat{\mu}_0)^2}{n + m}.$$

Hence (4.16) gives, after a monotonic transformation,

$$\begin{aligned} \lambda^{-2/(n+m)} = \left(\frac{\hat{\sigma}_0^2}{\hat{\sigma}^2} \right) &= \frac{\sum (z_i - \hat{\mu}_0)^2 + \sum (x_i - \hat{\mu}_0)^2}{\sum (z_i - \hat{\mu})^2 + \sum (x_i - \hat{\eta})^2} \\ &= \frac{\sum (z_i - \hat{\mu})^2 + n(\hat{\mu}_0 - \hat{\mu})^2 + \sum (x_i - \hat{\eta})^2 + m(\hat{\mu}_0 - \hat{\eta})^2}{\sum (z_i - \hat{\mu})^2 + \sum (x_i - \hat{\eta})^2}. \end{aligned}$$

Rewriting

$$n(\hat{\mu} - \hat{\mu}_0)^2 + m(\hat{\eta} - \hat{\mu}_0)^2$$

$$= n\left(\bar{z} - \frac{n\bar{z} + m\bar{x}}{n+m}\right)^2 + m\left(\bar{x} - \frac{n\bar{z} + m\bar{x}}{n+m}\right)^2$$

$$= \frac{nm^2}{(n+m)^2}(\bar{z} - \bar{x})^2 + \frac{mn^2}{(n+m)^2}(\bar{z} - \bar{x})^2$$

$$= \frac{nm}{n+m}(\bar{z} - \bar{x})^2,$$

we obtain

$$\lambda = \left(1 + \frac{t^2}{n+m-2}\right)^{-(n+m)/2}$$

after setting

$$t = \frac{(\bar{z} - \bar{x})/\sqrt{1/n + 1/m}}{s}$$

and

$$s^2 = \frac{\sum(z_i - \bar{z})^2 + \sum(x_i - \bar{x})^2}{n+m-2}.$$

Here, the term $\sum(z_i - \bar{z})^2$ is sampled from $\sigma^2 \chi^2_{n-1}$, and $\sum(x_i - \bar{x})^2$ is sampled from an independent $\sigma^2 \chi^2_{m-1}$. Therefore, their sum is $\sigma^2 \chi^2_{n+m-2}$, and the distribution of t is t_{n+m-2}. The acceptance region is of type $(-t_{\alpha/2}, t_{\alpha/2})$, differing from the situation in section 4 4.1 only in the degrees of freedom of the t distribution.

Remark 4.4.2 In the above discussion, it was essential to introduce the assumption of equal variance for the two populations, but in some situations this may not reasonably be believed to hold. Testing hypotheses (4.23) without this assumption gives rise to the so-called Behrens–Fisher problem, which does not lend itself to an exact solution, i e the null distribution of the test statistic cannot be determined exactly. Only asymptotic approximations are available. □

Remark 4.4.3 When the appropriateness of the equal variance assumption is unclear, one might think of the following scheme: first the equality of the variances is checked via a formal test procedure (cf. Exercise 4.4.8) and, should this hypothesis be accepted, the above test for equality of the means is performed.

This scheme is legitimate provided that the following observations are taken into account If the two distributions do indeed have equal mean and equal variance, and the two test procedures are performed at levels α_1 and α_2 respectively, then the probability that both equality hypotheses are accepted is $(1 - \alpha_1)(1 - \alpha_2)$, since it can be proved that the two test statistics are stochastically independent. The combined test procedure then has significance level $\alpha_1 + \alpha_2$, omitting the smaller-order term $\alpha_1\alpha_2$. Therefore, in this case, a combined test procedure of level α can be performed by setting, for instance, $\alpha_1 = \alpha_2 = \alpha/2$.

Generally, the cascade application of several test procedures produces a global significance level higher than those of the individual test procedures. In the problem considered here, the discussion has been dramatically simplified by the independence of the two test statistics, but this is an extremely special case, while in general the computation of the actual level of the combined test is a very difficult exercise, if indeed it is at all possible.

Often, in practical statistical work, several test statistics are applied to the same data, each followed by either formal acceptance or rejection of the hypothesis or computation of the p-value. In addition to the need to take care to avoid rigid attitudes in hypothesis testing, as mentioned in section 4.1.1, caution is necessary in interpreting numerical results when repeated testing occurs, because the associated probabilities are computed for each individual test statistic as if it were the only one applied to the data. Interpretation of formal tests or p-values becomes more and more troublesome when the number of test statistics involved is substantial. *

□

4.4.4 Student's t for paired data

A somewhat different situation occurs when each subject, from a group of size n, is observed on two occasions, before and after the treatment has been given. The advantage of this experimental design over that of section 4.4.3 is the increased precision of the treatment effect estimates achieved by eliminating the between-subjects variability component, as explained later. This scheme is an instance of a more general situation where measurements are taken from each subject at various successive times, producing **repeated measurements** or **longitudinal data**, a data structure suitable for analysing the effect of the treatment over time.

* Typical situations involving a very large number of replicated test procedures are automatic model selection techniques; cf section 5 5

Denote by x_i the observation from the ith subject before the treatment is given, and by z_i the value taken from the same subject after the treatment $(i = 1, \ldots, n)$. Assuming that x_i is sampled from an $N(\eta, \sigma^2)$ random variable, and that z_i is sampled from an $N(\mu, \sigma^2)$ random variable, we have to test the hypotheses

$$\begin{cases} H_0 : \delta = 0, \\ H_1 : \delta \neq 0, \end{cases} \tag{4.24}$$

where $\delta = \mu - \eta$.

Although the problem appears in many ways similar to the case of section 4.4.3, there is a major difference, in that both elements (x_i, z_i) are sampled from the same subject, and it is unrealistic to believe that they are sampled independently. Suppose, for instance, that the treatment under consideration should decrease the blood pressure of some patients; if the ith subject is above the population mean value before treatment, then it is likely that the same patient will be above the new population mean value after treatment. This lack of independence makes the method of section 4.4.3 inapplicable here.

A simple stochastic scheme for interpreting this situation is given by

$$x_i = \eta + \omega_i + \xi_i, \quad z_i = \mu + \omega_i + \varepsilon_i \quad (i = 1, \ldots, n),$$

where $\omega_i \sim N(0, \sigma_\omega^2)$, called the **subject effect**, represents the deviation between the ith subject mean level and the population mean level, while ξ_i and ε_i are normal variates associated to the measurement instances.

By considering the difference $y_i = z_i - x_i$, the component ω_i is removed, with a decrease in variability which is often substantial Moreover, the question of equal variance of the ξ_i, ε_i components vanishes. It suffices to apply the one-sample t test to the new data $(y_1, \ldots, y_n)$, which are sampled from an $N(\delta, \sigma_*^2)$ random variable, say.

4.4.5 One-way analysis of variance

The problem discussed in section 4.4.3 can be extended to the case of several populations as follows: m samples are available, where the ith sample $(y_{i1}, \ldots, y_{in})$ is taken from $N(\mu_i, \sigma^2)$ for $i = 1, \ldots, m$. Equivalently, we can write

$$y_{ij} = \mu_i + \varepsilon_{ij} \qquad (i = 1, \ldots, m; \; j = 1, \ldots, n)$$

where the ε_{ij} are independent $N(0,\sigma^2)$ random variables

The hypotheses to be tested are

$$\begin{cases} H_0 : \mu_1 = \mu_2 = \cdots = \mu_m, \\ H_1 : \text{at least one equality does not hold.} \end{cases}$$

which collapses to the case of section 4.4.3 if $m = 2$.

Because of the connection with Student's t, one might think of dealing with this problem by repeated use of two-sample t tests. If $m = 3$, for instance, the null hypothesis is given by combination of $\mu_1 = \mu_2$ and $\mu_2 = \mu_3$ Then, the above idea would be implemented by the use of the t test twice, once for each of the component hypotheses, followed by acceptance of the overall null hypothesis if both elementary null hypotheses were accepted. In fact, this approach cannot be pursued, for the reasons discussed in Remark 4 4.3 about repeated significance test procedures, with the additional complication that the sample from population 2 would enter both t statistics, hence inducing stochastic dependence between them.

We therefore drop this approach and apply the LRT criterion from scratch. By an immediate extension of previous results, it follows that the MLEs in Θ_0 are

$$\hat{\mu}_0 = \frac{\sum_i \sum_j y_{ij}}{nm} = \overline{y}, \quad \hat{\sigma}_0^2 = \frac{\sum_i \sum_j (y_{ij} - \hat{\mu}_0)^2}{nm}$$

and in Θ are

$$\hat{\mu}_i = \frac{\sum_j y_{ij}}{n} = \overline{y}_i \qquad (i = 1, \ldots, m),$$

$$\hat{\sigma}^2 = \frac{\sum_i \hat{\sigma}_i^2}{m} = \frac{1}{m} \sum_i \left(\frac{1}{n} \sum_j (y_{ij} - \overline{y}_i)^2 \right);$$

then

$$\begin{aligned} \lambda &= \frac{L(\hat{\mu}_0, \ldots, \hat{\mu}_0, \hat{\sigma}_0^2)}{L(\hat{\mu}_1, \ldots, \hat{\mu}_m, \hat{\sigma}^2)} = \left(\frac{\hat{\sigma}_0^2}{\hat{\sigma}^2} \right)^{-nm/2} \cdot \\ &= \left(\frac{\sum_i \sum_j (y_{ij} - \hat{\mu}_0)^2}{\sum_i \sum_j (y_{ij} - \hat{\mu}_i)^2} \right)^{-nm/2} \end{aligned}$$

$$= \left(\frac{\sum_i \sum_j [(y_{ij} - \hat{\mu}_i) + (\hat{\mu}_i - \hat{\mu}_0)]^2}{\sum_i \sum_j (y_{ij} - \hat{\mu}_i)^2} \right)^{-nm/2}$$

$$= \left(1 + \frac{n \sum_i (\hat{\mu}_i - \hat{\mu}_0)^2}{\sum_i \sum_j (y_{ij} - \hat{\mu}_i)^2} \right)^{-nm/2} .$$

The critical region is then given by sufficiently high values of

$$\frac{\sum_i (\bar{y}_i - \bar{y})^2 n}{\sum_i \sum_j (y_{ij} - \bar{y}_i)^2} = \frac{D_0}{D}.$$

To obtain the null hypothesis distribution of D_0/D, notice that the term D/σ^2 is the sum of m independent components, each sampled from a χ^2_{n-1} random variable, hence $D/\sigma^2 \sim \chi^2_{m(n-1)}$. The quantity

$$\frac{D_0}{\sigma^2} = \sum_{i=1}^{m} \left(\frac{\bar{y}_i - \bar{y}}{\sigma/\sqrt{n}} \right)^2$$

is the sample variance of $\bar{y}_1 \sqrt{n}/\sigma, \ldots, \bar{y}_m \sqrt{n}/\sigma$; its null distribution is χ^2_{m-1} Moreover, D_0 and D are mutually independent, since D_0 is a function of the sample means $\bar{y}_i$ only ($i = 1, \ldots, n$), while D is a function of the sample variances $\hat{\sigma}_i^2$ ($i = 1, \ldots, n$) only, and the two set of variates are independent. Then

$$F^* = \frac{D_0/(\sigma^2(m-1))}{D/(\sigma^2(m(n-1)))}$$

$$= \frac{n \sum_i (\bar{y}_i - \bar{y})^2/(m-1)}{\sum_i \sum_j (y_{ij} - \bar{y}_i)^2/(m(n-1))}$$

is sampled from an F random variable with $(m-1, m(n-1))$ degrees of freedom, and the critical region is (F_α, ∞), where F_α is the α-level upper quantile of the F distribution. Correspondingly, the observed significance level is given by the probability that an $F_{m-1, m(n-1)}$ random variable exceeds the sample value F^*.

4.4.6 Testing the variance of a normal distribution

In the same setting as section 4.4.1, consider the hypotheses

$$\begin{cases} H_0 : \sigma^2 = \sigma_0^2 \\ H_1 : \sigma^2 \neq \sigma_0^2, \end{cases}$$

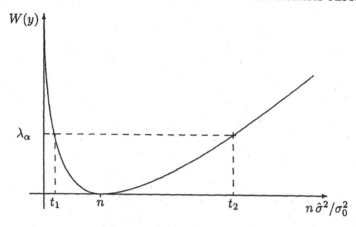

Figure 4 3 *Likelihood ratio statistic for testing the variance of a normal population, plotted as a function of $n\hat{\sigma}^2/\sigma_0^2$*

where σ_0^2 is a given value. The corresponding LRT statistic

$$W = -n \log \frac{\hat{\sigma}^2}{\sigma_0^2} + n \left(\frac{\hat{\sigma}^2}{\sigma_0^2} - 1 \right)$$

depends on the data only through $\hat{\sigma}^2$, or equivalently through $T = n\hat{\sigma}^2/\sigma_0^2$. If W is plotted as a function of T, the result looks like Figure 4.3

To identify precisely the rejection region of the LRT procedure, we should select a value λ_α achieving the required level, and reject the null hypothesis when T is outside the interval (t_1, t_2) associated with λ_α. In fact, the quantity whose distribution we can determine is T itself, this distribution is $\omega\chi_{n-1}^2$, where the multiplier $\omega = \sigma^2/\sigma_0^2$ is equal to 1 under the null hypothesis. Then we should find the roots t_1 and t_2 of the equations

$$W(t_1) = W(t_2), \qquad \Pr\{t_1 < T < t_2\} = 1 - \alpha$$

where W is considered as a function of T, and the probabilities are computed under the null distribution.

However, this scheme is not pursued in practice as it is too cumbersome, or at least it was when computers were not available. Instead, with the aid of χ^2 distribution tables, two values c_1 and c_2 are chosen so that

$$\Pr\{T < c_1\} = \Pr\{T > c_2\} = \tfrac{1}{2}\alpha$$

and the acceptance interval (c_1, c_2) is adopted. For large n, the substitutes c_1, c_2 are close to t_1, t_2, in a relative sense.

4.4.7 Testing the parameter of a binomial distribution

Practical problems where the outcome of the observation is of dichotomous or binary type, 'success' or 'failure' say, are perhaps the most common.

If the probability of 'success' in an individual observation is θ, then the observed frequency y of successes in n independent trials is sampled from $Y \sim Bin(n, \theta)$. Consider the testing problem

$$\begin{cases} H_0 : \theta = \theta_0, \\ H_1 : \theta \neq \theta_0, \end{cases}$$

where θ_0 is a specified value. It is easy to think of situations where $\theta_0 = \frac{1}{2}$, as this value represents a symmetry situation between success and failure. However, there are many cases where a different value of θ_0 is considered; for instance, the problem discussed in section 1.4.2 had $\theta_0 = 0.05$.

The LRT criterion leads to

$$\lambda(y) = \frac{\theta_0^y (1 - \theta_0)^{n-y}}{\hat{\theta}^y (1 - \hat{\theta})^{n-y}} = \frac{\theta_0^{n\hat{\theta}} (1 - \theta_0)^{n(1-\hat{\theta})}}{\hat{\theta}^{n\hat{\theta}} (1 - \hat{\theta})^{n(1-\hat{\theta})}}$$

where $\hat{\theta} = y/n$. The plot of $W(y) = -2 \ln \lambda(y)$ is qualitatively similar to Figure 4.3, apart from the discreteness of y. The minimum point occurs at $y = n\theta_0$, or some achievable value close to it, and the function increases asymmetrically away from $n\theta_0$, unless $\theta_0 = \frac{1}{2}$.

Therefore, the rejection region of the null hypothesis is given by y's in two intervals of the form $[0, t_1]$ and $[t_2, n]$, where t_1, t_2 are sufficiently distant from $n\theta_0$. The precise meaning of 'distant' is determined by the distribution of $\lambda(Y)$ under the null hypothesis, i e the critical points t_1, t_2 are such that

$$\Pr\{Y \leq t_1; \theta_0\} + \Pr\{Y \geq t_2; \theta_0\} = \alpha, \qquad \lambda(t_1) = \lambda(t_2),$$

for a given significance level α. In all but fortuitous cases, only approximate solutions of these equations are feasible due to the discreteness of Y.

There is the additional complication that the distribution of $\lambda(Y)$, or equivalently $W(Y)$, cannot be written in closed form. Hence, the distribution must be computed by enumeration, i e by

numerical summation of the probabilities of all relevant sample
points, as illustrated later in an example.

To avoid these computational complications, one can seek an
approximate solution based on asymptotic theory results. As $n \rightarrow$
∞, (4.18) provides the limit distribution, which can be used for
computing critical points and p-values

In the present testing problem, the other two test statistics, W_e
and W_u, give rise to solutions that are simpler, in a sense, than
the LRT. From (4.13) and (4.14), we obtain, up to an inessential
transformation,

$$W_e = \frac{(y - n\theta_0)^2}{n\,\hat{\theta}(1 - \hat{\theta})}, \quad W_u = \frac{(y - n\theta_0)^2}{n\theta_0(1 - \theta_0)}$$

respectively. Their numerator is explicitly based on the deviation
between observed and expected frequencies, y and $n\theta_0$. These two
statistics differ only in the normalization method of the denominator: W_u considers the mean value of $(Y - n\theta_0)^2$ computed at
$\theta = \theta_0$, while W_e sets $\theta = \hat{\theta}$. For both test statistics, the null
asymptotic distribution is χ_1^2 Since $E[W_u(Y); \theta_0] = 1$ exactly,
then the asymptotic distribution χ_1^2 has the precise mean value of
the statistic of W_u.

Example 4.4.4 Suppose that $n = 10, \theta_0 = 0.6, y_{\text{obs}} = 3$, so that
the sample value of

$$W(y) = 2\left(y\ln\frac{y/n}{\theta_0} + (n - y)\ln\frac{1 - y/n}{1 - \theta_0}\right).$$

is $w_{\text{obs}} = 3.676$. The computation of the p-value

$$\alpha_{\text{obs}} = \Pr\{W \geq w_{\text{obs}}; \theta_0\}$$

is performed by summing the probabilities of all values of Y associated with a value of W not smaller than w_{obs}. In computing $y\ln y$,
we set $0\ln 0 = 0$ since $\lim_{x\to 0} x\ln x = 0$. Table 4.1 summarizes
the outcome of such enumeration, marking with a $\sqrt{}$ the values of
W not smaller than w_{obs} By summing the relevant probabilities,
the observed significance level is obtained, namely $\alpha_{\text{obs}} = 0.1011$,
a value not in conflict with the null hypothesis. $\square$

4.4.8 Testing the parameter of a multinomial distribution

A direct generalization of the problem considered in section 4.4.7
occurs when the observations are of general qualitative type, in-

Table 4.1 *Computation of the p-value associated with the LRT statistic for the parameter of a binomial variate when $y = 3$, $n = 10$, $\theta_0 = 0.6$*

y	$W(y)$	$\{W(y) \geq w_{\text{obs}}\}$	$\Pr\{Y = y\}$
0	18.325	$\checkmark$	0 0001
1	11.013	$\checkmark$	0.0016
2	6.696	$\checkmark$	0.0106
3	3.676	$\checkmark$	0.0425
4	1.622		0.1115
5	0.408		0.2007
6	0.000		0.2508
7	0.432		0.2150
8	1.830		0 1209
9	4 526	$\checkmark$	0.0403
10	10.216	$\checkmark$	0.0060

stead of binary; an equivalent name for such data or variables is **categorical**. Denote by $A_0, A_1, \ldots, A_r$ the possible outcomes of the observations, called **levels** in this context. The observed frequencies $(n_1, \ldots, n_r)^{\mathsf{T}}$ are generated by a multinomial random variable

$$(N_1, \ldots, N_r)^{\mathsf{T}} \sim Bin_r(n, (\pi_1, \ldots, \pi_r)^{\mathsf{T}}),$$

where the remaining frequency is determined by the constraint on the overall number n of observations, and $\pi_1, \ldots, \pi_r$ are probabilities such that $\sum_{j=1}^{r} \pi_j \leq 1$. For a summary of the properties of the multinomial distribution, see section A.6.

The direct extension of the hypothesis on the binomial distribution parameter is

$$\begin{cases} H_0 : \pi_j = p_j, & \text{for } j = 1, \ldots, r, \\ H_1 : \pi_i \neq p_i & \text{for some } i \end{cases} \qquad (4.25)$$

where $p_1, \ldots, p_r$ are r given positive real numbers such that $\sum_{1}^{r} p_j \leq 1$. On setting

$$p_0 = 1 - \sum_{j=1}^{r} p_j, \quad \pi_0 = 1 - \sum_{j=1}^{r} \pi_j, \quad n_0 = n - \sum_{j=1}^{r} n_j,$$

the log-likelihood can be written as

$$\ell(\pi) = \sum_{j=0}^{r} n_j \ln \pi_j$$

$$= \sum_{j=1}^{r} n_j \ln \pi_j + n_0 \ln(1 - \pi_1 - \cdots - \pi_r)$$

with partial derivatives

$$\frac{\partial \ell}{\partial \pi_j} = \frac{n_j}{\pi_j} - \frac{n_0}{\pi_0} \quad (j = 1, \ldots, r),$$

implying that $\hat{\pi}_j / n_j = $ constant, and then

$$\hat{\pi}_j = n_j / n \quad (j = 1, \ldots, r),$$

since the sum of the $r+1$ probabilities must be 1. The LRT statistic is

$$W = 2\{\ell(\hat{\pi}) - \ell(p)\}$$

$$= 2 \sum_{j=0}^{r} n_j \ln(\hat{\pi}_j / p_j),$$

which in this case is often denoted by G^2. Since its exact distribution is virtually uncomputable, and it depends on the unknown probabilities anyhow, the distribution of W, or G^2, is usually approximated by χ_r^2.

For the present problem, we also consider the score test statistic W_u. Since

$$\mathrm{E}\left[-\frac{\partial^2 \ell}{\partial \pi_i \, \partial \pi_j}\right] = \delta_{ij} \frac{n}{\pi_j} + \frac{n}{\pi_0} \quad (i, j = 1, \ldots, r)$$

where δ_{ij} is the Kronecker delta, the expected Fisher information is

$$I(\pi) = (D + 1_r 1_r^{\mathsf{T}}) n / \pi_0$$

where 1_r is the $r \times 1$ vector with all elements equal to 1, and

$$D = \mathrm{diag}(\pi_0 / \pi_1, \ldots, \pi_0 / \pi_r).$$

To invert $I(\pi)$, make use of (A.26); then the (i, j)th entry of $(D + 1_r 1_r^{\mathsf{T}})^{-1}$ is

$$\frac{\delta_{ij} \pi_j}{\pi_0} - \left(\frac{\pi_i}{\pi_0}\right)\left(\frac{\pi_j}{\pi_0}\right) \frac{1}{1 + \sum_{j=1}^{r} \pi_j / \pi_0}.$$

Replacing the π_j by the p_j in $I(\pi)^{-1}$, we obtain

$$W_u = \frac{p_0}{n} \sum_{i=1}^{r} \sum_{j=1}^{r} \left(\frac{n_i}{p_i} - \frac{n_0}{p_0} \right) \left(\frac{\delta_{ij}p_i}{p_0} - \frac{p_ip_j}{p_0} \right) \left(\frac{n_j}{p_j} - \frac{n_0}{p_0} \right)$$

$$= \frac{1}{n} \left\{ \sum_{i=1}^{r} \left(\frac{n_i}{p_i} - \frac{n_0}{p_0} \right)^2 p_i - \sum_{i=1}^{r} \sum_{j=1}^{r} \frac{(n_ip_0 - n_0p_i)(n_jp_0 - n_0p_j)}{p_0^2} \right\}$$

$$= \frac{1}{n} \left\{ \sum_{j=1}^{r} \frac{n_j^2}{p_j} + \frac{n_0[n_0(1 + p_0) - 2np_0] - (p_0 n - n_0)^2}{p_0^2} \right\}$$

$$= \sum_{j=0}^{r} \frac{n_j^2}{np_j} - n$$

$$= \sum_{j=0}^{r} \frac{(n_j - np_j)^2}{np_j}.$$

In this context, W_u reduces to a test statistic introduced by Karl Pearson in 1900, and extremely widely used since then, namely

$$X^2 = \sum_{j=0}^{r} \frac{(n_j - n_j^*)^2}{n_j^*}$$

where $n_j^* = n\,p_j$ is called the **expected frequency** for the corresponding level $(j = 0, 1, \ldots, r)$.

Much effort has been devoted to establishing which of the two statistics, G^2 and X^2, is preferable. The summary conclusion is that X^2 is superior at least for faster convergence to the asymptotic null distribution, that is the actual distribution of X^2 is close to χ_r^2 even for small values of the expected frequencies $n\pi_j$; a rule of thumb is that they must be all larger than 2 However, G^2 requires substantially higher expected frequencies to achieve the same degree of accuracy of approximation See Cressie and Read (1989) for a review of the literature The above statements about preferability of G^2 to X^2 do not extend to a general superiority of W_u over W.

Example 4.4.5 Suppose that (u_i, v_i), $i = 1, \ldots, n$ is an s r s from a continuous bivariate (U, V) random variable whose support is the unit square $Q = (0, 1) \times (0, 1)$, and that the hypotheses

$$\begin{cases} H_0 : (U, V) \text{ is uniformly distributed over } Q \\ H_1 : (U, V) \text{ is not uniformly distributed over } Q \end{cases}$$

must be tested.

A practical problem leading to the above formulation occurs when testing the properties of generators of pseudo-random numbers. These generators are algorithms designed to simulate the behaviour of a sequence of independently sampled values from a $U(0,1)$ variate; given such a sequence, pseudo-random numbers with another distribution can be obtained, by suitably transforming the original sequence. A full check of the independence and uniformity of all d-dimensional vectors formed by d contiguous elements of the sequence, for any positive integer value of d, cannot be pursued, but some form of partial checking can be performed. Consider $d = 2$, and obtain pairs (u_i, v_i) of adjacent elements in the sequence of pseudo-random numbers $(i = 1, \ldots, n)$; if the generator behaves correctly, these pairs should be sampled uniformly in Q.

Another practical problem leading to the above hypothesis testing problem arises in biological contexts. Here Q is a portion of territory, and (u_i, v_i) denotes the position where a plant of a certain species has been found $(i = 1, \ldots, n)$. The objective of the study is to establish whether the members of the species are uniformly distributed over the territory, as otherwise this indicates some form of attraction or repulsion among members of the species.

Strictly speaking, the above hypotheses testing problem belongs to the province of nonparametric statistics, since the set of the distributions encompassed by H_0 and H_1 is given by all continuous distributions with support Q, and this set does not form a parametric class. We can reduce the above problem to a parametric one in the following way: partition Q into a number of subsets, for instance an $m \times m$ grid, such that each subset has probability $\pi_j = 1/m^2$ under H_0 $(j = 1, \ldots, m^2)$, and the problem is now in the form of (4.25), where $r = m^2 - 1$; hence, make use of G^2 or X^2.

We said that the problem can be 'reduced' from nonparametric to parametric type, but we cannot make it equivalent. The above procedure involves only the probabilities attached to the m^2 subsets of Q, and these probabilities are the only ones influencing the behaviour of X^2 and G^2 Therefore, if (U, V) had a totally nonuniform distribution within some of the little subsquares, but its probability $1/m^2$ was maintained, none of the test statistics could detect such non-uniformity, no matter how large the sample size. We could attempt to get around this problem by increasing the

Table 4.2 *A frequency table.*

	B_0	...	B_j	..	B_c	
A_0	n_{00}	..	n_{0j}	..	n_{0c}	n_{0+}
A_1	n_{10}	..	n_{1j}	..	n_{1c}	n_{1+}
..	.	...	.	..	...	...	..
A_i	n_{i0}	..	n_{ij}	.	n_{ic}	n_{i+}
.	..	...	....	......		.
A_r	n_{r0}	.	n_{rj}	..	n_{rc}	n_{r+}
	n_{+0}	...	n_{+j}	...	n_{+c}	$n = n_{++}$

'resolution', i e m, but this is possible only up to a certain point, for a given value of n: if m is too large, the expected frequencies fall, the approximation of the asymptotic distribution worsens, and the power of the test procedure decreases. □

4.4.9 Frequency tables

Instead of one variable, consider now the case that two categorical variables, A and B, are observed from each subject of a sample with n elements Denote the levels of A by A_0,   , A_r, and those of B by $B_0, \ldots, B_c$ Then Table 4.2 shows the frequency or **contingency** table giving the observed frequency n_{ij} for each of the $(r + 1) \times (c + 1)$ alternative outcomes of the observations. The marginal frequencies are obtained by summing all elements in the same row or column, and are denoted by replacing the corresponding row or column label by a + sign.

The purpose is to test the hypothesis that the $c + 1$ conditional distributions associated with each level of B are all equal. More specifically, if π_{ij} denotes the probability that A takes level A_i conditionally on B equal to B_j, the hypothesis under consideration is the equality of the $c + 1$ multinomial distributions, i e

$$\begin{cases} H_0 : \pi_{ij} = \pi_{i0} & \text{for all } i, j > 0 \\ H_1 : \{\text{some of the equalities do not hold}\} \end{cases}$$

By a simple extension of results of the previous subsection, we obtain that under the null hypothesis the MLE of the common value π_{i0} is n_{i+}/n, while under the alternative $\hat{\pi}_{ij} = n_{ij}/n_{+j}$.

Then it follows that

$$G^2 = W = 2 \left\{ \sum_{i=0}^{r} \sum_{j=0}^{c} n_{ij} \ln \frac{n_{ij}}{n_{+j}} - \sum_{i=0}^{r} n_{i+} \ln \frac{n_{i+}}{n} \right\}.$$

whose asymptotic null distribution is χ^2_{rc}. The corresponding expression of Pearson's X^2 test statistic is

$$X^2 = \sum_{i=0}^{r} \sum_{j=0}^{c} \frac{(n_{ij} - n_{i+}\, n_{+j}/n)^2}{n_{i+}\, n_{+j}/n},$$

but to derive this expression one must introduce the form of W_u under constraint on the parameters, which we shall not develop

Remark 4.4.6 The null hypothesis considered here deals with the two categorical variables asymmetrically. In our exposition, A plays the role of the **response**, while B is called a **factor** The purpose was to establish whether the probabilities of $A_0, \ldots, A_r$ vary with the levels of the factor B. To be specific, A could represent the performance of pupils at a school, and B the social class of the family. The implication is that, in the above discussion, the column totals $n_{+0}, \ldots, n_{+c}$ have been regarded as fixed. It can immediately be seen that, interchanging the roles of A and B, and then fixing the row totals, the corresponding test function is the same, as is the formal test procedure, although the subject-matter implications of statistical significance are very different.

Finally, there is also the possibility of treating symmetrically the categorical variables. From the formal viewpoint, this means that only the total number of observations n is regarded as fixed. Again, the test statistic is unchanged, but not the subject-matter implications.

These points will be discussed further in Chapter 6. For a systematic treatment of categorical data, see Agresti (1990).   □

Remark 4.4.7 Some of the above results are relevant also in a slightly different context. If a homogeneous Markov chain with states $S_0, S_1, \ldots, S_m$ is observed for $n + 1$ time units, the table of transition frequencies is similar to Table 4.2, with $r = c = m$.

To estimate the matrix of transition probabilities (π_{ij}) whose generic entry gives the probability of transition from state S_i to state S_j, we can reuse part of the results of section 4.4.8, arguing as follows. If, for a given state S_i, we consider all transitions with that initial state, these give rise to a likelihood of the form considered in section 4.4.8 with π_j replaced by π_{ij}, since the Markov assumption makes the transitions mutually independent. Therefore, the MLE of the transition probability is $\hat{\pi}_{ij} = n_{ij}/n_{i+}$ (for $i, j = 0, 1, \ldots, m$).

However, the distribution theory for the MLE for dependent data is different from that for the independent case. For a discussion of this point, see for instance Basawa and Prakasa Rao (1980, Chapters 4, 7). □

4.5 Interval estimation

In estimation problems, the mere production of an estimate, $\hat{\theta}$ say, is seldom a satisfactory conclusion. It is almost always necessary to obtain an assessment of the degree of accuracy of the estimate, and correspondingly a set of plausible values of the parameter θ, including $\hat{\theta}$ as one of its elements. When θ is a scalar, this set of plausible values is typically an interval, which is then called an **interval estimate**.

Although logically distinct, the interval estimation problem and the hypothesis testing problem have a strong formal connection, as will be explained soon; this is why we consider interval estimation in this chapter.

4.5.1 Pivotal quantities

Suppose there exists a function $T(y, \theta)$ from $\mathcal{Y} \times \Theta$ to $\mathbb{R}^m$ for some positive integer m, such that $T(Y, \theta)$ has a distribution not depending on θ, when θ is the true parameter value; we then say that $T(y, \theta)$ is a **pivotal quantity**. Notice that $T(y, \theta)$ is not a statistic.

Then, for any subset B of $\mathbb{R}^m$,

$$\Pr\{T(Y, \theta) \in B; \theta\} \tag{4.26}$$

is a value depending on B but not on θ. Choose one set B such that the above probability is $1 - \alpha$ and write the event $\{y : T(y, \theta) \in B\}$ as $\{y : C(y) \ni \theta\}$ where $C(y)$ is a subset of Θ. Then

$$\Pr\{C(Y) \ni \theta; \theta\} = 1 - \alpha \tag{4.27}$$

which must be read as 'the probability that the random set $C(Y)$ includes the true parameter θ is equal to $1 - \alpha$, for any possible value of θ'. We then say that $C(y)$ is a **confidence** set of level $1 - \alpha$ for θ.

Notice that statement (4.27) refers to the random set $C(Y)$, not to the observed set $C(y)$; in other words, statement (4.27) holds a priori, not a posteriori. Therefore, once the sample has been taken, and a specific confidence interval (c_1, c_2) is computed, it is not appropriate to say 'interval (c_1, c_2) contains θ with probability 0 90'. Since (c_1, c_2) is a fixed interval, either it contains the true parameter value, or it does not contain it: there is no randomness in this situation; hence, there is no probability to talk about What it is legitimate to say is: 'the interval (c_1, c_2) is a specific instance of a procedure for selecting intervals, and this procedure generates intervals which 90% of the times contain θ', a sentence which is conventionally summarized as '(c_1, c_2) is a 90% confidence interval' However, the logical involution of the end conclusion is one of the points against this approach.

If Y is a real continuous random variable, then a pivotal quantity always exists. Denoting by $F(y; \theta)$ the distribution function of Y, we have that $F(Y; \theta)$ is uniformly distributed over (0,1), recalling the results of section A 2.2.

Example 4.5.1 If $\overline{y}$ is the arithmetic mean of an s r s of size n from an $N(\theta, 1)$ random variable, then $\overline{Y} - \theta \sim N(0, 1/n)$ is a pivotal quantity, and

$$
\begin{aligned}
1 - \alpha &= \Pr\left\{(\overline{Y} - \theta)\sqrt{n} \in (-z_{\alpha/2}, z_{\alpha/2})\right\} \\
&= \Pr\left\{-z_{\alpha/2}/\sqrt{n} < \theta - \overline{Y} < z_{\alpha/2}/\sqrt{n}\right\} \\
&= \Pr\left\{\overline{Y} - z_{\alpha/2}/\sqrt{n} < \theta < \overline{Y} + z_{\alpha/2}/\sqrt{n}\right\} \\
&= \Pr\left\{(\overline{Y} - z_{\alpha/2}/\sqrt{n}, \overline{Y} + z_{\alpha/2}/\sqrt{n}) \ni \theta\right\}
\end{aligned}
$$

if $\Phi(-z_{\alpha/2}) = \alpha/2$. Therefore $(\overline{y} - z_{\alpha/2}/\sqrt{n}, \ \overline{y} + z_{\alpha/2}/\sqrt{n})$ is a confidence interval of level $1 - \alpha$. In this case, the width of the interval does not depend on the data, but only on n and α, and it is a decreasing function of both. $\qquad\square$

4.5.2 Neyman's approach

The previous subsection has provided a general framework, but it does not say how to obtain pivotal quantities, or how to move from statement (4.26) to (4.27), or how to choose among pivotal quantities, when more than one is known. As for the last point, notice, for instance, that in Example 4.5.1 another pivotal quantity is $\tilde{m} - \theta$, if $\tilde{m}$ is the sample median.

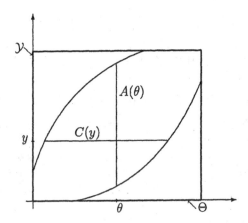

Figure 4 4 *Schematic representation of the construction of a confidence interval according to Neyman's criterion*

In response to the above problem, Jerzy Neyman has proposed a general criterion for selecting a pivotal quantity which is optimal in given sense. A confidence set $C(y)$ is regarded as 'optimal' if the probability that $C(Y)$ contains a parameter value different from the true one is as small as possible. To elaborate, the requirement is

$$\Pr\{C(Y) \ni \theta_*; \theta_*\} = 1 - \alpha,$$
$$\Pr\{C(Y) \ni \theta; \theta_*\} = \text{minimum, for all } \theta \neq \theta_* \tag{4 28}$$

for all $\theta_* \in \Theta$ There is nothing in (4.28) directly implying that the geometric area of $C(y)$, or its length in the scalar case, is the smallest possible among all confidence sets of the same level $1 - \alpha$, although in a large number of examples it turns out that this is indeed the case.

Suppose that a test procedure of level α is available to test the hypothesis that the parameter is equal to a given value, against a two-sided alternative. If θ is the value specified by the null hypothesis, the test procedure identifies an acceptance region, which depends on θ itself, $A(\theta)$ say. On the other hand, if y is given, there exists a set $C(y)$ of θ values for which the null hypothesis is accepted. The logical equivalence, illustrated by Figure 4.4,

$$\{y \in A(\theta)\} \quad \Longleftrightarrow \quad \{\theta \in C(y)\} \tag{4.29}$$

must hold, and then

$$\Pr\{Y \in A(\theta); \theta\} = \Pr\{\theta \in C(Y); \theta\} = 1 - \alpha.$$

The underlying pivotal quantity is the indicator function of the set $\{y : y \in A(\theta)\}$. If the test procedure which defines $A(\theta)$ has maximum power, then

$$\Pr\{Y \in A(\theta); \theta_*\} = \Pr\{\theta \in C(Y); \theta_*\} = \text{minimum}$$

satisfying Neyman's optimality criterion. The practical rule for building a confidence set is then as follows.

Select a test procedure for testing the hypothesis $\theta = \theta_0$ against the alternative $\theta \neq \theta_0$, at level α. The set of all values θ_0 for which the null hypothesis would be accepted is a confidence set of level $1 - \alpha$.

Example 4.5.2 In section 4.4.1, we concluded that the set of sample values leading to acceptance of the null hypothesis (4.22) at level α is such that

$$-t_{\alpha/2} < \frac{(\overline{y} - \mu_0)\sqrt{n}}{s} < t_{\alpha/2}.$$

By proceeding similarly to Example 4.5.1, the corresponding confidence interval for μ is

$$\left(\overline{y} - t_{\alpha/2}\frac{s}{\sqrt{n}}, \ \overline{y} + t_{\alpha/2}\frac{s}{\sqrt{n}}\right).$$

In this case, the width of the interval is random, as it is proportional to s.   □

Example 4.5.3 To obtain a $(1 - \alpha)$-level confidence interval for the parameter σ^2 of an $N(\mu, \sigma^2)$ random variable, recall the results of section 4.4.6. Since $T = n\hat{\sigma}^2/\sigma^2 \sim \chi^2_{n-1}$ is a pivotal quantity, we have

$$\begin{aligned}
1 - \alpha &= \Pr\{c_1 < T < c_2\} \\
&= \Pr\left\{\frac{c_1}{n\hat{\sigma}^2} < \frac{1}{\sigma^2} < \frac{c_2}{n\hat{\sigma}^2}\right\} \\
&= \Pr\left\{\frac{n\hat{\sigma}^2}{c_2} < \sigma^2 < \frac{n\hat{\sigma}^2}{c_1}\right\},
\end{aligned}$$

implying that the confidence interval for σ^2 is $(n\hat{\sigma}^2/c_2, \ n\hat{\sigma}^2/c_1)$.
                                                                    □

Example 4.5.4 Given a value y sampled from a Poisson random variable Y, a confidence interval for its mean value θ is required. To test the hypothesis that θ is equal to θ_0, the LRT statistic is

$$\lambda = \frac{e^{-\theta_0}\theta_0^y/y!}{e^{-\hat{\theta}}\hat{\theta}^y/y!} = e^{y-\theta_0}\left(\frac{\theta_0}{y}\right)^y$$

if $y > 0$. This ratio is a function of y, first increasing and then decreasing; hence the acceptance region is of the form

$$\{y : y_1(\theta_0) < y < y_2(\theta_0)\}$$

where y_1, y_2 are two values, depending on θ_0, chosen so that the level of the test procedure is α.

To perform the inversion required by (4.29), it is convenient to refer to Figure 4.4, even though in our case both Θ and $\mathcal{Y}$ are unbounded, and $\mathcal{Y}$ is discrete; however, this does not invalidate the sort of argument to follow. Of the two curves of Figure 4.4, consider first the top one, associated with the inequality $y < y_2(\theta_0)$ in the specification of the rejection region. Taking into account (A.13), write

$$\Pr\{Y < y\} = \sum_{k=0}^{y-1}\frac{e^{-\theta}\theta^k}{k!} = \Pr\{V_y > \theta\}$$

where V_y is a gamma random variable with index y and scale parameter 1. Setting the above probability equal to $1 - \alpha/2$, since this is the probability of falling in the right-hand part of the rejection region, it follows that the top curve of Figure 4.4, considered as a function of y, gives the $(\alpha/2)$-level upper percentile of the gamma distribution with index y. Since the observed y is discrete, the equality of the above probability to $1 - \alpha/2$ will be only approximate, in general.

By arguing similarly for the bottom curve of Figure 4.4, associated with the inequality $y_1(\theta_0) < y$, we obtain

$$\alpha/2 = \Pr\{Y \le y\} = \sum_{k=0}^{y}\frac{e^{-\theta}\theta^k}{k!} = \Pr\{V_{y+1} > \theta\}$$

where V_{y+1} is a gamma random variable with index $y+1$ and scale parameter 1. Therefore the bottom curve of Figure 4.4 represents the $(1-\alpha/2)$-level percentile of the gamma distribution with index $y+1$.

Since tables of the χ^2 quantiles are more widely available that those of the gamma distribution, it is convenient to rewrite our conclusions using χ^2 quantiles. Then the $(1-\alpha)$-level confidence interval for θ is

$$\left(\tfrac{1}{2}c_{2y}, \ \tfrac{1}{2}c_{2(y+1)}\right)$$

where c_{2y} is the $(\alpha/2)$-level of the χ^2_{2y} distribution and $c_{2(y+1)}$ is the $1 - \alpha/2$ level quantile of the $\chi^2_{2(y+1)}$ distribution.

The treatment of the case $y = 0$ is left to the reader as an exercise.

If n independent observations from the same Poisson random variable are available, let y denote their sum, which is sampled from a *Poisson* $(n\,\theta)$ random variable. The method described above provides a confidence interval for $n\,\theta$, which is immediately transformed into an interval for θ.   □

4.5.3 Approximate methods

In practice, we are not often in the fortunate situation obtaining in the examples of the previous subsection, where the exact distribution of the likelihood ratio is known, and in addition the equivalence (4.29) has an explicit representation. Moreover, even when we are, impracticalities may quickly arise, as illustrated by the apparently innocuous Example 4.5.4.

To make the problem more tractable, a simplification is obtained, without attempting to invert (4.29) in closed form, by applying the rule on page 144 for each specific data set and problem. Moreover, since the exact distribution of the test statistic cannot usually be computed, the asymptotic distribution of the statistic has to be employed.

By using the LRT as the reference test statistic, and working with its asymptotic distribution, the set of values for which the null hypothesis is accepted is

$$\{\theta : 2(\ell(\hat{\theta}) - \ell(\theta)) < c_\alpha\} \qquad (4.30)$$

where c_α is the $(1 - \alpha)$-level quantile of χ^2_k where $k = \dim(\theta)$.

Consider for instance Figure 4.5, representing a log-likelihood function for a scalar parameter θ. A confidence interval with approximate level $1 - \alpha$ is (h_1, h_2), where h_1 and h_2 are obtained by solving the two equations

$$\ell(h_1) = \ell(h_2) = \ell(\hat{\theta}) - c_\alpha/2, \qquad (4.31)$$

where c_α refers to the χ^2_1 distribution. Notice that a different choice of α would lead to a non-connected confidence set, due to the presence of two local maxima.

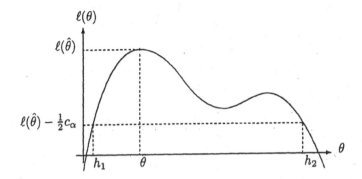

Figure 4.5 *Confidence interval computed via the log-likelihood function*

For general k, application of (4.30) involves finding all values of θ which solve the equation

$$\ell(\theta) = \ell(\hat{\theta}) - c_\alpha/2,$$

a task which is commonly tackled with the aid of a computer. Moreover, there is a practical difficulty in representing graphically the outcome when $k > 2$. In regular circumstances, the above equation determines a bounded connected set of Θ as a confidence region but, again, this does not hold in all cases.

Numerical examples of the implementation of (4.30) will be presented in section 4.5.4. Before that, we describe a further simplification which is commonly used. For a one-dimensional parameter θ, asymptotic results of Chapter 3 state that

$$\Pr\left\{-z_{\alpha/2} < (\hat{\theta} - \theta)\sqrt{I(\theta)} < z_{\alpha/2}\right\} \to 1 - \alpha \qquad (4.32)$$

as $n \to \infty$; here $\Phi(-z_{\alpha/2}) = \alpha/2$. Formula (4.32) applies in all cases for which (3.19) holds; the latter has been proved only for a special case, but its validity is in fact much broader, even though not universal.

From (4.32) we have that, for finite n, the corresponding confidence interval is

$$(\hat{\theta} - z_{\alpha/2}\hat{I}(\theta)^{-1/2}, \; \hat{\theta} + z_{\alpha/2}\hat{I}(\theta)^{-1/2}), \qquad (4.33)$$

with approximate confidence level $1 - \alpha$. The quantity $\hat{I}(\theta)$ is an estimate of $I(\theta)$; we can use either $I(\hat{\theta})$ or $\mathcal{I}(\hat{\theta})$ – the latter is preferable for the reasons presented in section 3.4.8. Moreover, $\mathcal{I}(\hat{\theta})$ has the important practical advantage of being more easily obtained since it does not involve an expected value; in addition, $\mathcal{I}(\hat{\theta})$ can even be computed when $\ell(\theta)$ is too complicated to differentiate explicitly, and numerical differentiation is then used.

If we use $I(\hat{\theta})$ for $\hat{I}(\theta)$ in (4.33), the corresponding confidence interval can be regarded as the outcome of applying the rule on page 144 using the Wald test as reference test procedure.

The quantity $\hat{I}(\theta)^{-1/2}$, called the **standard error**, regulates the width of the confidence interval, and therefore acts as an indicator of the accuracy of the estimate. The term 'standard error' is not necessarily used, however, in connection with the MLE, but in general it denotes the estimated standard deviation of an estimator. It is strongly recommended that, when an estimated value is reported, its standard error is *always* quoted too.

Formula (4.33) has the advantage, compared to (4.30), of being simpler to implement, since it does not require the solution of equations (4.31), which are usually nonlinear. However, there are also disadvantages, as illustrated in Examples 4.5.5 and 4.5.6 below. Moreover, (4.30) enjoys an equivariance property with respect to reparametrizations; cf. Exercise 4 18.

Since the asymptotic theory results are based on the approximation of the log-likelihood by a parabola, it follows that the resulting approximate distributions of the test statistics will perform poorly when the actual shape of the log-likelihood is very different from a parabola. Figure 4.5 illustrates an unfavourable situation in this respect, especially so because the log-likelihood has two local maxima. In other cases, especially if the log-likelihood maintains the property of concavity, the approximation provided by asymptotic theory can be improved by a suitable reparametrization from θ to ψ, as sketched in section 3.3.5 and in Example 3.3.5. The asymptotic theory results are employed on the ψ scale to construct a confidence interval, and this is then converted back to the original θ scale.

Example 4.5.5 If y is a value sampled from a $Bin(n, \theta)$ random variable, formula (4.33) leads to

$$\hat{\theta} \pm z_{\alpha/2} \sqrt{\frac{\hat{\theta}\,(1 - \hat{\theta})}{n}}$$

where $\hat{\theta} = y/n$. If, for instance, $n = 30$ and $y = 5$, then $\hat{\theta} = 0.167$

with standard error 0.068, and the corresponding 95% confidence interval is (0.033, 0.300). At the 99% level, the same formula gives $(-0.009, 0.342)$ part of which interval is inadmissible; in this case, the negative values must be discarded, but it is certainly unpleasant to make such adjustments to fix absurd values produced by the formula.

This situation is caused by the poor degree of approximation provided by (4.33) in the present case. One solution is to go back to (4.30), which cannot produce values outside Θ. However, for the present example, consider the reparametrization via the logit transformation, namely

$$\omega = \ln \frac{\theta}{1 - \theta} = \text{logit}(\theta)$$

which varies in $(-\infty, \infty)$. The asymptotic distribution of the MLE of ω can be obtained using the asymptotic distribution of $\hat{\theta}$ and Corollary A.8.10 on setting $f(\theta) = \text{logit}(\theta)$. It turns out that

$$\sqrt{n}\,(\hat{\omega} - \omega) \xrightarrow{d} N\left(0, \frac{1}{\theta(1 - \theta)}\right)$$

as $n \to \infty$. This asymptotic distribution can be used to construct confidence intervals on the ω scale, and then to convert them back to the θ scale by the inverse logit transformation

$$\theta = \frac{e^{\omega}}{1 + e^{\omega}}.$$

For the data considered earlier, this procedure gives the new intervals (0.071, 0.343) and (0.053, 0.414) at confidence levels 95% and 99%, respectively.

Besides always producing admissible intervals, it can be shown that the use of the logit reparametrization gives more accurate results than the direct use of the θ scale, in the sense that the actual confidence level of the new intervals is closer to the nominal level. $\square$

When the parameter of interest is one component of a multi-dimensional parameter, the standard error is obtained from the appropriate diagonal element of the inverse information matrix.

Example 4.5.6 In Example 3.1.13, we obtained the estimates $\hat{\alpha} = 0.439$ and $\hat{\beta} = 0.862$. From the inverse Hessian matrix of the log-likelihood, the square roots of the diagonal elements give the standard errors, 0.316 and 0.576, respectively. Using (4.33) with a confidence level of 0.95, so that $z_{\alpha/2} = 1.96$, we obtain the confidence intervals

$$(-0.181, \ 1.059), \qquad (-0.267, \ 1.99).$$

Since the data available are quite limited, the standard errors are relatively large, and thus the confidence intervals are wide.

These intervals are 95% confidence sets for α and β separately, and they cannot be combined directly to obtain a confidence region for the pair (α, β). To obtain such a region, (4.30) must be used with $\theta = (\alpha, \beta)$; this procedure will be illustrated in section 4.5.4.

$\square$

As remarked earlier, the inappropriateness of the asymptotic theory results for finite sample sizes is related to departures of the log-likelihood from the ideal quadratic shape. To detect and circumvent such difficulties, the recommended practice is to inspect the graph of $\ell(\theta)$ itself, at least when dim(θ) is 1 or 2. When dim(θ) is higher, but the parameter of interest is still of small dimension, the profile log-likelihood can be considered.

The plot of $\ell(\theta)$ over a reasonable range of θ, particularly when we are dealing with a distribution which does not belong to the exponential family, allows various features to be detected, such as multiple local maxima or non-concave shape, so that appropriate action can be taken, for instance when (4.33) is better replaced by (4.30).

4.5.4 Numerical examples

Example 4.5.7 For a summary illustration of the discussion of the previous subsection, consider the data in the following table, reporting the outcomes of a medical experiment carried out to evaluate the degree of improvement in the state of a group of patients after they have been given a certain therapy.

Improvement after therapy		
None	Mild	Marked
26	22	8

The observed frequencies, denoted by (n_0, n_1, n_2), are sampled from a trinomial distribution, where one of the components, n_0 say, is determined by the other two and the total number n of

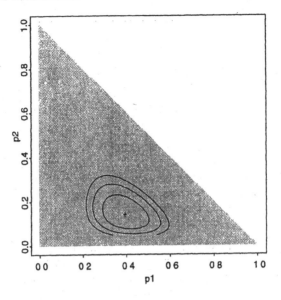

Figure 4.6 *Confidence regions of levels 75%, 95%, 99% for the parameter* (π_1, π_2) *of a trinomial random variable   The MLE is marked by* +

observations. Then the parent random variable is $Bin_2(n, (\pi_1, \pi_2))$, where $n = n_0 + n_1 + n_2$, $0 \le \pi_i \le 1 (i = 1, 2)$ and $\pi_0 = 1 - \pi_1 - \pi_2 \ge 0$. The log-likelihood function is

$$\ell(\pi_1, \pi_2) = \sum_{i=0}^{2} n_i \ln \pi_i$$

with corresponding MLE

$$\hat{\pi}_i = n_i / n \qquad (i = 0, 1, 2)$$

as obtained in section 4.4.8.

Figure 4.6 shows a plot of the log-likelihood, shifted so that it takes value 0 at the MLE, marked by a + symbol, i e this function is the relative log-likelihood. The plot contains three contour curves of ordinate $-\frac{1}{2}c_\alpha$ where c_α is the upper α-level quantile of the χ_2^2 distribution. The portion of the plane inside one contour curve is then a $1 - \alpha$ confidence region for the parameter (π_1, π_2) by rule (4.30).

For the confidence interval of a single parameter component, π_2

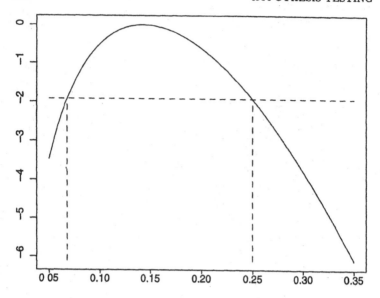

Figure 4 7 *Relative profile log-likelihood versus π_2, and a 95% confidence interval for the parameter π_2 of a binomial random variable*

for instance, consider the profile log-likelihood

$$\ell^*(\pi_2) = \ell(\hat{\pi}_1(\pi_2), \pi_2),$$

where

$$\hat{\pi}_1(\pi_2) = \frac{n_1(1 - \pi_2)}{n_0 + n_1}$$

provides the maximum of ℓ with respect to π_2 for each fixed value of π_1 After some algebra, we obtain

$$\ell^*(\pi_2) = c + n_2 \ln \pi_2 + (n - n_2) \ln(1 - \pi_2)$$

which is the same function as we would have obtained if, from the beginning, the first two outcome types had been merged into a single class, enabling us to work with a binomial distribution.

The plot of $\ell^*(\pi_2)$, in relative form, is provided by Figure 4.7, where a 95% confidence interval is also shown. This interval is again obtained using (4.30) and (4.31), except that c_α is provided by the χ_1^2 distribution.

□

Table 4.3 *Annual mean of the Wolf number, 1749–1979 (read data across) (Data from Andrews and Herzberg (1985), reproduced with permission.)*

8.99	9.13	6.90	6.91	5.54	3.49	3.09	3.19	5.69	6.89
7.34	7.92	9.26	7.82	6.71	6.02	4.57	3.37	6.15	8.35
10.30	10.04	9.03	8.15	5.89	5.53	2.64	4.45	9.61	12.42
11.22	9.20	8.25	6.20	4.77	3.18	4.90	9.10	11.49	11.44
10.87	9.48	8.16	7.74	6.84	6.40	4.61	4.00	2.52	2.01
2.60	3.80	5.83	6.71	6.56	6.89	6.49	5.30	3.17	2.85
1.59	0.00	1.19	2.22	3.49	3.73	5.95	6.76	6.40	5.48
4.89	3.95	2.56	2.00	1.33	2.92	4.07	6.02	7.05	8.00
8.18	8.42	6.91	5.24	2.92	3.63	7.54	11.02	11.76	10.16
9.26	7.94	6.06	4.91	3.26	3.87	6.32	7.84	9.92	11.15
9.79	8.15	8.03	7.36	6.24	4.53	2.59	2.07	4.77	7.40
9.68	9.78	8.78	7.68	6.63	6.85	5.52	4.03	2.69	6.10
8.59	11.79	10.55	10.08	8.14	6.68	4.13	3.36	3.50	1.83
2.44	5.68	7.36	7.72	7.98	7.96	7.22	5.03	3.61	2.59
2.50	2.65	5.96	8.54	9.21	8.83	7.99	6.46	5.12	5.16
3.48	3.07	1.65	2.24	4.93	6.47	7.96	7.33	7.87	6.96
6.62	4.31	2.38	1.89	1.20	3.09	6.88	7.55	10.19	8.97
7.97	6.13	5.11	3.77	2.40	4.08	6.65	7.99	8.30	8.82
8.06	5.97	4.60	3.33	2.37	2.95	6.00	8.92	10.70	10.47
9.42	8.23	6.89	5.53	4.04	3.09	5.75	9.61	12.24	11.67
11.62	9.16	8.35	5.60	3.72	2.10	6.16	11.90	13.78	13.59
12.60	10.60	7.34	6.13	5.28	3.19	3.88	6.84	9.67	10.29
10.27	10.23	8.16	8.30	6.17	5.86	3.93	3.54	5.24	9.61
12.47									

Example 4.5.8 The Wolf number is an index which measures sunspot activity; it reflects both the total number of sunspots and the number of sunspot clusters observed at any given time on the sun's surface. Table 4.3 contains the annual mean of the Wolf number, obtained by averaging monthly data reported by Andrews and Herzberg (1985, pp. 67–74). This famous time series has been analysed in detail by several authors, and some very elaborate formulations have been proposed to model the observed data. However, we shall confine our discussion to the simplest approach.

The data are plotted in Figure 4.8 after a square root transformation has been applied. This transformation is monotonic for

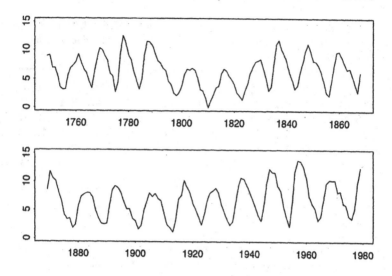

Figure 4 8 *Plot of square root of the Wolf number, 1749–1979*

positive values like the entries of Table 4.3, hence it does not change the actual information contained in the table, and it has the advantage of making the marginal distribution of the data closer to normality, which simplifies the subsequent analysis. If we denote by y_t the square root of the tth datum, it is the series $(y_1, \ldots, y_n)$ which we shall regard as the variable of interest to us, rather than the original data. In Figure 4.8, pairs of successive observations have been joined by a straight line to improve readability.

First of all, notice that Figure 4.8 exhibits an obvious oscillating pattern, but the distances between adjacent peaks and the heights of the peaks are not constant, so that the series cannot be interpolated by a periodic function. Pseudo-periodic behaviour, similar to that observed in this time series, can be produced by a second-order autoregressive process of the form

$$Y_t - \mu = \alpha_1(Y_{t-1} - \mu) + \alpha_2(Y_{t-2} - \mu) + \varepsilon_t$$

where Y_t represents the random variable generating the value observed at time t, μ is the mean value, and α_1, α_2 are the autoregressive parameters. Equivalently, we write

$$Z_t = Y_t - \mu$$

$$Z_t = \alpha_1 Z_{t-1} + \alpha_2 Z_{t-2} + \varepsilon_t,$$

where the process $\{Z_t\}$ has the same correlation structure as $\{Y_t\}$ but zero mean value

We shall now digress to discuss some general properties of second-order autoregressive processes. The notion of the autoregressive process has been presented in Example 2.4.5 with special emphasis on the first-order case However, we shall not describe in detail all the technical aspects of the second-order case, leaving a number of these as an exercise; the interested reader could consult the literature on time series, for instance Priestley (1981, in particular pp. 169–173).

In Example 2 4.5, to make sure that all the components of a first-order autoregressive process have the same marginal distribution, a particular value was chosen for the variance of Y_1. A similar condition is imposed here, involving this time the first two components, and requiring that

$$\begin{pmatrix} Y_1 \\ Y_1 \end{pmatrix} \sim N_2 \left(\begin{pmatrix} \mu \\ \mu \end{pmatrix}, \sigma^2 \begin{pmatrix} 1 - \alpha_2^2 & -\alpha_1(1 + \alpha_2) \\ -\alpha_1(1 + \alpha_2) & 1 - \alpha_2^2 \end{pmatrix}^{-1} \right)$$

where the autoregressive parameters α_1, α_2 must satisfy the constraints

$$|\alpha_2| < 1, \quad \alpha_1 + \alpha_2 < 1, \quad \alpha_2 - \alpha_1 < 1,$$

which determine a triangular region, T say. Therefore, the parameter space of $\theta = (\mu, \alpha_1, \alpha_2, \sigma^2)$ is given by $\mathbb{R} \times T \times \mathbb{R}^+$.

Given an observed realization $(y_1, \ldots, y_n)$ of the process, the corresponding likelihood can be obtained by multiplying the density function for the first two components by the conditional distribution of the subsequent ones, schematically

$$L(\mu, \alpha_1, \alpha_2, \sigma^2) \propto f_2(y_1, y_2) \prod_{t=3}^{n} f_c(y_t | y_{t-1}, y_{t-2}),$$

where f_2 is the above-described density of (Y_1, Y_2), and

$$f_c(y_t | y_{t-1}, y_{t-2}) = \frac{1}{\sqrt{2\pi\sigma^2}} \exp\left(-\frac{1}{2\sigma^2}(z_t - \alpha_1 z_{t-1} - \alpha_2 z_{t-2})^2 \right)$$

is the conditional density of Y_t $(t = 3, \ldots, n)$ given the past values, and $z_j = y_j - \mu$ $(j = 1, \ldots, n)$.

In time-series methodology, it is customary to use an approximated form of likelihood, ignoring the f_2 term in the above expression for L. This approximation produces an expression which lends itself to a much simpler algebraic treatment. Moreover, it can be shown that the contribution to L from the f_2 term is small compared to the contribution from the product of the f_c terms, when n is large and (α_1, α_2) is not close to the boundary of the parameter space.

However, we shall retain the exact form of L, which is usable for any choice of n, α_1, α_2 To write down the likelihood function in compact form, define

$$\alpha = \begin{pmatrix} -1 \\ \alpha_1 \\ \alpha_2 \end{pmatrix}, \qquad N = \begin{pmatrix} n & n-1 & n-2 \\ n-1 & n-2 & n-3 \\ n-2 & n-3 & n-4 \end{pmatrix},$$

$$D_1 = \begin{pmatrix} d_0 & d_1 & d_2 \\ d_0 & d_1 & d_2 \\ d_0 & d_1 & d_2 \end{pmatrix}, \qquad D_2 = \begin{pmatrix} d_{00} & d_{01} & d_{02} \\ d_{10} & d_{11} & d_{12} \\ d_{20} & d_{21} & d_{22} \end{pmatrix},$$

where d_r and d_{rs} are defined as in Example 2.4.5; notice that D_2 is a symmetric matrix. Simple but tedious algebra shows that twice the log-likelihood is

$$2\,\ell(\mu, \alpha_1, \alpha_2, \sigma^2) = -n \log \sigma^2 + \log((1+\alpha_2)^2\{(1-\alpha_2)^2 - \alpha_1^2\})$$
$$- \frac{1}{\sigma^2}\,\alpha^\top (D_2 - 2\mu\,D_1 + \mu^2\,N)\alpha.$$

Removing duplicated components in D_1 and D_2, we conclude that the minimal sufficient statistic is nine-dimensional.

A closed-form expression for the MLE is not feasible. However, partial explicit maximization of ℓ is possible, by maximizing the log-likelihood for each given value of α_1, α_2 with respect to the other two parameters, obtaining

$$\hat{\mu}(\alpha_1, \alpha_2) = \frac{\alpha^\top D_1 \alpha}{\alpha^\top N \alpha},$$

$$\hat{\sigma}^2(\alpha_1, \alpha_2) = \frac{1}{n}\alpha^\top (D_2 - 2\,\hat{\mu}(\alpha_1, \alpha_2)\,D_1 + \hat{\mu}(\alpha_1, \alpha_2)^2\,N)\alpha$$
$$= \frac{1}{n}\left(\alpha^\top D_2 \alpha - \frac{(\alpha^\top D_1 \alpha)^2}{\alpha^\top N \alpha} \right),$$

When these expressions are substituted in ℓ, a profile log-likelihood function

$$\ell^*(\alpha_1, \alpha_2) = \ell(\hat{\mu}(\alpha_1, \alpha_2), \alpha_1, \alpha_2, \hat{\sigma}^2(\alpha_1, \alpha_2))$$

is defined, which is more easily handled since it involves only two parameters instead of four.

Armed with these theoretical results, we return to the sunspots time series. The observed values of the non-redundant components of D_1 and D_2 are

$$(1494 \quad 1472 \quad 1454), \qquad \begin{pmatrix} 11\,570 & & \\ 11\,105 & 11\,333 & \\ 10\,300 & 10\,903 & 11\,157 \end{pmatrix},$$

respectively. For these values of the sufficient statistic, the relative profile log-likelihood function $\ell^*(\alpha_1, \alpha_2) - \ell^*(\hat{\alpha}_1, \hat{\alpha}_2)$ is plotted in Figure 4.9 using a contour-level representation The maximum of this function occurs at $\alpha_1 = 1.394, \alpha_2 = -0.700$, represented in the contour plot by a + sign; the associated values of the other two parameters are $\mu = 6.493, \sigma^2 = 1.426$.

The log-likelihood function descends quite rapidly away from the MLE point in the (α_1, α_2) plane, indicating a sharp preference for the selected point. However, the rate of decrease of the function values is not the same in the various directions, as indicated by the elliptical shapes of the contour curves. Moreover, the negative slope of the main axes of the ellipsoids indicates negative correlation between the estimates.

The levels of the contour plot in Figure 4.9 have been chosen equal to one half of the upper percentage points of the χ_2^2 distribution; we are thus proceeding exactly as in Example 4.5.7. Notice, however, a major difference with respect to the other case: the autoregressive nature of the process generating the data induces serial correlation among successive observations.

The theoretical framework of the present example is, then, not covered by the asymptotic theory developed in the last two chapters. As already mentioned, however, the actual validity of the asymptotic results obtained extends to much wider situations. In particular, it covers autoregressive and similar models used in time-series analysis; see, for instance, Anderson (1971, especially sections 5.5 5.6). For this reason, we can retain the regions depicted in Figure 4.9 as a valid confidence regions at the quoted confidence levels.

As a simple check of the adequacy of the fitted model, some simulated data are plotted in Figure 4.10 These have been generated with the aid of a computer sampling a sequence from an autoregressive process with the same parameters as our estimates $(\hat{\mu}, \hat{\alpha}_1, \hat{\alpha}_2, \hat{\sigma}^2)$. It is apparent that the simulated series reproduces to a moderate extent the features of the observed series in Fig-

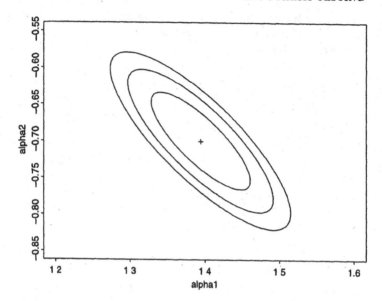

Figure 4.9 *Confidence regions of levels 75%, 95%, 99% for the parameter* (α_1, α_2) *of a second-order autoregresive model. The MLE is marked by* $+$

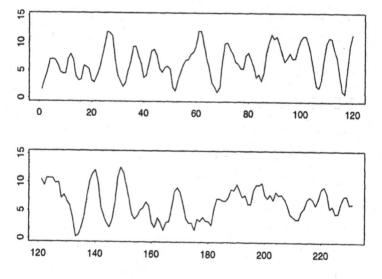

Figure 4.10 *Simulated data from the fitted autoregressive process*

ure 4.8, but it is also quite clear that the simulated series lacks certain aspects of the real data, such as its more distinct regularity of fluctuation and the peculiar asymmetry of the peaks, which rise more rapidly than they fall.

More elaborate modelling of the data is therefore called for. However, we shall not develop this analysis any further, and refer the reader to the literature quoted by Andrews and Herzberg (1985, pp. 67–74) and by Priestley (1988, pp. 69–72 and p. 81).   □

Exercises

4.1 Remark 4.2.5 states that Wald test statistic is not invariant with respect to reparametrizations, while the score function and the LRT statistics enjoy this property. Prove these assertions.

4.2 Obtain the power function of the test procedure discussed in Remark 4.3.1.

4.3 Suppose that observations are taken from independent random variables, all of negative exponential type but with different parameter. Specifically, the ith distribution has scale parameter x_i/θ $(i = 1, \ldots, n)$, where $(x_1, \ldots, x_n)$ are known positive reals and θ is an unknown positive parameter.

(a) Construct a test procedure for the hypotheses

$$\begin{cases} H_0 : \theta = \theta_0, \\ H_1 : \theta > \theta_0. \end{cases} \tag{4.34}$$

(b) Construct a 95% confidence interval for θ.

4.4 If $(y_1, \quad, y_n)$ is an s r s from a beta random variable with parameters $(\theta, 1)$, derive a test procedure for the hypotheses

$$\begin{cases} H_0 : \theta = 1, \\ H_1 : \theta \neq 1. \end{cases}$$

Show that the LRT criterion rejects the null hypothesis if

$$Q = -\sum_{i=1}^{n} \ln y_i$$

is sufficiently large or sufficiently small. Show that the null distribution of $2Q$ is χ^2_{2n}, and determine the power function of the test procedure.

4.5 The density function

$$f(y; \mu, \sigma) = \begin{cases} \dfrac{1}{\sqrt{2\pi}\sigma y} \exp\left[-\dfrac{1}{2\sigma^2}(\ln y - \mu)^2\right] & \text{if } y > 0, \\ 0 & \text{otherwise,} \end{cases}$$

where $\sigma > 0$, is obtained by applying the $\exp(\cdot)$ transformation to an $N(\mu, \sigma^2)$ random variable. Although nearly everybody agrees that f should be called the 'antilognormal' or 'exp-normal' distribution, it is commonly called 'log normal'. Assume that an s r s from this distribution is available, and develop a test procedure for hypotheses (4.22), stating whether or not it is an exact level test.

4.6 Let $(Y_1, \ldots, Y_n)$ be independent normal random variables with unit variance and mean values

$$\mathrm{E}[Y_i] = \theta_1 + \theta_2 t_i \quad (i = 1, \ldots, n),$$

where $(t_1, \ldots, t_n)$ are known constants such that $\sum t_i = 0$, and θ_1, θ_2 are unknown parameters. Obtain the LRT procedure for the hypotheses

$$\begin{cases} H_0 : \theta_1 = \theta_2 = 0, \\ H_1 : \text{at least one parameter is not 0.} \end{cases}$$

Compute the exact distribution of the test statistic, both under the null and the alternative hypothesis.

4.7 Assume that an s r s from $N(\eta, \sigma^2)$ and an independent s r s from $N(\mu, \sigma^2)$ are available. Derive the LRT procedure for

$$\begin{cases} H_0 : \mu \le \eta, \\ H_1 : \mu > \eta. \end{cases}$$

4.8 Assume that an s r s from $N(\mu_1, \sigma_1^2)$ and an independent s r s from $N(\mu_2, \sigma_2^2)$ are available. Obtain the LRT procedure for

$$\begin{cases} H_0 : \sigma_1^2 = \sigma_2^2, \\ H_1 : \sigma_1^2 \ne \sigma_2^2. \end{cases}$$

4.9 (Generalization of Exercise 4.8 to several populations.) From m normal populations with mean values $\mu_1, \ldots, \mu_m$ and variance $\sigma_1^2, \ldots, \sigma_m^2$, respectively, samples of size $n_1, \ldots, n_m$ are drawn. Consider the LRT statistic to test the hypothesis that all variances are equal against the alternative that

$\sigma_i^2 \neq \sigma_j^2$ for some i and j, and show that it is a function of

$$\prod_{i=1}^n \left(\frac{s_i^2}{s_a^2}\right)^{n_i/2}$$

for suitable quantities $s_1^2, \ldots, s_m^2, s_a^2$. Determine the form of the rejection region, and compute the asymptotic distribution of $W(Y)$ when all sample sizes $n_1, \ldots, n_m$ go to infinity at the same rate.

4.10 Let $(y_1, \ldots, y_n)$ be an s r s from $N_k(\mu, \Omega)$ as in Exercise 2.14. Using the LRT criterion, obtain a test statistic for the null hypothesis stating that Ω is a diagonal matrix against the alternative hypothesis that it is not diagonal. Obtain the distribution of the test statistic.

4.11 Given a positive real μ_0, obtain the LRT procedure for the hypotheses

$$\begin{cases} H_0 : \mu = \mu_0, \\ H_1 : \mu \neq \mu_0, \end{cases}$$

about the mean value μ of an exponential random variable assuming that an s r s is available. Check by direct computation that the null asymptotic distribution of $W = -2\ln\lambda$ is χ_1^2

In reliability experiments, units are activated and allowed to work until a failure occurs; the time y_i at which the ith unit fails is recorded $(i = 1, \ldots, n)$. Often, to avoid the experiment lasting too long, only the first r failure times, i e the first r order statistics, $y_{(1)} < y_{(2)} < \cdots < y_{(r)}$, are recorded, and then the experiment is stopped. Assuming that the failure times are distributed exponentially, obtain a test procedure for the above hypothesis H_0 about the mean μ of the distribution when only the censored sample $(y_{(1)}, y_{(2)}, \ldots, y_{(r)})$ is available.

4.12 Extend the analysis of variance method to the case of unequal sample sizes among the various populations.

4.13 Verify that, when $r = 1$, the test function W_u of section 4.4.8, namely X^2, collapses to the test function W_u of section 4.4.7.

4.14 Verify that X^2 in section 4.4.9 becomes W_u from section 4.4.8, if $r = 1$.

4.15 Compute $E[X^2]$ when X^2 is defined as in section 4.4.8.

4.16 In the setting of Example 4.5.4, obtain a confidence interval for θ when $y = 0$ is observed.

4.17 If an s r s from a random variable with distribution function $F(\cdot)$ is available, obtain an interval estimate for $F(1)$. [Note that this is a problem in nonparametric statistics.]

4.18 Show that the confidence intervals produced by rule (4.30) are equivariant with respect to reparametrizations, while those produced by (4.33) are not.

CHAPTER 5

Linear models

5.1 Relationships among variables

Whatever the context (science, technology or business) and the type of study (observational or experimental) under consideration, only infrequently are we interested in a single variable, or even in several variables considered separately. Far more often, we wish to examine the relationships linking the variables under examination. It is not too hazardous to state that most empirical studies are motivated by the study of relationships among variables. Examples are so obvious that an explicit list seems unnecessary.

In the empirical analysis of the relationships among variables, statistical methods play a central role, and the present chapter is entirely dedicated to introducing suitable techniques for such an analysis. This account is limited to somewhat specialized techniques; however, experience has taught us that these techniques are those with greatest practical relevance, and in addition they provide tools for developing more general and elaborate techniques.

The first restriction that we introduce is the presence of a **response variable**, y say, and one or more **explanatory variables**,* say $x_1, \ldots, x_p$. Therefore the variables are treated asymmetrically, as the response variable is considered to vary according to variations in the explanatory variables, as the term itself implies.

In the previous chapters we have already discussed some cases, albeit simple, containing an explanatory and a response variable. In the logistic regression framework of Example 2.4.4, the variable x represents the dose of a certain substance which is given to individuals and plays the role of explanatory variable; the variable y

---

* The terms **concomitant variable** and **covariate** are synonyms of the term 'explanatory variable' The term **dependent variable** is sometimes used instead of 'response variable'; similarly, explanatory variables are sometimes called **independent variables** However, the use of the terms 'independent' and 'dependent' is not recommended, at the very least because the 'independent' variables can be quite strongly correlated

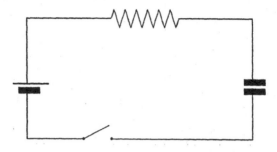

Figure 5.1 *RC circuit*

is a binary variable denoting the presence or absence of a certain event and plays the role of the response variable. Another example is provided by the contingency tables in section 4.4.9, where one factor is treated as the response and the other factor as explanatory. Notice that we do not distinguish here between quantitative and qualitative (also called factor) variables.

Our second restriction is that the variable y represents the sum of two terms, namely

$$y = r(x_1, \ldots, x_p) + \varepsilon. \tag{5.1}$$

Here, the term $r(x_1, \ldots, x_p)$ is called the **systematic component** and represents the contribution to y from the explanatory variables. The term ε is called the **erratic component** or **error term**; ε is the random discrepancy between y and $r(x_1, \ldots, x_p)$, free from any connection with the explanatory variables. Expression (5.1) is said to define a **regression model**. Notice that, although quite general, this model does not include either of the above-mentioned examples, i.e. logistic regression and frequency tables.

There are cases, especially in experimental settings, where both y and the x_i variables are quantitative, and the form of the function $r(\cdot)$ is known from subject-matter considerations, except that its expression depends upon certain unknown characteristic values, i e parameters. Our target is to make inferences on these parameters from a given set of observations.

Consider for instance the electric RC circuit in Figure 5.1. If one measures the voltage V at the capacitor terminals starting when

the circuit is closed, it is known that the following relationship
holds (using suitable measurement units):

$$V = E\{1 - \exp(-t/T)\}, \qquad (5.2)$$

where E denotes the voltage of the power supply unit, t is the time
since closure of the circuit, and $T = RC$ is a constant characteristic
value of the circuit, depending on R and C, the characteristic values
of the resistor and of the capacitor, respectively.

In this example, we can regard V as the response variable, t
as an explanatory variable, E and T as parameters which must
be evaluated. The function $r(\cdot)$ is given by the right-hand side of
(5.2).

Let us now close the circuit Figure 5.1 and take some measurements at successive time points, obtaining various pairs of values (t, V). If the measuring instruments were error-free, the read-
ings from the instruments were totally accurate and the circuit
behaviour were ideal (without voltage ripples in the power supplied or other irregularities), then two pairs of values, (t_1, V_1) and
(t_2, V_2) say, would be sufficient to identify the values of E and T
by solving two nonlinear equations.

In practice, the circuit behaviour is not expected under ideal conditions and the observed pairs of values (t, V) satisfy a relationship
of the form

$$V = E\{1 - \exp(-t/T)\} + \varepsilon \qquad (5.3)$$

where ε is a term which includes all sources of departure from the
ideal behaviour. Unless some additional information is available,
the ε term will be regarded as a random component.

Under scheme (5.3), we cannot simply solve two nonlinear equations to identify the parameters E and T, and a different approach
must be pursued.

We shall deal with a special case of (5.1), where $r(\cdot)$ is a **linear
function of the parameters**, namely

$$y = \beta_1 x_1 + \cdots + \beta_p x_p + \varepsilon, \qquad (5.4)$$

where $\beta_1, \ldots, \beta_p$ are the unknown parameters.

As a special case of (5.1), the class of models defined by (5.4)
excludes some kinds of relationship, such as (5.3). Nevertheless, in
spite of its apparent limitations, model (5.4) turns out to be useful
in a very large number of real data analyses. The reasons for the
practical success of model (5.4) rest both in its logical and math-

ematical structure as well as in its flexibility in handling practical situations: therefore is it worth explicitly listing its features.

- First of all, (5.4) is extremely simple from the logical viewpoint.

- There are cases where the relationship among the relevant variables is indeed of type (5.4). For instance, if we take the height of a boy as the response variable and the height of his father as an explanatory variable, then we observe an approximately linear relationship, although with a substantial error component.*

- Sometimes, (5.4) holds for transformed variables of the original variables. For instance, consider the steady-state equation of a perfect gas (using appropriate measurement units),

$$pV = RT,$$

where p represents pressure, V volume, R a universal constant and T the absolute temperature. Suppose that we wish to check the above equation experimentally, and that at the same time we wish to estimate the value of R; assume that the experiment was set up so as precisely to control the temperature and volume of the gas, while the pressure was determined by the other two variables. After logarithmic transformation, we have

$$\log p = \log R + \log T - \log V$$

which is of type (5.4), except for the ε term. Since the components of the new model are directly related to the original variables, then conclusions can be transferred back again.

- Model (5.4) does not require that the relationship is linear with respect to the variables, just with respect to the parameters. Therefore an expression such as

$$y = \beta_1 + \beta_2 z_1 + \beta_3 z_1^3 + \beta_4 z_1 z_2 + \beta_5 \cos(z_2 + 1) + \varepsilon$$

is included as it can be written as (5.4), with $p = 5$, on setting $x_1 = 1$, $x_2 = z_1$, $x_3 = z_1^3$, $x_4 = z_1 z_2$, $x_5 = \cos(z_2 + 1)$.

- Even when it is not entirely appropriate, as in (5 3), (5.4) can

---

* This phenomenon was noticed by F Galton Further, since the slope of the line was less than 1, meaning that the son of a very tall father is not as tall as the father and the son of a very short father is not as short as the father, he introduced the term 'regression to the mean' to describe this tendency of returning towards the central value of the distribution of points Later on, the term 'regression' came into common use to denote any form of linear relationship

provide an approximation to the 'real' expression connecting
the variables. In a number of cases, one is interested mainly in
the local behaviour of the relationship in a neighbourhood of a
nominated point, and a regular function can be locally approximated via a Taylor series expansion. For instance, expanding
(5.3) up to the second-order term in a neighbourhood of t_0, we
have

$$V = \beta_0 + \beta_1 x_1 + \beta_2 x_2 + \varepsilon \qquad (5\,5)$$

where

$$\begin{aligned}
\beta_0 &= E(1 - e^{-t_0/T})/T, \\
\beta_k &= (-1)^{k-1} E e^{-t_0/T}/T^k, \quad x_k = (t - t_0)^k/k! \quad \text{for } k = 1, 2
\end{aligned}$$

When there is no available theory supporting a specific form for
the function $r(\cdot)$, then it is plausible to use (5.4) as a working
approximation.

- Model (5.4) lends itself to a relatively simple mathematical treatment of the associated statistical techniques. This feature is related to but distinct from the first remark of the present list.

- One can include in the model both quantitative and qualitative
  explanatory variables, as described later.

The purpose of the rest of this chapter is the analysis of a certain number, n say, of observations sampled from model (5.4). The
main aim of the analysis is to make inferences on the values of the
parameters $\beta_1, \ldots, \beta_p$. Since the error component ε is considered
to have been generated by a random mechanism, y also has to be
regarded in the same way; therefore the stochastic properties of ε
must be specified before proceeding.

5.2 Second-order hypotheses

5.2.1 The least-squares criterion

Suppose that n observations $(y_1, \ldots, y_n)$ are available, where each
component is generated by (5.4). These observations are sampled
from random variables $(Y_1, \ldots, Y_n)$ satisfying

$$Y_i = \beta_1 x_{i1} + \cdots + \beta_p x_{ip} + \varepsilon_i \quad (i = 1, \ldots, n), \qquad (5.6)$$

where Y_i is the ith component of the response variable, x_{i1} is the
ith value of the explanatory variable x_1, and so on. Using matrix

notation for compactness, we write

$$Y = X\beta + \varepsilon \tag{5.7}$$

where $Y = (Y_1, \ldots, Y_n)^\top$ is the random vector with the n components of the response variable, $X = (x_{ij})$ is an $n \times p$ matrix, called the **design matrix** or **regression matrix**, containing the values of the p explanatory variables $(n \geq p)$, $\varepsilon = (\varepsilon_1, \ldots, \varepsilon_n)^\top$ is a vector containing the n components of the error term, and $\beta = (\beta_1, \ldots, \beta_p)^\top$ is the vector containing the **regression parameters**. Assume that

- $E[\varepsilon] = 0$;

- $\text{var}[\varepsilon] = \sigma^2 I_n$ for some unknown positive σ^2;

- X is a matrix of non-random elements with full rank p.

We remind the reader of the operators $E[\cdot]$, $\text{var}[\cdot]$ for random vectors, defined in section A.4.2.

If (5.7) and the related assumptions hold, we say that (5.7) defines a linear regression model or, briefly, a **linear model**. The above set of assumptions concerning ε is said to represent the **second-order hypotheses** since only moments up to second-order are involved.

Model (5 7) and related assumptions are particularly suited for the mathematical description of problems arising from controlled experiments, where the experimenter can control the values taken by relevant factors to examine the corresponding value of the response variable. In this setting, X contains the values of the experimental factors, which are non-stochastic since they are chosen by the experimenter. The error term is due to measuring errors (which explains the name) and, if the instruments are not biased, then it follows that $E[\varepsilon] = 0$. Finally, if the various experiments are conducted in such a way as not to influence each other, then the stochastic independence assumption is satisfied, implying uncorrelated errors.

Using y ($y \in \mathbb{R}^n$), the observed vector of the multivariate random variable Y, we aim to make inferences on the set of $p + 1$ parameters $\beta_1, \ldots, \beta_p, \sigma^2$ (although σ^2 is only a nuisance parameter). It follows immediately that

$$E[Y] = X\beta, \quad \text{var}[Y] = \sigma^2 I_n, \tag{5.8}$$

but we do not have enough ingredients to write down the probability distribution of Y, hence not even the likelihood function for the observed vector y.

To follow the likelihood principle, it would be necessary to introduce additional hypotheses on the distribution of ε, from which we can obtain the distribution of Y. For the moment, we prefer not to do so; instead, we introduce an alternative criterion in place of the likelihood principle. A reasonable proposal is to choose β so as to minimize

$$\|y - \mu\|,$$

the Euclidean distance between the observed vector and its mean value, $\mu = \mathrm{E}[Y]$, under the proposed linear model; this mean value is a function of β. Equivalently, we can minimize the square of this distance,

$$
\begin{aligned}
Q(\beta) &= \|y - \mu\|^2 \\
&= (y - \mu)^\top (y - \mu) \\
&= (y - X\beta)^\top (y - X\beta).
\end{aligned}
\tag{5.9}
$$

Obviously, here we regard β as a variable quantity, not the 'true value', similarly to maximum likelihood estimation.

With the present choice, we are establishing the **least-squares criterion (or method)**, whose name reminds us that it operates by minimizing the sum of squared differences between the components of y and the corresponding components of $\mu = \mathrm{E}[Y]$.

One must not get the impression that the least-squares criterion is relevant only to linear models (5.7). In fact, it applies to the more general form (5.1), assuming that $r(\cdot)$ depends on some parameter β. The only difference is that, in (5.9), μ will no longer be a linear function of β. Therefore, in such a case, we use the terms **nonlinear model** and **nonlinear least squares**. However, we shall not develop this aspect, and restrict ourselves to (5.7).

Since, for any variable (both response and explanatory), all that we know is the set of observed values included in its corresponding vector, it is common practice to describe such a vector as 'variable'.

5.2.2 Least-squares geometry

In this section, we consider the various entities from the geometrical point of view, ignoring for the moment the statistical and probabilistic aspects. Regard vectors $y, x_1, \ldots, x_p$, containing the

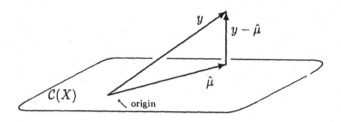

Figure 5 2 *Projection of y onto $\mathcal{C}(X)$*

response and the explanatory variables respectively, as elements of the vector space $\mathbb{R}^n$. As β varies in $\mathbb{R}^p$, the expression $X\beta = \beta_1 x_1 + \cdots + \beta_p x_p$ represents a linear combination of the columns $x_1, \ldots, x_p$ of X with coefficients β, i e it gives the parametric equation of the linear subspace of $\mathbb{R}^n$ generated by the columns of X. This subspace, which we shall denote by $\mathcal{C}(X)$, is a linear subspace, with dimension p, of $\mathbb{R}^n$. In fact, if $X\beta \in \mathcal{C}(X)$ and $a \in \mathbb{R}$, then $a(X\beta) = X(a\beta) \in \mathcal{C}(X)$; moreover, if $X\beta$ and $Xb \in \mathcal{C}(X)$, then also $X\beta + Xb = X(\beta + b) \in \mathcal{C}(X)$.

Model (5.7) states that $\mu = \mathrm{E}[Y]$ lies in $\mathcal{C}(X)$ and the least-squares criterion chooses the element of $\mathcal{C}(X)$ which minimizes the Euclidean distance between the vector y and the space $\mathcal{C}(X)$ Denote this element of $\mathcal{C}(X)$ by $\hat{\mu} = X\hat{\beta}$, which is uniquely identified by the coefficients $\hat{\beta} \in \mathbb{R}^p$. Figure 5.2 illustrates the situation.

To identify explicitly $\hat{\beta}$, let us minimize the function $Q(\beta)$ by making use of differentiation rules for matrix expressions, namely

$$\frac{d}{dx}Ax = A^\top, \quad \frac{d}{dx}x^\top Bx = 2Bx$$

if B is a symmetric matrix. By differentiating $Q(\beta)$ with respect to β, we obtain

$$
\begin{aligned}
\frac{d}{d\beta}Q(\beta) &= \frac{d}{d\beta}(y - X\beta)^\top(y - X\beta) \\
&= \frac{d}{d\beta}(y^\top y - 2y^\top X\beta + \beta^\top X^\top X\beta)
\end{aligned}
$$

$$= 2(X^\mathsf{T} X \beta - X^\mathsf{T} y)$$

and by equating this derivative to 0, we deduce that the required minimum point satisfies

$$X^\mathsf{T} X \beta = X^\mathsf{T} y, \qquad (5\ 10)$$

called the **normal equations**.

Inversion of the matrix $X^\mathsf{T} X$ is legitimate since the condition that the rank of X is equal to p implies also that $X^\mathsf{T} X$ has rank p Therefore, we conclude that the minimum of $Q(\beta)$ is attained when β equals

$$\hat{\beta} = (X^\mathsf{T} X)^{-1} X^\mathsf{T} y. \qquad (5\ 11)$$

The estimates $\hat{\beta}$ of the regression parameters are also called **regression coefficients**.

To check that (5 11) does indeed give a minimum point for $Q(\beta)$, consider the matrix of second derivatives

$$\frac{d^2}{d\beta\, d\beta^\mathsf{T}} Q(\beta) = 2\, X^\mathsf{T} X$$

which is a positive definite matrix, ensuring that $Q(\beta)$ is a convex function.

Equation (5.10) also can be obtained by a geometric rather than an algebraic argument: using well-known results from geometry, we can say that the vector $\hat{\mu} \in \mathcal{C}(X)$ which minimizes the distance from y is such that

$$(y - X\hat{\beta}) \perp \mathcal{C}(X)$$

which requires orthogonality of $(y - \hat{\mu})$ and the base vectors of $\mathcal{C}(X)$, i e we need

$$(y - X\hat{\beta})^\mathsf{T} X = 0,$$

which is equivalent to (5.10).

Associated with $\hat{\beta}$ there is the **projection** vector of y onto $\mathcal{C}(X)$, called the vector of **fitted values**, namely

$$\begin{aligned} \hat{\mu} &= X\hat{\beta} \\ &= X(X^\mathsf{T} X)^{-1} X^\mathsf{T} y \\ &= Py \end{aligned} \qquad (5.12)$$

where $P = X(X^\mathsf{T} X)^{-1} X^\mathsf{T}$ is called the **projection matrix** on $\mathcal{C}(X)$. This matrix effectively defines an operator, associated with the matrix X, whose role is in fact to project any vector $y \in \mathbb{R}^n$, transforming it into another vector, $Py \in \mathcal{C}(X)$, having minimum distance from y. It is easily checked that P is symmetric and $P^2 = P$, i e it is idempotent, which implies $Py = P^2y$; the geometric meaning of this formal fact is that projecting a projection has no effect For later use, notice that the above properties imply that the rank of P is

$$\text{rk}(P) = \text{tr}(P) = \text{tr}((X^\top X)^{-1} X^\top X) = p.$$

We can decompose y into two terms: its projection $\hat{\mu}$ onto $\mathcal{C}(X)$ and its **residual** component, given by the difference

$$y - \hat{\mu} = y - X(X^\top X)^{-1} X^\top y = (I_n - P)y. \qquad (5.13)$$

These two terms are orthogonal. More generally, $y - \hat{\mu}$ is orthogonal to all elements of $\mathcal{C}(X)$; in fact, any element of $\mathcal{C}(X)$ can be written as Xa for some vector of coefficients a, and

$$
\begin{aligned}
(Xa)^\top (y - \hat{\mu}) &= (Xa)^\top (y - Py) \\
&= a^\top X^\top (y - X(X^\top X)^{-1} X^\top y) \\
&= 0.
\end{aligned}
$$

Even $I_n - P$ is a projection matrix: it projects the elements of $\mathbb{R}^n$ onto the space orthogonal to $\mathcal{C}(X)$. Proceeding as for the rank of P, we obtain $\text{rk}(I_n - P) = n - p$.

Orthogonality of the projection vector Py and the residual vector has a straightforward corollary: by explicitly writing the squared norm of $\hat{\mu} + (y - \hat{\mu})$ we obtain

$$\|y\|^2 = \|\hat{\mu}\|^2 + \|y - \hat{\mu}\|^2 \qquad (5.14)$$

which is a special case of Pythagoras's theorem.

5.2.3 Least-squares statistics

In this section, we examine from a statistical viewpoint the quantities introduced in the previous section. Therefore we now regard the observations y and the error component as generated by random variables We then have

$$
\begin{aligned}
\mathrm{E}\left[\hat{\beta}\right] &= \mathrm{E}\left[(X^\top X)^{-1} X^\top Y\right] \\
&= (X^\top X)^{-1} X^\top \mathrm{E}[Y] \\
&= (X^\top X)^{-1} X^\top X\beta \\
&= \beta, \qquad (5.15)
\end{aligned}
$$

showing that $\hat{\beta}$ in an unbiased estimate of β. Moreover, we have that

$$\mathrm{E}[\,\hat{\mu}\,] = \mu.$$

The dispersion matrix of the estimates is

$$
\begin{aligned}
\mathrm{var}\left[\hat{\beta}\right] &= (X^{\mathsf{T}}X)^{-1}X^{\mathsf{T}}\mathrm{var}[Y]\left((X^{\mathsf{T}}X)^{-1}X^{\mathsf{T}}\right)^{\mathsf{T}} \\
&= (X^{\mathsf{T}}X)^{-1}X^{\mathsf{T}}(\sigma^2 I_n)X(X^{\mathsf{T}}X)^{-1} \\
&= \sigma^2(X^{\mathsf{T}}X)^{-1}
\end{aligned}
\tag{5.16}
$$

and

$$
\begin{aligned}
\mathrm{var}[\,\hat{\mu}\,] &= X\mathrm{var}\left[\hat{\beta}\right]X^{\mathsf{T}} \\
&= \sigma^2 X(X^{\mathsf{T}}X)^{-1}X^{\mathsf{T}} \\
&= \sigma^2 P.
\end{aligned}
$$

Another important statistical property is consistency. From (5.15) and (5.16), consistency holds if and only if the diagonal elements of $V = (X^{\mathsf{T}}X)^{-1}$ converge to 0 as $n \to \infty$.

The fulfilment of this condition depends on the matrix X, or, more precisely, on the sequence of row vectors of X, whose numbers of rows must increase as $n \to \infty$. It is therefore impossible to assess the consistency of $\hat{\beta}$ in general terms, without making assumptions on the asymptotic behaviour of X.

However, from the following informal argument, it is quite plausible that $V \to 0$ in the majority of practical cases. Denoting by $\tilde{x}_i^{\mathsf{T}}$ the ith *row* of X for $i = 1, \ldots, n$, we can write

$$X^{\mathsf{T}}X = \sum_{i=1}^{n} \tilde{x}_i\,\tilde{x}_i^{\mathsf{T}}\,.$$

The right-hand side suggests that $X^{\mathsf{T}}X$ increases, and correspondingly $V \to 0$, as $n \to \infty$, unless the sequence $\tilde{x}_1, \tilde{x}_2, \ldots$ converges sufficiently rapidly to the null vector.

Remark 5.2.1 In the above computations, we have regarded the matrix X as non-stochastic. However, it is very frequently the case that all the variables under consideration, response and explanatory, are determined simultaneously; therefore, we are dealing stochastic explanatory variables. The most typical instance of this kind of situation occurs when two or more variables are observed on the same statistical unit, such as weight (y) and height (x). In this example, the explanatory variable is actually of the same kind as the response variable (in fact we could interchange the roles of the two variables), i e they are non-random.

Consideration of non-stochastic X is supported by the following argument. In most common cases, the distribution of the explanatory variables does not contain any information on the relationship with the response variable; remember that we are interested in making inferences on this relationship, not on the distribution of the explanatory variables. Therefore, we examine the variables *conditionally* on the values taken by X. In other words, we operate within the conditionality principle (cf. section 3.3.7).   □

Remark 5.2.2 Linear models assume that the error component perturbs the value of the response variable, i e it makes its 'true' value unobservable, but the explanatory variable must not be affected by a similar problem. These assumptions are in agreement with the idea mentioned above that the explanatory variables are chosen by the experimenter in a controlled experimental framework.

If even the explanatory variables were observed with error, we would have to deal with a situation such as

$$
\begin{aligned}
x &= x_* + \eta \\
y &= \beta x_* + \varepsilon
\end{aligned}
$$

where x_* is the 'true' explanatory variable, which is directly related to the response variable, while x is the observed explanatory variable. Models of the above form are said to define a **structural relationship**; they represent an evolution of the regression models, but their treatment is beyond the scope of this book. An extensive account on this sort of problem is given by Fuller (1987).   □

So far we have dealt with estimation of β only. Although to a lesser degree than β, we are interested in estimation of σ^2 too. The least-squares criterion does not say how to tackle the new problem, however.

Since the generic term ε_i is such that $\mathrm{E}\left[\varepsilon_i^2\right] = \sigma^2$, it is reasonable to estimate σ^2 via the arithmetic mean of the $\hat{\varepsilon}_i^2$, where $\hat{\varepsilon}_i$ is the generic component of the residual vector

$$
\hat{\varepsilon} = y - \hat{\mu}.
$$

Therefore we consider

$$
\hat{\sigma}^2 = \frac{\sum_i \hat{\varepsilon}_i^2}{n} = \frac{\|\hat{\varepsilon}\|^2}{n} \tag{5.17}
$$

as an estimate of σ^2 Expression (5.17) can be rewritten in a number of alternative forms, making use of the following equalities:

$$
\begin{aligned}
\|\hat{\varepsilon}\|^2 &= Q(\hat{\beta}) \\
&= (y - \hat{\mu})^\top (y - \hat{\mu}) \\
&= y^\top (I_n - P)^\top (I_n - P) y \\
&= y^\top (I_n - P) y = \varepsilon^\top (I_n - P) \varepsilon \\
&= y^\top y - y^\top X \hat{\beta}.
\end{aligned}
$$

For computing the mean value of (5.17), then, we have

$$
\begin{aligned}
\mathrm{E}[n\hat{\sigma}^2] &= \mathrm{E}[y^\top (I_n - P) y] \\
&= \mu^\top (I_n - P) \mu + \mathrm{tr}((I_n - P)\sigma^2 I_n) \\
&= \sigma^2 (n - p), \qquad\qquad (5.18)
\end{aligned}
$$

taking into account Lemma A.4.5. The above term $\mu^\top (I_n - P)\mu$ vanishes since $I_n - P$ projects onto the orthogonal space of $\mathcal{C}(X)$ which contains μ, hence

$$
(I_n - P)\mu = (I_n - X(X^\top X)^{-1} X^\top) X \beta = 0.
$$

Therefore $\hat{\sigma}^2$ is a biased estimate, although the bias vanishes as $n \to \infty$, for fixed p.

To obtain an unbiased estimate of σ^2, we consider the corrected residual variance

$$
s^2 = \hat{\sigma}^2 \frac{n}{n - p} = \frac{Q(\hat{\beta})}{n - p}. \qquad\qquad (5.19)
$$

which is a direct extension of (3.7).

Example 5.2.3 The simplest case that we can consider has $p = 1$ and $X = 1_n$, the vector of 1's, so that

$$
Y = 1_n \beta + \varepsilon
$$

where β is a scalar. This is just an unusual way of introducing a model where all components of Y have the same mean and the same variance, and are uncorrelated, a situation already considered several times under the normality assumption, using a slightly different formulation. Using (5.11), we have

$$
(X^\top X)^{-1} = 1/n, \qquad X^\top y = \sum_{i=1}^{n} y_i
$$

and

$$
\hat{\beta} = \frac{\sum y_i}{n},
$$

i e simply the arithmetic mean, with corresponding variance

$$\text{var}\left[\hat{\beta}\right] = \sigma^2/n.$$

The estimate of σ^2 using (5.19) turns out to coincide with the usual corrected sample variance for normal data (3.7).   □

Example 5.2.4 Consider now the case $p = 2$, with the first column of X equal to 1_n, and the second one equal to $x = (x_1, \ldots, x_n)^\top$, so that

$$X = \begin{pmatrix} 1 & x_1 \\ 1 & x_2 \\ \vdots & \vdots \\ 1 & x_n \end{pmatrix}.$$

In this case the regression model, called **simple linear regression**, interpolates the values of the y variable using a straight line of the form

$$y = \beta_1 + \beta_2 x + \varepsilon,$$

called the **regression line**. We have

$$X^\top X = \begin{pmatrix} n & \sum_i x_i \\ \sum_i x_i & \sum_i x_i^2 \end{pmatrix},$$

$$(X^\top X)^{-1} = \frac{1}{n\sum_i x_i^2 - (\sum_i x_i)^2} \begin{pmatrix} \sum_i x_i^2 & -\sum_i x_i \\ -\sum_i x_i & n \end{pmatrix},$$

$$X^\top y = \begin{pmatrix} \sum_i y_i \\ \sum_i x_i y_i \end{pmatrix},$$

$$\hat{\beta} = \frac{1}{n\sum_i x_i^2 - (\sum_i x_i)^2} \begin{pmatrix} \sum_i y_i \sum_i x_i^2 - \sum_i x_i \sum_i x_i y_i \\ n\sum_i x_i y_i - \sum_i x_i \sum_i y_i \end{pmatrix}$$

$$= \begin{pmatrix} \bar{y} - \dfrac{s_{xy}}{s_{xx}}\bar{x} \\ \dfrac{s_{xy}}{s_{xx}} \end{pmatrix},$$

where

$$\bar{x} = \frac{\sum_i x_i}{n}, \quad \bar{y} = \frac{\sum_i y_i}{n},$$

$$s_{xx} = \sum_i (x_i - \bar{x})^2, \quad s_{xy} = \sum_i (x_i - \bar{x})(y_i - \bar{y}).$$

The above expressions simplify considerably if $\bar{x}=0$, a condition which can always be satisfied on changing the parametrization from

$$y = 1_n\beta_1 + x\beta_2 + \varepsilon$$

to

$$y = 1_n(\beta_1 + \beta_2\bar{x}) + (x - 1_n\bar{x})\beta_2 + \varepsilon$$

which has intercept $\alpha = \beta_1 + \beta_2\bar{x}$ and slope β_2 as before. The new explanatory variable $z = x - 1_n\bar{x}$ has $\sum_i z_i = 0$ as required.

Besides speeding up our computations, this reparametrization offers the advantage of uncorrelated estimates $\hat{\alpha}, \hat{\beta}_2$, i e α and β_2 are orthogonal parameters, as can be checked by considering $(X^T X)^{-1}$, obviously replacing x by z

An additional advantage of the reparametrization is the numerical stability of the estimates, i e they are less sensitive to rounding errors in computations.

An important statistic associated with the simple linear regression context is the **sample correlation**

$$r = \frac{s_{xy}}{\sqrt{s_{xx}\, s_{yy}}}, \tag{5.20}$$

where $s_{yy} = \sum_i(y_i - \bar{y})^2$. This quantity is used as a descriptive index to quantify the degree of closeness of the observed data points to the fitted line. In fact, the range of r is $[-1, 1]$, and $r = \pm 1$ means that the points lie exactly on the regression line, with the sign of r in accordance with the sign of $\hat{\beta}_2$. On the other hand, value $r = 0$ denotes a poor fit of the line to the points, but the cause of this situation needs to be examined in each specific case, since it can arise in a number of different ways.

A closely connected quantity is the **coefficient of determination**

$$r^2 = \frac{s_{xy}^2}{s_{xx}\, s_{yy}} = 1 - \frac{\|y - \hat{\mu}\|^2}{\|y - \bar{y}1_n\|^2}$$

which represents the fraction of variability of the y_i which can be imputed to the linear component of the relationship.

The present example offers the opportunity to stress a feature of least-squares estimation. The method works by minimizing $\|y - \mu\|^2$ with respect to the Euclidean metric of $\mathbb{R}^n$. However, when we consider the points (x_i, y_i) as elements of $\mathbb{R}^2$, the distances between the observed values y_i and the fitted values $\hat{\mu}_i$ on the regression line are measured vertically, hence they do *not* correspond to the *geometric* distance between the points and the regression line. Figure 5.3 illustrates this point: a scatter plot and the corresponding

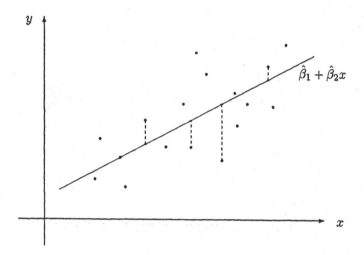

Figure 5.3 *Scatter plot and regression line*

regression line are represented; this line is chosen by minimizing the sum of squared lengths measured parallel to the vertical axis, as indicated (for a few points) by the dashed lines.   □

Remark 5.2.5 One of the 'second-order hypotheses' is that the error component in (5.7) must have null mean value. In a large number of cases, this assumption does not really introduce any restriction, as it is automatically satisfied

Suppose that one column of X, the first say, is equal to 1_n, a very frequent occurrence in practice. Denote by μ_ε the mean value of ε_i, and by $\tilde{X}$ and $\tilde{\beta}$ the remaining columns of X and the remaining rows of β after deleting the first, respectively. Then we can write (5.7) as

$$
\begin{aligned}
Y &= 1_n\beta_1 + \tilde{X}\tilde{\beta} + \varepsilon \\
  &= 1_n(\beta_1 + \mu_\varepsilon) + \tilde{X}\tilde{\beta} + (\varepsilon - 1_n\mu_\varepsilon)
\end{aligned}
$$

where now the error component $\varepsilon - 1_n\mu_\varepsilon$ has null mean value, having included $\mathrm{E}[\varepsilon]$ in the intercept. Therefore the restriction on the mean value of ε is not really there. We just need to be careful in interpreting the value of the intercept; luckily, in most cases interest focuses on the other parameters, and the problem vanishes. □

Theorem 5.2.6 (Gauss–Markov) *Under second-order hypotheses for model (5.7), the least-squares estimator $\hat{\beta}$ is such that*

$$\text{var}\left[\hat{\beta}\right] \leq \text{var}[T] \qquad (5.21)$$

for any estimator T of the form $T = C^{\top}y$, where C is an $n \times p$ matrix such that $\mathrm{E}[T] = \beta$ for all β.

Proof. Since the equality $\mathrm{E}\left[C^{\top}Y\right] = C^{\top}X\beta = \beta$ must hold for all $\beta \in \mathbb{R}^{p}$, it follows that $C^{\top}X = I_{p} = X^{\top}C$. Taking into account (5.16) and $\text{var}[T] = \sigma^{2}C^{\top}C$, the statement of the theorem is equivalent to

$$C^{\top}C - (X^{\top}X)^{-1} \geq 0,$$

that is

$$C^{\top}C - C^{\top}X(X^{\top}X)^{-1}X^{\top}C = C^{\top}(I_{n} - P)C \geq 0.$$

This relationship is true since $I_{n} - P$ is symmetric and idempotent. Therefore, we have

$$\begin{aligned} a^{\top}C^{\top}(I_{n} - P)Ca &= a^{\top}C^{\top}(I_{n} - P)^{\top}(I_{n} - P)Ca \\ &= \|(I_{n} - P)Ca\|^{2} \geq 0 \end{aligned}$$

for any $a \in \mathbb{R}^{p}$. □

Remark 5.2.7 Inequality (5 21) must be taken in the ordering among matrices. An implication of this fact is the following: consider a linear combination of the parameters, often called the **contrast**, of the form

$$\psi = a^{\top}\beta$$

where $a \in \mathbb{R}^{p}$ is a vector of given constants. It follows that $\hat{\psi} = a^{\top}\hat{\beta}$ is an unbiased estimate of ψ and its variance is the minimum achievable variance among unbiased estimates of ψ. In particular, on choosing a equal to the null vector except for the jth component equal to 1 ($j \in \{1, \ldots, p\}$), we have $\psi = \beta_{j}$ and $\hat{\psi} = \hat{\beta}_{j}$. Therefore, any single component of $\hat{\beta}$ is the best linear estimate of the corresponding component of β □

Remark 5.2.8 The practical meaning of the Gauss–Markov theorem is as follows: within the class of unbiased linear estimators of β, the least-squares estimator has the minimum variance (or minimum variance matrix). Therefore $\hat{\beta}$ is the 'best' estimator, adopting the usual criterion of ordering estimators according to their variance

The Gauss–Markov theorem provides a formal support for the use of the least-squares criterion, which so far has been justified only qualitatively. The result is particularly relevant if we consider that it does not involve assumptions on ε beyond second-order moments.   □

5.2.4 Sum-of-squares decomposition

Examine decomposition (5.14) with the observed vector y replaced by the random variable Y, leading to

$$\|Y\|^2 = \|PY\|^2 + \|(I_n - P)Y\|^2 \tag{5.22}$$

i e

$$Y^\top I_n Y = Y^\top PY + Y^\top(I_n - P)Y.$$

These three quadratic forms are associated with the symmetric idempotent matrices

$$I_n, \quad P, \quad I_n - P$$

whose ranks are

$$n, \quad p, \quad n - p,$$

respectively. The norms in (5.22) are commonly called the **total, regression and residual sum of squares** (SS), respectively. The ratios of each of these terms to the rank of the corresponding matrix are called **mean squares** (MS). In fact, even if the number of elements added is n for all three SS terms, the vectors in (5.22) belong to subspaces of lower dimensions which represent the 'effective' vector components.

The study of the probability distribution of the sum of squares in (5.22) plays a fundamental role in statistics Clearly, such an analysis is possible only after the distribution of Y has been specified, but we can already compute the mean values. The mean value of the last term in (5.22) has been obtained in (5.18); similar computations also yield the other mean values. The results are summarized in Table 5.1.

We are often interested in deciding whether vector μ, or equivalently β, satisfies a certain condition; the simplest example is the condition $\mu = 0$. The entries of the column headed E[MS] in Table 5.1 vary according to whether the condition $\mu = 0$ is

Table 5.1 *Sum-of-squares decomposition*

Source	SS	rank	E[MS]
Regression	$\|PY\|^2$	p	$\sigma^2 + \mu^{\mathsf{T}} P\mu/p$
Residual	$\|(I_n - P)Y\|^2$	$n - p$	σ^2
Total	$\|Y\|^2$	n	$\sigma^2 + \mu^{\mathsf{T}}\mu/n$

satisfied; therefore their sample values can be used for deciding about $\mu = 0$. In particular, the comparison of the regression and of the residual sum of squares is relevant for this purpose, although a formal procedure can be defined only after the distribution of the quadratic form has been established, which is one of the goals of the following sections.

5.2.5 Constrained estimates

Consider now the problem of estimating β when there exist linear constraints among the components of β, i e β is such that

$$H\beta = 0 \qquad (5.23)$$

where H is a $q \times p$ matrix ($q \le p$) of rank q whose elements are given constants. The solution of this constrained estimation problem will be particularly useful for testing hypotheses about β (which we shall tackle later on), but it is also an interesting problem in its own right

First of all, consider the geometrical meaning of condition (5.23), which constrains β to lie in a certain subset of the linear space $\mathcal{C}(X)$; it can be checked that this subset represents a linear subspace of $\mathcal{C}(X)$, $\mathcal{C}_0(X)$ say, of dimension $p - q$. The situation is illustrated by Figure 5.4.

The constrained least-squares estimate (LSE) is obtained by minimizing $Q(\beta)$ under the condition required. Therefore, under condition (5.23), we must minimize

$$f(\alpha, \beta) = (y - X\beta)^{\mathsf{T}}(y - X\beta) + 2(H\beta)^{\mathsf{T}}\alpha,$$

where α is a vector of Lagrange multipliers. On differentiating f with respect to α, β and equating these expressions to 0, we obtain

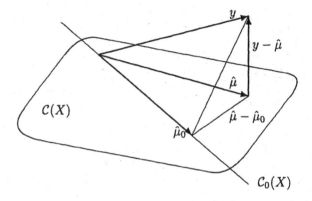

Figure 5.4 *Projection of y onto C(X) and onto the subspace $C_0(X)$*

the equations

$$\begin{cases} X^\top X \beta + H^\top \alpha = X^\top y, \\ H\beta = 0. \end{cases}$$

Multiplying the first equation on the left by $(X^\top X)^{-1}$, we obtain

$$\beta = \hat{\beta} - (X^\top X)^{-1} H^\top \alpha,$$

which leads to

$$\alpha = \{H(X^\top X)^{-1} H^\top\}^{-1} H\hat{\beta}.$$

Then, the minimum of f is attained for β equal to

$$\hat{\beta}_0 = \hat{\beta} - (X^\top X)^{-1} H^\top K H \hat{\beta}. \qquad (5.24)$$

where

$$K = \{H(X^\top X)^{-1} H^\top\}^{-1}. \qquad (5.25)$$

The corresponding projection of y onto $C_0(X)$ is given by

$$\begin{aligned} \hat{\mu}_0 &= X\hat{\beta}_0 \\ &= \hat{\mu} - X(X^\top X)^{-1} H^\top K H\hat{\beta} \\ &= (P - P_H)y = P_0 y \end{aligned}$$

having set

$$\begin{aligned} P_H &= X(X^\top X)^{-1} H^\top K H(X^\top X)^{-1} X^\top, \qquad (5.26) \\ P_0 &= P - P_H. \end{aligned}$$

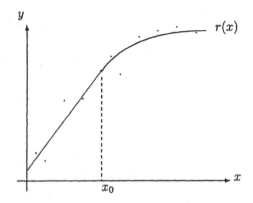

Figure 5.5 *Piecewise linear/quadratic curve*

Example 5.2.9 To illustrate the above results, consider the data represented in Figure 5.5 and suppose that they must be interpolated by a function $r(x)$, which must be linear up to the given point x_0, and quadratic to the right of x_0, with the additional condition that $r(x)$ is continuous with continuous derivative at x_0. In other words, consider the model

$$y = r(x) + \varepsilon$$

where

$$r(x) = \begin{cases} \beta_1 + \beta_2 x & \text{for } x \leq x_0, \\ \beta_3 + \beta_4 x + \beta_5 x^2 & \text{for } x > x_0, \end{cases}$$

under the constraints

$$\beta_1 + \beta_2 x_0 - (\beta_3 + \beta_4 x_0 + \beta_5 x_0^2) = 0$$
$$\beta_2 - (\beta_4 + 2\beta_5 x_0) = 0$$

to ensure continuity of $r(x)$ and $r'(x)$ at x_0. Clearly, $r(\cdot)$ is a nonlinear function with respect to x, but it is a linear function of the parameters, as we shall see in a moment. We can proceed in two steps: first, we estimate the five parameters ignoring the constraints; then, we apply (5.24) to obtain the final estimate.

To compute the unconstrained estimates, we build an X matrix whose first m rows are of the form

$$(1 \quad x_i \quad 0 \quad 0 \quad 0)$$

where the x_i are values not greater that x_0, while the remaining

$n - m$ rows of X are of the form

$$(0 \quad 0 \quad 1 \quad x_i \quad x_i^2)$$

where $x_i > x_0$. Matrix X contains two blocks of 0s such that estimation of the first two components of β is completely separated from estimation of the remaining three components. After obtaining the unconstrained vector $\hat{\beta}$, we must convert it into a vector $\hat{\beta}_0$ satisfying the constraints. The appropriate matrix H to represent the above constraints is

$$H = \begin{pmatrix} 1 & x_0 & -1 & -x_0 & -x_0^2 \\ 0 & 1 & 0 & -1 & -2x_0 \end{pmatrix}$$

which has to be inserted into (5.24) to transform $\hat{\beta}$ into $\hat{\beta}_0$ so that $r(x)$ satisfies the regularity conditions.   □

Consider the conclusions reached so far. We have a new projection matrix P_0 which projects any vector of $\mathbb{R}^n$ onto the subspace $C_0(X)$. By applying this projection matrix to y we obtain $\hat{\mu}_0$, which is the element of $C_0(X)$ with smallest distance from y. Moreover, the following properties hold.

- The vector $y - \hat{\mu}_0$ is orthogonal to all elements of $C_0(X)$ In fact, let Xc be any element of $C(X)$ such that $Hc = 0$; then we have $(y - \hat{\mu}_0)^\top Xc = 0$. In particular, we have

$$(y - \hat{\mu}_0) \perp \hat{\mu}_0$$

- The projection of $y - \hat{\mu}_0$ onto $C(X)$ is $P(y - \hat{\mu}_0) = \hat{\mu} - \hat{\mu}_0$ such that

$$\hat{\mu} - \hat{\mu}_0 \perp \hat{\mu}_0.$$

To conclude, we have established the decomposition

$$y = \hat{\mu}_0 + (\hat{\mu} - \hat{\mu}_0) + (y - \hat{\mu}),$$

where the three terms on the right-hand side are orthogonal to each other. Therefore, we can write

$$\|y\|^2 = \|\hat{\mu}_0\|^2 + \|\hat{\mu} - \hat{\mu}_0\|^2 + \|y - \hat{\mu}\|^2, \qquad (5.27)$$

which extends (5.14).

5.3 Normal theory

5.3.1 Likelihood

So far we have dealt with linear models without introducing any assumptions on the error distribution besides those on the first- and second-order moments.

It is, however, quite natural to ask whether we can introduce a more detailed specification of the distribution of ε, and then make use of the methods developed in the previous chapters. This further step is not only possible but in fact necessary for proceeding to hypothesis testing about the parameter values.

The simplest and most common hypothesis about the distribution of ε is normality. Therefore, from now on, we shall add the assumption that ε has *multivariate normal* distribution. Taking into account assumptions already made, we then have

$$\varepsilon \sim N_n(0, \sigma^2 I_n). \tag{5.28}$$

From (5.8) and well-known properties of normal random variables, it follows that

$$Y \sim N_n(X\beta, \sigma^2 I_n).$$

Using (A 27), we can write the log-likelihood of $\theta = (\beta^\top, \sigma^2)^\top$ for a given vector y, namely

$$
\begin{aligned}
\ell(\theta) &= -\tfrac{1}{2} n \ln(\sigma^2) - \tfrac{1}{2}\sigma^{-2}\|y - X\beta\|^2 \qquad (5.29)\\
&= -\tfrac{1}{2} n \ln(\sigma^2) - \tfrac{1}{2}\sigma^{-2}(y^\top y - 2y^\top X\beta + \beta^\top X^\top X\beta).
\end{aligned}
$$

This log-likelihood clearly has an exponential structure with minimal sufficient statistics $(y^\top X, y^\top y)$ of dimension $p+1$ like θ; therefore, it represents a regular exponential family if θ is free to vary over the whole admissible set $\mathbb{R}^p \times \mathbb{R}^+$.

It is clear that the MLE of β is obtained by minimizing $Q(\beta)$ as defined by (5.9), which is the same quantity used by the least-squares method. This remark implies that the maximum of (5.29) coincides with the minimum of $Q(\beta)$, so we conclude that the MLE of β coincides with the LSE (5.11) and explains why in the previous sections we denoted the LSE using the same notation used so far for MLE. It is clear that this connection between MLE and LSE relies on normality assumption (5.28), and does not hold in general.

By a computation similar to that of Example 3.1.10, we obtain that the MLE of σ^2 coincides with (5.17). However, for the reasons explained earlier, the usual estimate of σ^2 is s^2, defined by (5.19).

Consider now the implications of (5.28) on decomposition (5.22) and the corresponding Table 5.1. From Fisher–Cochran theorem A.5.5 and its Corollary A.5.6, it follows that

- $SS_{tot} = Y^T Y \sim \sigma^2 \chi_n^2(\delta)$, where $\delta = \mu^T \mu / \sigma^2 = \beta^T X^T X \beta / \sigma^2$;
- $SS_{reg} = Y^T P Y \sim \sigma^2 \chi_p^2(\delta)$;
- $SS_{res} = Y^T (I_n - P) Y \sim \sigma^2 \chi_{n-p}^2$ with non-centrality parameter equal to 0 since $\mu^T (I_n - P) \mu = 0$;
- the random variables SS_{reg} and SS_{res} are independent.

5.3.2 A first example of the use of the LRT

Consider the problem of deciding whether $\mu = 0$, which is equivalent to $\beta = 0$. In the notation of Chapter 4, we have to test the hypotheses

$$\begin{cases} H_0 : \beta = 0, \\ H_1 : \beta \neq 0. \end{cases} \tag{5.30}$$

To make use of the LRT, we find the MLE under the null hypothesis and under the general hypothesis; the maximum value of (5.29) under H_0 is achieved at

$$\theta = \begin{pmatrix} 0 \\ \hat{\sigma}_0^2 \end{pmatrix}, \quad \hat{\sigma}_0^2 = \|y\|^2 / n$$

and the unconstrained maximum value is at

$$\theta = \begin{pmatrix} \hat{\beta} \\ \hat{\sigma}^2 \end{pmatrix}.$$

The LRT criterion leads to the test statistic

$$\begin{aligned} \lambda = \lambda(y) &= \frac{(\hat{\sigma}_0^2)^{-n/2} \exp(-\frac{1}{2} \hat{\sigma}_0^{-2} \|y\|^2)}{(\hat{\sigma}^2)^{-n/2} \exp(-\frac{1}{2} \hat{\sigma}^{-2} \|y - \hat{\mu}\|^2)} \\ &= \left(\frac{\|y - \hat{\mu}\|^2}{\|y\|^2} \right)^{n/2}, \end{aligned}$$

which can replaced by the monotonic function

$$\lambda^* = \lambda^{-2/n} - 1 = \frac{\|\hat{\mu}\|^2}{\|y - \hat{\mu}\|^2}. \tag{5.31}$$

We reject the null hypothesis if λ is sufficiently small, i e if λ^* is sufficiently large. To determine the distribution of λ^*, notice that the right-hand side terms of (5.31) are the same as in Table 5.1, they have χ^2 distribution up to a multiplicative constant σ^2, and they are mutually independent. Therefore, the quantity of

$$F = \lambda^* \frac{n-p}{p} = \frac{\text{MS}_{reg}}{\text{MS}_{res}} \qquad (5.32)$$

is distributed as an F random variable with $(p, n-p)$ degrees of freedom and non-centrality parameter $\delta = \beta^\top X^\top X \beta / \sigma^2$ which vanishes under H_0

To conclude, the operational procedure is as follows: MS_{reg} and MS_{res} are obtained as ratios of the SS terms and the associated ranks; the ratio (5.32) is computed and compared with the critical point of the F distribution with $(p, n-p)$ degrees of freedom. Alternatively, the area to the right of the observed F statistic under the $F_{p,n-p}$ distribution gives the p-value.

5.3.3 Linear hypotheses

Clearly, the hypotheses (5.30) are a very special case which has been dealt with separately to ease presentation. Our final goal, however, is to consider more general testing problems; namely, we wish to deal with the hypotheses

$$\begin{cases} H_0 : H\beta = 0, \\ H_1 : H\beta \neq 0, \end{cases} \qquad (5.33)$$

where H is a $q \times p$ matrix $(q \leq p)$, of rank q, whose elements are known constants.

The null hypothesis (5.33) introduces a number of linear constraints on the components of the parameters; therefore, it is said to be a **linear hypothesis**. By applying the LRT criterion to hypotheses (5.33) and using (5.24) for computing the MLE under the constraint $H\beta = 0$, we obtain the test statistic

$$\begin{aligned} \lambda &= \frac{L(\hat{\theta}_0)}{L(\hat{\theta})} \\ &= \frac{(\hat{\sigma}_0^2)^{-n/2} e^{-n/2}}{(\hat{\sigma}^2)^{-n/2} e^{-n/2}} \\ &= \left(\frac{Q(\hat{\beta})}{Q(\hat{\beta}_0)} \right)^{n/2} \end{aligned}$$

where

$$\hat{\sigma}_0^2 = \frac{Q(\hat{\beta}_0)}{n} = \frac{\|(I_n - P_0)y\|^2}{n}.$$

The test statistic λ can be replaced by the monotonic transformation already introduced,

$$\begin{aligned}
\lambda^* = \lambda^{-2/n} - 1 &= \frac{Q(\hat{\beta}_0) - Q(\hat{\beta})}{Q(\hat{\beta})} \\
&= \frac{\|y - \hat{\mu}_0\|^2 - \|y - \hat{\mu}\|^2}{\|y - \hat{\mu}\|^2} \qquad (5.34) \\
&= \frac{\|\hat{\mu} - \hat{\mu}_0\|^2}{\|y - \hat{\mu}\|^2}.
\end{aligned}$$

Similarly, to test function (5.31), we shall reject the null hypothesis for sufficiently large values of λ^*. To obtain the distribution of λ^*, or equivalently of its monotonic transformation (5.32), notice that (5.27) gives, when written for random variables,

$$\|Y\|^2 = \|P_0 Y\|^2 + \|PY - P_0 Y\|^2 + \|Y - PY\|^2,$$

where the quadratic forms of Y are associated with the matrices

$$I_n, \quad P_0, \quad P - P_0, \quad I_n - P$$

with respective ranks

$$n, \quad p - q, \quad q, \quad n - p$$

as can easily be checked. From the Fisher–Cochran theorem, we conclude, in addition to what we have already established at the end of section 5.3.1, that SS_{reg} can be decomposed as

$$\begin{aligned}
SS_{reg} &= \|P_0 Y\|^2 + \|(P - P_0)Y\|^2 \\
&= SS_{reg(C_0)} + SS_{reg(\perp C_0)}
\end{aligned}$$

where

$$SS_{reg(C_0)} \sim \sigma^2 \chi_{p-q}^2(\delta_1), \qquad SS_{reg(\perp C_0)} \sim \sigma^2 \chi_q^2(\delta_2)$$

are independent of SS_{res}. We have set

$$\begin{aligned}
\delta_1 &= (\mu^\top P_0 \mu)/\sigma^2, \\
\delta_2 &= (\mu^\top P_H \mu)/\sigma^2 \\
&= (\beta^\top H^\top K H \beta)/\sigma^2,
\end{aligned}$$

where K is defined by (5.25); if $H\beta = 0$, then $\delta_2 = 0$. For computing the non-centrality parameters, or equivalently $\mathrm{E}[\mathrm{MS}_{reg(C_0)}]$

Table 5.2 *Decomposition of the sum of squares into three components*

Component	SS	d.f	E[MS]
Regression[1]	$\|P_0Y\|^2$	$p-q$	$\sigma^2(1+\delta_1/(p-q))$
Regression[2]	$\|(P-P_0)Y\|^2$	q	$\sigma^2(1+\delta_2/q)$
Residual	$\|(I_n-P)Y\|^2$	$n-p$	σ^2
Total	$\|Y\|^2$	n	$\sigma^2+\mu^\mathsf{T}\mu/n$

[1] Independent of H_0

[2] Departure from H_0

and $E\big[MS_{reg(\perp C_0)}\big]$, we proceed as for (5.18). Then we conclude that

$$
\begin{aligned}
F &= \lambda^* \frac{n-p}{q} \\
&= \frac{\|\hat\mu - \hat\mu_0\|^2/q}{\|Y-\hat\mu\|^2/(n-p)} \\
&= \frac{MS_{reg(\perp C_0)}}{MS_{res}} \\
&= \frac{(H\hat\beta)^\mathsf{T} K (H\hat\beta)/q}{s^2}
\end{aligned}
\tag{5.35}
$$

is distributed as an F random variable with $(q, n-p)$ degrees of freedom, and the F distribution is central or non-central, depending on whether the null or the alternative hypothesis is true. Therefore, having chosen a significance level α, one accepts the null hypothesis if the sample value of F is smaller than the critical value F_α which leaves on the right a probability α of an F distribution with $(q, n-p)$ degrees of freedom. Alternatively, the p-value is given by the area between the observed F value and ∞ under the curve of an F density function with $(q, n-p)$ degrees of freedom. It is convenient to arrange the sum-of-squares decomposition as in Table 5.2.

A slight generalization of (5.33) is given by

$$
\begin{cases}
H_0 : H\beta = h, \\
H_1 : H\beta \neq h,
\end{cases}
\tag{5.36}
$$

where h is a $q \times 1$ vector of given constants. Beside arising as a genuine hypothesis testing problem in some situations, the new version is relevant for the construction of **confidence regions**, by means of the rule stated on page 144.

By a simple extension of the results of section 5.2.5, the new constrained estimate is given by

$$\hat{\beta}_0 = \hat{\beta} - (X^\mathsf{T} X)^{-1} H^\mathsf{T} K (H\hat{\beta} - h) \qquad (5.37)$$

and the corresponding extension of (5.35) is

$$F = \frac{(H\hat{\beta} - h)^\mathsf{T} K (H\hat{\beta} - h)/q}{s^2}. \qquad (5.38)$$

Example 5.3.1 An important special case of (5.38) arises when we wish to test the null hypothesis that a specific component of β is equal to a given real number h, i e we are considering the hypotheses

$$\begin{cases} H_0 : \beta_r = h, \\ H_1 : \beta_r \neq h, \end{cases}$$

where $1 \leq r \leq p$; in many practical cases, $h = 0$. We can write these hypotheses in the standard form (5.36) with

$$H = (0 \quad \dots \quad 0 \quad 1 \quad 0 \quad \dots \quad 0),$$

where the non-null term is in the rth position. Using (5.26), the numerator of (5.38) becomes

$$\|\hat{\mu} - \hat{\mu}_0\|^2 = (\hat{\beta}_r - h)^2/v_{rr}$$

where v_{rr} is the entry at position (r, r) of $V = (X^\mathsf{T} X)^{-1}$, i e it is $\mathrm{var}\left[\hat{\beta}_r\right]$ up to the scale factor σ^2. Then, (5.38) becomes

$$F = \frac{(\hat{\beta}_r - h)^2}{s^2 \, v_{rr}} \qquad (5.39)$$

with $(1, n - p)$ degrees of freedom.

From the connection between the t and F distributions, the square root of (5.39),

$$t = \frac{\hat{\beta}_r - h}{s \sqrt{v_{rr}}},$$

has a t_{n-p} distribution. This test statistic can be used in the same manner as in section 4.4.1: comparison of the sample value of $|t|$ with a suitable percentage point t' of the t_{n-p} distribution defines a test procedure equivalent to that above based on the F test statistic, but with the practical advantage of indicating the direction of departure from the null value through the sign of $\hat{\beta}_r - h$.

From the rule on page 144, a $(1 - \alpha)$-level confidence interval for β_r is given by all values of h such that the above t statistic is less than t', where t' is the value exceeded with probability $\alpha/2$ by a Student's random variable. The set of all h such that

$$\left| \frac{\hat{\beta}_r - h}{s\sqrt{v_{rr}}} \right| < t'$$

is given by the interval

$$(\hat{\beta}_r - t' s \sqrt{v_{rr}}, \ \hat{\beta}_r + t' s \sqrt{v_{rr}}). \tag{5.40}$$

□

Remark 5.3.2 Expression (5.34) shows that the likelihood ratio depends on the sample values only via

$$Q(\hat{\beta}) = \|y - \hat{\mu}\|^2, \quad Q(\hat{\beta}_0) = \|y - \hat{\mu}_0\|^2.$$

Each of these two quantities is called the **deviance** or **residual sum of squares** of the corresponding linear model. More precisely, λ^* is given by the difference between the two deviances divided by $Q(\hat{\beta})$, whose role is essentially to provide an estimate of σ^2 up to a multiplicative constant. In fact, where σ^2 is known, λ^* reduces to the mere difference of deviances; see Exercise 5.20. □

5.4 Some important applications

In this section, we shall develop some important applications of the general results obtained so far. Therefore, this section plays a role analogous to the corresponding section of Chapter 4. In fact, some of the problems we shall tackle have already been treated earlier; this sort of repetition is deliberate, to present two different ways of handling the same problem.

5.4.1 Multiple regression

In Example 5.2 4, we considered a simple regression model suitable for interpolating n pairs of data by means of a straight line. This scheme can be extended in several directions.

By adding a second explanatory variable, we interpolate the values of the response variable y by means of a **regression plane** of the form

$$y = \beta_1 + \beta_2 x_1 + \beta_3 x_2$$

say. By adding further explanatory variables, we define a **regression hyperplane**.

In fact, what we are describing is the same framework as in section 5.2, except that the first column of matrix X is 1_n, and the values taken by the other variables are of continuous type. In such a case, we refer to the above model as a regression model in strict sense, or **multiple regression**, where the term 'multiple' is in contrast to the term 'simple' of Example 5.2.4.

Hypothesis testing and confidence intervals for the individual components of β can be tackled with the formulae developed in Example 5.3.1. In some cases, a global test statistic is required to test for significance of the whole set of explanatory variables. This is similar to testing problem (5.30) but, in the present case, the test statistic must ignore component β_1, since in many cases this is known to be non-zero.

The hypothesis $\beta_2 = \ldots = \beta_p = 0$ can be written in standard form (5.33) using

$$H = (0, I_{p-1}),$$

a $(p-1) \times p$ matrix with 0s in the first column, which is associated with β_1. There is no loss of generality in assuming that 1_n is orthogonal to the other columns, since otherwise we can replace the values of each variable by the deviations from their own arithmetic mean, as at the end of Example 5.2.4. Then write

$$X = (1_n, \tilde{X})$$

where $\tilde{X}$ is $n \times (p-1)$. The condition $1_n \perp \tilde{X}$ implies

$$(X^\top X)^{-1} = \begin{pmatrix} 1/n & 0 \\ 0 & \tilde{V} \end{pmatrix},$$

where $\tilde{V} = (\tilde{X}^\top \tilde{X})^{-1}$. By substituting these expressions in (5.24) and (5.26), we obtain, after some algebra,

$$H(X^\top X)^{-1} X^\top = \tilde{V} \tilde{X},$$
$$P_H = \tilde{X} \tilde{V} \tilde{X}^\top,$$
$$\|P_H y\|^2 = \|\tilde{X} (\hat{\beta}_2, \ldots, \hat{\beta}_p)^\top\|^2$$

and the test statistic (5.35) is

$$F = \frac{\|\tilde{X}(\hat{\beta}_2,\ldots,\hat{\beta}_p)^{\mathsf{T}}\|^2/(p-1)}{s^2}$$

whose null distribution is $F_{(p-1),(n-p)}$.

A particularly important special case of the above test statistic arises when $p = 2$. We are then testing the null hypothesis that the slope parameter of the regression line of Example 5.2.4 is 0. The above matrix $\tilde{X}$ reduces to vector z defined in Example 5.2.4 and the above test statistic becomes

$$F = \frac{\left((\sum z_i^2)\,\hat{\beta}_2\right)^2}{s^2}.$$

Similarly to Example 5.3.1, we can consider the signed square root statistic

$$t = \frac{\hat{\beta}_2\sqrt{\sum_i(x_i - \bar{x})^2}}{s}$$

whose null distribution is t_{n-2}. Further, by using (5.40), a $(1-\alpha)$-level confidence interval for β_2 is given by

$$\hat{\beta}_2 \pm t'\,\frac{s}{\sqrt{\sum_i(x_i - \bar{x})^2}} \tag{5.41}$$

where t' is the value exceeded with probability $\alpha/2$ by a t_{n-2} random variable.

For general p, the coefficient of determination introduced in Example 5.2.3 can be generalized to

$$R^2 = 1 - \frac{\|y - \hat{\mu}\|^2}{\|y - \bar{y}\mathbf{1}_n\|^2}$$

which can be interpreted as the fraction of variability due to the linear relationship between the response variable and the set of the explanatory variables other than $\mathbf{1}_n$.

Example 5.4.1 For a numerical illustration of multiple regression, consider Table 5.3, containing data from a sample of black cherry trees. For each sample unit, three measurements are given, namely

D: diameter of the tree measured at a given height from the ground (in inches),

H: height of the tree (in feet),

V: volume of timber (in cubic feet).

Table 5 3 *Black cherry tree data, from Ryan, Joiner and Ryan (1985)*
For a set of 31 trees, three measurements are given· diameter (inches),
height (feet) and volume (cubic feet)

Diameter	Height	Volume	Diameter	Height	Volume
8 2	70	10.3	12.9	85	33.8
8.6	65	10.3	13.3	86	27.4
8.8	63	10.2	13.7	71	25.7
10.5	72	16.4	13.8	64	24.9
10 7	81	18.8	14.0	78	34 5
10.8	83	19.7	14.2	80	31.7
11.0	66	15.6	14 5	74	36 3
11.0	75	18 2	16.0	72	38.3
11.1	80	22.6	16.3	77	42.6
11 2	75	19.9	17.3	81	55.4
11.3	79	24.2	17.5	82	55 7
11.4	76	21 0	17.9	80	58 3
11.4	76	21.4	18 0	80	51 5
11 7	69	21 3	18 0	80	51 0
12 0	75	19.1	20.6	87	77.0
12.9	74	22.2			

These data are taken from the second edition of the *Minitab Handbook* (Ryan, Joiner and Ryan 1985, p.278), but were also included
in the first edition, and have since been analysed by a number of
authors. For instance, they are discussed throughout the book by
Atkinson (1985).

Of the three variables, V is measured after the tree has been
pulled down, while D and H are measured beforehand. The purpose of the analysis is to find a simple rule relating V to D and H,
to be used for predicting the volume of timber of unfelled trees on
the basis of measurements of D and H. Since D is measured more
quickly and accurately than H, a rule based on D only would be
preferred.

As a preliminary stage of the data analysis, examine the scatter
plots of the (D, V) and (H, V) pairs, shown in Figure 5 6. These
plots exhibit a clear monotonicity between variables, loosely similar to a linear relationship. There is, however, a mild curvature
in the scatter plot of the (D, V) pair, and an apparent increase
of variability of the response variable in the (H, V) plot, as H in-

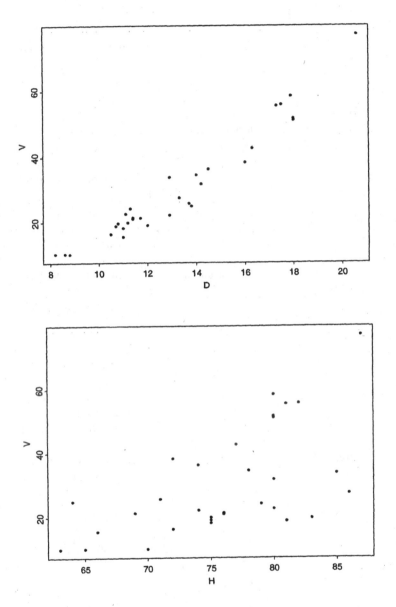

Figure 5.6 *Cherry tree data: scatter plots of (D, V) and (H, V)*

creases. With the aid of a computer and suitable software, further insight can be gained by a three-dimensional dynamic plot of the triplet (D, H, V).

Simple geometric considerations suggest that a plausible relationship among the variables is

$$V = \gamma_0\, D^{\gamma_1}\, H^{\gamma_2}$$

where $\gamma_0, \gamma_1, \gamma_2$ are constants to be estimated. Further, we can conjecture that γ_1 is close to 2 and γ_2 is close to 1. The coefficient γ_0 is more vaguely located, but a value close to $\pi/4$ or $\pi/12$ would indicate that the shape of the tree trunk is close to a cylinder or a cone, respectively. In fact, these values for γ_0 apply if the three measurements are expressed using homogeneous units; if the original units are maintained, they must be divided by 12^2.

The above type of relationship among D, H, V also has the advantage of accommodating the observed curvature between V and D

After taking logarithms of all variables, we formulate the linear model

$$y = \alpha + \gamma_1\, x_1 + \gamma_2\, x_2 + \varepsilon$$

where

$$y = \log V, \quad \alpha = \log \gamma_0,$$
$$x_1 = \log D, \quad x_2 = \log H,$$

and the error term ε is tentatively assumed to be $N(0, \sigma^2)$. Scatter plots of the transformed variables are given in Figure 5.7, showing a form of dependence closer to linearity and constant variability than for the original variables.

After defining the 31×3 design matrix X containing a column of 1's, a column with the values of $\log(D)$ and a third column of $\log(H)$, numerical computations give

$$(X^\top X)^{-1} = \begin{pmatrix} 96.613 & 3.1340 & -24.171 \\ 3.1340 & 0.8437 & -1.223 \\ -24.171 & -1.2228 & 6.308 \end{pmatrix},$$

$$X^\top y = \begin{pmatrix} 101.5 \\ 263.0 \\ 439.9 \end{pmatrix},$$

whose multiplication gives the regression coefficients. For their

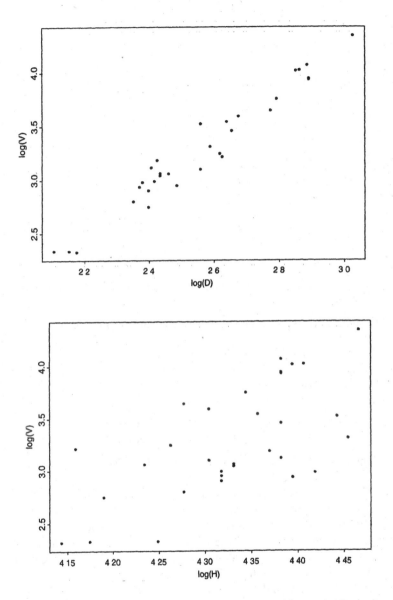

Figure 5.7 *Cherry tree data: scatter plots of* $(\log D, \log V)$ *and* $(\log H, \log V)$

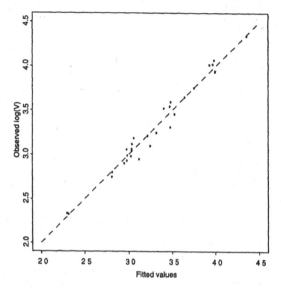

Figure 5.8 *Cherry tree data: observed values of* $\log V$ *versus fitted values*

standard errors, we take the square root of the product of

$$s^2 = \frac{\sum_i (y_i - \hat{y}_i)^2}{28} = (0.08172)^2$$

by the diagonal elements of $(X^{\mathsf{T}} X)^{-1}$. Finally, we obtain

$$\hat{\alpha} = -6.620 \quad \text{(standard error 0.803)},$$
$$\hat{\gamma}_1 = 1.976 \quad \text{(standard error 0.075)},$$
$$\hat{\gamma}_2 = 1.119 \quad \text{(standard error 0.205)}.$$

The t test statistic for nullity of γ_2 is $1.119/0.205 = 5.45$, which is extremely significant when compared with the percentage points of the t_{28} distribution; hence, a prediction rule based on D alone does not seem appropriate.

As a partial check of the adequacy of the fitted model, Figure 5.8 plots the observed versus the fitted values of $\log V$, and the identity line. The points lie along the identity line, with no signs of systematic departure from it.

To obtain a confidence region for (γ_1, γ_2), let

$$H = \begin{pmatrix} 0 & 1 & 0 \\ 0 & 0 & 1 \end{pmatrix}$$

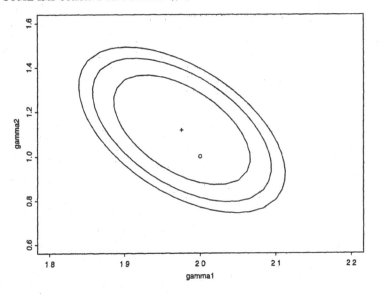

Figure 5.9 *Cherry tree data: confidence regions for (γ_1, γ_2), at confidence levels 75%, 90% and 95%; the + sign marks the MLE point, o shows the (2,1) point*

and then K, defined by (5.25), is

$$K = \begin{pmatrix} 1.648 & 0.320 \\ 0.320 & 0.221 \end{pmatrix}$$

Next, compute the F statistic (5.38) for a range of possible choices of $h = (h_1, h_2)^{\top}$. Figure 5.9 shows the contour curves of this statistic at levels chosen to be equal to the 75%, 90% and 95% points of the $F_{2,28}$ distribution; the region inside each curve is then a confidence region of the corresponding confidence level. We are then proceeding similarly to the construction of Figure 4.6, except for the different (but equivalent) test statistic.

In Figure 5.9, the + sign marks the MLE point, while o corresponds to point $(2, 1)$, i.e. the pair of values originally conjectured for the parameters. The o symbol is well inside all the confidence regions, hence providing evidence in support of the conjecture.

As for γ_0, we obtain $\hat{\gamma} = 0.001333$ by exponentiation of $\hat{\alpha}$. This number can be written as

$$\hat{\gamma}_0 \approx 0\,244 \frac{\pi}{4 \times 144}$$

showing better agreement with the 'cone shape' than the 'cylinder shape'. A more careful analysis of this point is left as an exercise to the reader.   □

5.4.2 Polynomial regression

An important special case arises when we wish to interpolate the values of the variable y using a function of the form

$$y = \beta_1 + \beta_2 x + \beta_3 x^2 + \cdots + \beta_{m+1} x^m + \varepsilon \qquad (5.42)$$

where the explanatory variables are in fact powers of the same variable x.

For ease of explanation, consider the case $m = 2$. If the variable x has been observed at values $(x_1, \ldots, x_n)$, then the regression model becomes

$$y = \begin{pmatrix} 1 & x_1 & x_1^2 \\ 1 & x_2 & x_2^2 \\ \vdots & \vdots & \vdots \\ 1 & x_n & x_n^2 \end{pmatrix} \begin{pmatrix} \beta_1 \\ \beta_2 \\ \beta_3 \end{pmatrix} + \varepsilon$$

which is called a **quadratic regression** scheme; here the term 'quadratic' refers to the kind of relationship between y and x in (5.42), but the model remains linear in the parameters. Obviously, we can add a further power of x, leading to cubic regression, and so on. The general term used for this sort of model is **polynomial regression**.

It is always possible to define a linear transformation of the columns of matrix X in order to to make them orthogonal, using Gram–Schmidt orthogonalization. This operation is often used when the x values are equally spaced, a rather frequent situation.

Continuing with the quadratic regression case, if for instance $n = 5$ and $x_i = i$ (for $i = 1, \ldots, 5$), then the two matrices

$$X = \begin{pmatrix} 1 & 1 & 1 \\ 1 & 2 & 4 \\ 1 & 3 & 9 \\ 1 & 4 & 16 \\ 1 & 5 & 25 \end{pmatrix}, \qquad Z = \begin{pmatrix} 1 & -2 & 2 \\ 1 & -1 & -1 \\ 1 & 0 & -2 \\ 1 & 1 & -1 \\ 1 & 2 & 2 \end{pmatrix}$$

span the same column space, i e $\mathcal{C}(X) = \mathcal{C}(Z)$. Therefore, the two

Table 5.4 *Analysis of variance table for quadratic regression*

Component	SS	d.f.
Constant	$(\sum y_i)^2/n$	1
Linear	$(\sum y_i z_{i2})^2/\sum z_{i2}^2$	1
Quadratic	$(\sum y_i z_{i3})^2/\sum z_{i3}^2$	1
Residual	$\sum_i y_i^2 - \sum_j (\hat{\gamma}_j^2 \sum_i z_{ij}^2)$	$n-3$
Total	$\sum_i y_i^2$	n

linear models

$$y = X\beta + \varepsilon, \qquad y = Z\gamma + \varepsilon$$

are perfectly equivalent from a geometric viewpoint. There are, however, important statistical differences:

- matrix $(Z^{\mathsf{T}}Z)^{-1}$ is diagonal, implying that estimates of γ have a simple expression, namely

$$\hat{\gamma}_i = \frac{z_i^{\mathsf{T}} y}{\|z_i\|^2}$$

where z_i is the ith column of Z;

- as $(Z^{\mathsf{T}}Z)^{-1}$ is diagonal, the components of $\hat{\gamma}$ are uncorrelated random variables, which makes assessment of their significance easier;

- the term $\|PY\|^2$ of Table 5.1 can be decomposed into mutually independent χ_1^2 components, as summarized in Table 5.4.

For generic values of n and of the degree m of the polynomial $(m < n)$, the column vectors of Z are obtained from the values of certain polynomials, which are in fact called **orthogonal polynomials**. There exist tables containing such values; see for instance *Biometrika Tables* (Pearson and Hartley, 1970–72).

5.4.3 Two-sample t test

Consider the same framework and problem as in section 4 4.3 except that we now denote the two population means by β_1, β_2 instead of μ, η. Combine the two samples into a single response vector y, i e let

$$y = (z_1, \ldots, z_n, x_1, \ldots, x_m)^{\mathsf{T}}$$

of length $n + m$; further, define

$$
X = \begin{pmatrix} 1 & 0 \\ 1 & 0 \\ \vdots & \vdots \\ 1 & 0 \\ 0 & 1 \\ 0 & 1 \\ \vdots & \vdots \\ 0 & 1 \end{pmatrix} = \begin{pmatrix} 1_n & 0 \\ 0 & 1_m \end{pmatrix}, \qquad \beta = \begin{pmatrix} \beta_1 \\ \beta_2 \end{pmatrix}.
$$

Notice that the origin of a sample element from one or other population has been represented by means of **indicator variables**, taking values 1 when the sample elements belong to the associated population, 0 otherwise. Therefore, we have expressed an intrinsically qualitative variable such as group membership by means of quantitative variables.

Using (5.11) and (5.17), we have that

$$
X^\top X = \begin{pmatrix} n & 0 \\ 0 & m \end{pmatrix}, \quad X^\top y = \begin{pmatrix} \sum_i z_i \\ \sum_i x_i \end{pmatrix}, \quad \hat{\beta} = \begin{pmatrix} \hat{\beta}_1 \\ \hat{\beta}_2 \end{pmatrix} = \begin{pmatrix} \bar{z} \\ \bar{x} \end{pmatrix},
$$

and it is easy to check that the MLEs of μ, η, σ^2 are as in Chapter 4. Testing hypotheses (4.23) is equivalent to considering (5.33) with

$$
H = (1 \quad -1)
$$

of dimension 1×2. By applying (5.24) we have

$$
\begin{aligned}
\hat{\beta}_0 &= \begin{pmatrix} \bar{z} \\ \bar{x} \end{pmatrix} - \begin{pmatrix} 1/n & 0 \\ 0 & 1/m \end{pmatrix} \begin{pmatrix} 1 \\ -1 \end{pmatrix} (1/n + 1/m)^{-1} (1 \quad -1) \begin{pmatrix} \bar{z} \\ \bar{x} \end{pmatrix} \\
&= \begin{pmatrix} \bar{z} - \dfrac{\bar{z} - \bar{x}}{n(1/m + 1/m)} \\ \bar{x} - \dfrac{\bar{z} - \bar{x}}{m(1/m + 1/m)} \end{pmatrix} \\
&= \frac{n\bar{z} + m\bar{x}}{n + m} \begin{pmatrix} 1 \\ 1 \end{pmatrix}.
\end{aligned}
$$

We could have obtained $\hat{\beta}_0$ in a simpler way, by noticing that under the null hypothesis y is a sample with all $n + m$ elements taken from the same population, but we preferred to use the general formulae exactly for demonstration purposes.

Therefore the terms required for computing (5.35) are

$$SS_{reg(\perp C_0)} = n\left(\bar{z} - \frac{n\bar{z} + m\bar{x}}{n+m}\right)^2 + m\left(\bar{x} - \frac{n\bar{z} + m\bar{x}}{n+m}\right)^2$$

$$= \frac{(\bar{z} - \bar{x})^2}{1/n + 1/m},$$

$$SS_{res} = \sum_{i=1}^{n}(y_i - \hat{\beta}_1)^2 + \sum_{i=n+1}^{n+m}(y_i - \hat{\beta}_2)^2$$

$$= \sum_{i=1}^{n}(z_i - \bar{z})^2 + \sum_{i=1}^{m}(x_i - \bar{x})^2$$

and finally

$$F = \frac{(\bar{z} - \bar{x})^2/(1/n + 1/m)}{SS_{res}/(n+m-2)}$$

which is equal to t^2 if t is defined as in section 4 4.3. The distribution of F is $F_{1,n+m-2}$. We reject the null hypothesis for large values of the F test statistic, or equivalently for large values of $|t|$, as in section 4.4 3, and the connection between the two probability distributions ensures that the two procedures are in agreement.

5.4.4 One-way analysis of variance

Consider the same framework as in section 4.4.5, except that we replace the symbol μ_i by β_i $(i = 1, \ldots, m)$; further, let

$$y = (y_{11}, \ldots, y_{1n}, \cdots, y_{i1}, \ldots, y_{in}, \ldots, y_{m1}, \ldots, y_{mn})^\top,$$

$$X = \begin{pmatrix} 1_n & 0 & \cdots & 0 \\ 0 & 1_n & \cdots & 0 \\ \vdots & \vdots & \ddots & \vdots \\ 0 & 0 & \cdots & 1_n \end{pmatrix}$$

of dimensions $nm \times 1$ and $nm \times m$, respectively. The qualitative variable denoting membership of one of the m possible populations is often called a **factor** and its possible 'values' are called **levels**. In this case, the factor has been represented by means of m indicator variables, although we shall see later on that one of these variables is in fact redundant.

Then we have the linear model

$$y = X\beta + \varepsilon$$

where $\beta = (\beta_1, \ldots, \beta_m)^\top$ is the vector of the mean values. It can easily be checked algebraically that

$$X^\top X = nI_m, \qquad \hat{\beta} = (\bar{y}_1, \ldots, \bar{y}_m)^\top,$$

where the elements of $\hat{\beta}$ are the arithmetic means defined in section 4.4.5.

Matrix H representing equality of the m mean values is

$$H = \begin{pmatrix} 1 & -1 & 0 & \cdots & 0 & 0 \\ 1 & 0 & -1 & \cdots & 0 & 0 \\ \vdots & \vdots & \vdots & \ddots & \vdots & \vdots \\ 1 & 0 & 0 & \cdots & -1 & 0 \\ 1 & 0 & 0 & \cdots & 0 & -1 \end{pmatrix}$$

of dimension $(m-1) \times m$. This matrix is such that $H\beta$ is the vector of the $(m-1)$ differences $(\beta_1 - \beta_2), \ldots, (\beta_1 - \beta_m)$; therefore, testing the null hypothesis $H\beta = 0$ is equivalent to testing $\beta_1 = \beta_2 = \cdots = \beta_m$. To invert

$$H(X^\top X)^{-1}H^\top = \frac{1}{n}(I_{m-1} + 1_{m-1}1_{m-1}^\top),$$

we make use of (A.26) and obtain

$$K = (H(X^\top X)^{-1}H)^{-1} = n\left(I_{m-1} - \frac{1}{m}1_{m-1}1_{m-1}^\top\right).$$

Taking into account (5.26), write

$$\begin{aligned}
\|\hat{\mu} - \hat{\mu}_0\|^2 &= y^\top P_H y \\
&= n\hat{\beta}^\top H^\top \left(I_{m-1} - \frac{1}{m}1_{m-1}1_{m-1}^\top\right)H\hat{\beta} \\
&= n\left(\|H\hat{\beta}\|^2 - \frac{1}{m}\|1_{m-1}^\top H\hat{\beta}\|^2\right) \\
&= n\left(\sum_{i=2}^{m}(\bar{y}_i - \bar{y}_1)^2 - \frac{1}{m}\left((m-1)\bar{y}_1 - \sum_{i=2}^{m}\bar{y}_i\right)^2\right) \\
&= n\sum_{i=1}^{m}(\bar{y}_i - \bar{y})^2,
\end{aligned}$$

where $\bar{y}$ is the overall mean of the elements of y.

Again, even in this case, we could have obtained $SS_{reg(\perp c_0)}$ in a simpler way, noticing that under the null hypothesis the constrained estimate of β is an $m \times 1$ vector with all elements equal to $\bar{y}$. Then, for a generic element y_{ij}, the two estimates of $E[y_{ij}]$, under the general and the alternative hypothesis, are $\bar{y}_i$ and $\bar{y}$ respectively; we then obtain

$$SS_{reg(\perp C_0)} = \sum_{i=1}^{m}\sum_{j=1}^{n}(\bar{y}_i - \bar{y})^2,$$

which equals the expression obtained previously.

Remark 5.4.2 Equality of the m mean values can be expressed in many other ways, besides the condition $H\beta = 0$, for instance by the condition $\tilde{H}\beta = 0$, where

$$\tilde{H} = \begin{pmatrix} 1 & -1 & 0 & \cdots & 0 & 0 \\ 0 & 1 & -1 & \cdots & 0 & 0 \\ \cdot & \cdot & \vdots & \ddots & \vdots & \vdots \\ 0 & 0 & 0 & \cdots & 1 & -1 \end{pmatrix},$$

of dimension $(m-1) \times m$, is such that $\tilde{H}\beta$ gives the vector of the $(m-1)$ differences $(\beta_1 - \beta_2), \ldots, (\beta_{m-1} - \beta_m)$. Therefore testing $\tilde{H}\beta = 0$ is equivalent to testing $\beta_1 = \beta_2 = \cdots = \beta_m$. It can be checked that replacing matrix H by $\tilde{H}$ does not modify the sum-of-squares decomposition.   □

It is very common, especially in applied work, to adopt an alternative parametrization of the model. Instead of the representation

$$y_{ij} = \beta_i + \varepsilon_{ij},$$

it is often preferred, for easier interpretation of the quantities involved, to write

$$y_{ij} = \bar{\mu} + \alpha_i + \varepsilon_{ij} \tag{5.43}$$

where α_i represents the mean deviation of the ith population from the overall mean $\bar{\mu} = \sum_i \beta_i/m$.

The reason for introducing (5.43) is clearer if we interpret the row index of y_{ij} as the label attached to some **treatment** applied to the units or individuals, and the column index as a label to identify a unit within a give **treatment group**. Then α_i is regarded as the **treatment effect**, intended as the mean difference between a generic observation with a given treatment and the reference value $\bar{\mu}$. Schematically, we have the following decomposition

(observation) = (grand mean) + (treatment effect) + (error).

Obviously, (5.43) is only a different way of representing the same quantities we considered earlier; therefore it must be true that

$$\beta_i = \bar{\mu} + \alpha_i \qquad (i = 1, \ldots, m). \tag{5.44}$$

From the fact that $\sum \beta_i = \sum (\bar{\mu} + \alpha_i)$ it follows that

$$\sum_{i=1}^{m} \alpha_i = 0,$$

implying that one of the α_i can be written as a function of the others, for instance

$$\alpha_m = -\alpha_1 - \cdots - \alpha_{m-1}.$$

In fact, this equality (or another one equivalent to it) is necessary, to avoid non-identifiability problems with the model.

The corresponding linear model is

$$y = Z\gamma + \varepsilon \tag{5.45}$$

where

$$Z = \begin{pmatrix} 1_n & 1_n & 0 & \cdot & 0 & 0 \\ 1_n & 0 & 1_n & \cdots & 0 & 0 \\ \vdots & \vdots & \vdots & \ddots & \vdots & \vdots \\ 1_n & 0 & 0 & \cdots & 1_n & 0 \\ 1_n & 0 & 0 & \cdots & 0 & 1_n \\ 1_n & -1_n & -1_n & \cdots & -1_n & -1_n \end{pmatrix}, \qquad \gamma = \begin{pmatrix} \bar{\mu} \\ \alpha_1 \\ \vdots \\ \alpha_{m-1} \end{pmatrix},$$

whose dimensions are $nm \times m$ and $m \times 1$, respectively. Since this results in the equality

$$X(X^{\mathsf{T}} X)^{-1} X^{\mathsf{T}} = Z(Z^{\mathsf{T}} Z)^{-1} Z^{\mathsf{T}},$$

we have that

$$X\hat{\beta} = Z\hat{\gamma}$$

for all possible choices of y.

The pattern of Z shows that, as already mentioned, only $m - 1$ indicator variables are in fact necessary to represent a factor with m levels.

The identifiability condition $\sum_i \alpha_i = 0$ could be replaced by some other constraint which imposes nullity of a linear combination of the α_i; this other condition would simply imply that a certain weighted mean of the α_i is 0. This alternative choice of the constraint would change the estimates $\hat{\alpha}_i$, but not the projection $\hat{\mu}$ and the test function discussed shortly.

Table 5.5 *One-way analysis of variance table*

Component	SS	d.f.
Treatments	$n \sum_i (\bar{y}_i - \bar{y})^2$	$m - 1$
Error	$\sum_i \sum_j (y_{ij} - \bar{y}_i)^2$	$(m - 1)n$
Total	$\sum_i \sum_j (y_{ij} - \bar{y})^2$	$nm - 1$

Testing the hypothesis of equality of the m mean values β_j is now equivalent to testing the hypothesis $H_* \gamma = 0$, where

$$H_* = \begin{pmatrix} 0 & 1 & 0 & \cdots & 0 & 0 \\ 0 & 0 & 1 & \cdots & 0 & 0 \\ \vdots & \vdots & \vdots & \ddots & \vdots & \vdots \\ 0 & 0 & 0 & \cdots & 1 & 0 \\ 0 & 0 & 0 & \cdots & 0 & 1 \end{pmatrix}$$

is of dimension $(m - 1) \times m$ It is left to the reader to check that the corresponding sum-of-squares decomposition is the same as obtained in the previous parameterization.

We now develop the special form taken by Table 5.2 in the present case. Since generally one is not interested in testing a hypothesis on the overall mean $\bar{\mu}$, it is common practice to subtract the contribution of this term, $nm\bar{y}^2$, from the total sum of squares, leading to

$$\sum_i \sum_j y_{ij}^2 - nm\bar{y}^2 = \sum_i \sum_j (y_{ij} - \bar{y})^2$$

which is, up to a multiplicative constant, the 'total variance' of the data. This total variance is decomposed into a variance component due to the differences among the means of the populations and a residual variance due to the error component. Table 5.5 lists all these quantities.

Independence of the treatment sum of squares from the overall mean sum of squares, $nm\bar{y}^2$, follows from orthogonality of the first column of Z, associated with $\bar{\mu}$, and the remaining $m - 1$ columns, associated with the α_i. The test statistic for testing nullity of the α_i is obviously the same F statistic obtained in section 4.4.5.

Notice that $H_* \gamma = 0$ is exactly the hypothesis testing problem considered in section 5.4.1, but here we are taking advantage of the special pattern of the design matrix Z.

5.4.5 Two-way analysis of variance

The analysis of variance scheme considered in the previous subsection is the basis for a vast family of methods. We do not intend to move deeply into this territory, although it represents an important area of statistics. Our sole purpose is to sketch an initial extension, even dropping some of the technical details; the problem is too broad and difficult to be developed appropriately here. For a proper presentation of this topic, the reader is referred to more specialized treatments. In particular, the classic accounts are those of Scheffé (1959) and Cochran and Cox (1950); these two texts are different in style and coverage, and are to be regarded as complementary to each other.

Decomposition (5.43) is meaningful if the experimental units are essentially homogeneous and are given the treatment under the same conditions. In such a case, the observed differences among the treatment groups can safely be interpreted as being caused by the treatments themselves. It often happens, however, that such an ideal situation cannot be achieved, due to the difficulty either of repeating the experiment under constant conditions, or of finding sufficiently homogeneous experimental units or for some other reason. In these cases we may run into the problem of **confounding**, a term used to indicate that we are observing the effect of the treatment superimposed on some other effect; clearly, when confounding is present, the whole experiment may become ineffective, due to the difficulty or impossibility of meaningful interpretation of the results. This sort of problem must be taken into account both when the experiment is designed and when the data are analysed.

To illustrate these points, suppose that the treatments under consideration are represented by some agricultural fertilizers and the response variable is the amount of crop produced over a unit area of land. It may be difficult to perform all trials on soil of constant type, with the same chemical and geological properties, and indeed this would not even be desirable, since then the conclusions of the study would be applicable only to that specific kind of soil. On the other hand, it would be misleading to perform the experiment and the subsequent data analysis ignoring the soil properties,

Table 5.6 *Two-way analysis of variance*

Source	SS	d.f.
Fertilizer	$n \sum_i (\bar{y}_{i+} - \bar{y})^2$	$m - 1$
Soil type	$m \sum_j (\bar{y}_{+j} - \bar{y})^2$	$n - 1$
Error	$\sum_i \sum_j (y_{ij} - \bar{y}_{i+} - \bar{y}_{+j} + \bar{y})^2$	$(m-1)(n-1)$
Total	$\sum_i \sum_j (y_{ij} - \bar{y})^2$	$nm - 1$

since we would run into a case of confounded effects, namely those of treatments and soil type, which would be undistinguishable.

To overcome the problem, we can design the experiment so that each treatment is tested on each soil type (assuming that soils can be classified into a limited set of types). Let y_{ij} represent the yield produced with the ith fertilizer on soil type j, for $i = 1, \ldots, m$ and $j = 1, \ldots, n$. Assuming additivity of the various components, we can write

$$y_{ij} = \bar{\mu} + \alpha_i + \gamma_j + \varepsilon_{ij}$$

where γ_j is the effect of the soil type. For reasons similar to those leading to $\sum_i \alpha_i = 0$, we have the constraint

$$\sum_j \gamma_j = 0.$$

In this problem, it is likely that the parameters α_i relative to treatments are regarded as parameters of interest, while often the parameters associated with soil effect are nuisance parameters.

We shall not develop the resulting algebra in full as we did for section 5.4 4, and report only that one obtains an analysis of variance table like Table 5.6, where now a new source of variability due to soil type is present. In this table, $\bar{y}_{i+}$, $\bar{y}_{+j}$ denote the arithmetic means of the ith row and of the jth column of the data matrix, respectively.

We stress the essential role played by the assumption of additivity of effects: we are saying that the use of fertilizer i increases or decreases the yield always by the same mean amount, independent of the type of soil to which it is applied, and an analogous statement is valid for the soil effects. If additivity did not hold, we should write

$$y_{ij} = \bar{\mu} + \delta_{ij} + \varepsilon_{ij}$$

where δ_{ij} denotes the effect of treatment i on soil of type j. We then say that there is interaction between the two factors, treatment and soil type.

Unfortunately, we have now run into a non-identifiability problem, due essentially to the fact that δ_{ij} and ε_{ij} are not distinguishable. There is no escape from this trap, once the data have been collected: a different experimental design is necessary. We shall not deal with this problem, and refer the reader to the literature mentioned earlier.

5.4.6 Analysis of covariance

In section 5.4.1, we assumed that the columns of the design matrix X contained only values taken by quantitative variables. In sections 5.4.3–5.4.5, matrix X contained one or more indicator variables to represent the levels assumed by a qualitative variable. Consider now the mixed case, where both sorts of variables are present.

To avoid a somewhat abstract description, it is convenient to link it to a specific problem. Consider two groups of subjects: members of one group are given treatment A, while those of the other group are given treatment B. The purpose of the study is to investigate the difference, if any, in effect between the two treatments; in contrast to section 5.4.3, we now wish to take into account the presence of a quantitative explanatory variable in order to conduct a more accurate comparison between the treatments.

Consider for instance a situation where A and B are two different procedures for annealing metal, and y is a measure of the strength of the metal. If we perform a certain number of trials on each of the two treatments, making sure that the two treatments are treated exactly in the same way as far as experimental conditions are concerned, then Student's t test of section 5.4.3 is the appropriate tool for our purposes. It might be the case, however, that some relevant experimental factor cannot be controlled precisely by the experimenter, and this factor may play a relevant role in the experimental outcome. It could be the case that metal purity x cannot be kept constant across the various trials, but that the value of x influences the strength of the product. Therefore, we wish to compare A and B *adjusting for the differences in the values of* x, since otherwise we could interpret differences between groups as due to the treatments, while they might be due to uncontrollable variations – causing unbalancing – in the x values.

Introduce the simplifying assumption that the effect of x on y is proportional to x, and that the proportionality constant γ is the same for both groups; this implies that equalities of the form

$$y = \bar{\mu} + \alpha + \gamma x + \varepsilon,$$
$$y = \bar{\mu} - \alpha + \gamma x + \varepsilon,$$

hold true for the first and second group, respectively. The intercepts of the two lines have been parametrized similarly to (5.44).

Suppose now that we have obtained n measurements from each of the two groups, we can write the linear model

$$
\begin{pmatrix} y_1 \\ \vdots \\ y_{2n} \end{pmatrix}
=
\begin{pmatrix}
1 & 1 & x_1 \\
\vdots & \vdots & \vdots \\
1 & 1 & x_n \\
1 & -1 & x_{n+1} \\
\vdots & \vdots & \vdots \\
1 & -1 & x_{2n}
\end{pmatrix}
\begin{pmatrix} \bar{\mu} \\ \alpha \\ \gamma \end{pmatrix}
+ \varepsilon,
$$

where the first n rows of the response vector belong to the first group, and the remaining rows to the second group. Notice that the first two columns of the design matrix are orthogonal, a property which would not hold if we replaced '-1' by '0', for instance.

Clearly, the hypothesis of equal group sizes is made just for the sake of simplicity. In practical cases, this condition often fails, and the corresponding mathematical development is slightly more complicated, but analogous.

For ease of notation, denote the ith element of vector x by x_{iA}, and the $(i+n)$th element of x by x_{iB}; we use similar notation for the elements of vector y. This results

$$
X^\top X =
\begin{pmatrix}
2n & 0 & \sum x_i \\
0 & 2n & \sum x_{iA} - \sum x_{iB} \\
\sum x_i & \sum x_{iA} - \sum x_{iB} & \sum x_i^2
\end{pmatrix},
$$

$$
X^\top y =
\begin{pmatrix}
\sum y_i \\
\sum y_{iA} - \sum y_{iB} \\
\sum x_i y_i
\end{pmatrix}
$$

where summations are extended to all elements of the vectors. It can then be checked that the solutions of the normal equations satisfy

$$
\hat{\gamma} = \frac{\sum (x_{iA} - \bar{x}_A) y_{iA} + \sum (x_{iB} - \bar{x}_B) y_{iB}}{\sum (x_{iA} - \bar{x}_A)^2 + \sum (x_{iB} - \bar{x}_B)^2},
$$

$$\hat{\alpha} \;=\; \frac{1}{2}\left(\bar{y}_A - \bar{y}_B - \hat{\gamma}(\bar{x}_A - \bar{x}_B)\right),$$

$$\hat{\mu} \;=\; \bar{y} - \hat{\gamma}\bar{x},$$

where a bar denotes the arithmetic mean of the corresponding vector.

In this case, we are particularly interested in testing the hypotheses

$$\begin{cases} H_0 : \alpha = 0, \\ H_1 : \alpha \neq 0 \end{cases}$$

since 2α represents the mean difference between the two treatments, commonly called the **treatment effect**. Testing the null hypothesis above is possible using the F statistic which is now of type (5.39), namely

$$F = \frac{\hat{\alpha}^2\, 2n\{\sum(x_{iA} - \bar{x}_A)^2 + \sum(x_{iB} - \bar{x}_B)^2\}}{s^2\,\{\sum x_i^2 - (\sum x_i)^2/(2n)\}}$$

with $(1, 2n - 3)$ degrees of freedom.

One could criticize the assumption that the two regression lines are parallel, introducing instead two completely distinct regression lines for the two groups, with distinct slopes, γ_A, γ_B say.

Such an approach is certainly possible, and in fact appropriate in some cases. The problem then becomes substantially more complicated, not so much from the mathematical viewpoint as from the viewpoint of interpreting the results and numerical quantities involved. In this framework, it no longer makes sense to talk about treatment effect as a given value. In fact, if we choose two sample elements, one for each group, with the same x value, it turns out that $E[y_{iA}] - E[y_{iB}]$ is no longer constant, but depends on the value of x, namely

$$E[y_{iA}] - E[y_{iB}] = 2\alpha + (\gamma_A - \gamma_B)x$$

which varies with x. We then say that there is **interaction** between the treatment and the variable x. We shall not expand this topic further

5.5 On model selection

Our entire development of statistical inference theory has been based on the firm assumption that there exists a given statistical model on which any inferences can be based. According to our exposition in section 1.3, this model must satisfy two main conditions, namely:

(a) it is defined as a class of probability distributions, including the one which actually generates the data;

(b) this class is specified before the data are collected or, at least, before the data are analysed.

These requirements are easily fulfilled when the mechanism which generates the data is simple, or at least is perfectly understood. The clearest case arises in replicating an experiment with two possible outcomes. If the various replicates of the experiment are performed under well-controlled conditions which we can safely consider to be constant, then the distribution of the number of successes is binomial. In this case, knowing the physical mechanism generating the data translates immediately into a statement on the probabilistic properties of the phenomenon.

The previous example represents a sort of ideal case. Although it is not the only one allowing such a simple treatment, it remains true, however, that most practical cases are far more complicated, especially when we are dealing with problems involving the study of relationships among variables, as in the case of linear models or those to be described in Chapter 6.

When one analyses the relationships among variables, the number of variables to be considered is often quite high and the possible ways in which dependence may occur are extremely varied, making the problem of specifying statistical models far more complicated. In these more complex cases, formulating a statistical model fulfilling requirements (a) and (b) requires the existence of a solid theory for the phenomenon under consideration. Unfortunately, it is not often the case that such a theory is available; far more often, the model which is being fitted to the data does not exist before the data are produced, and it is built in the light of the data themselves.

Therefore, we have to face the problem of **model choice** (or **model selection**), a broad term which embraces all aspects of the issue discussed above, and gives rise to a number of more specialized subproblems: selection of relevant variables, choice of variables transformations, choice of the mathematical relationship among the variables, and so on.

Clearly, the whole issue is a very complex one, and a full treatment is beyond the scope of this book. We shall restrict ourselves to some general remarks.

There are various, quite distinct, ways of tackling the model selection problem. Some of them are essentially within the realm of parametric statistics, while others are outside this domain. The following discussion is very sketchy; in reading it, one must bear in mind that not all approaches deal exactly with the same sorts of problem; so a direct comparison cannot be carried out.

The first approach, **model extension**, operates within parametric statistics, by enlarging a linear model so as to include some form of nonlinearity.

For ease of explanation, we start with a very simple scheme, still within the linear model framework. If x is an explanatory variable whose effect on the response variable is suspected to be nonlinear, a common technique is to include in the linear model, besides x itself, a function such as x^2 or $\log(x)$ or some other nonlinear function, and then to evaluate the significance of the corresponding regression coefficient. This procedure is not adopted because one really believes that the nonlinear component, if present, is of the form x^2 or $\log(x)$, but simply as a means to capture at least in an approximate way the nonlinearity of the relationship.

The above methodology is rather crude (nevertheless, often quite effective!), but it can be improved and adapted to different situations. A common problem is the choice of the scale on which to express the relationship between y and $X\beta$, i e. the choice of the transformation $T(y)$ for which the linear model holds. A technique developed for this purpose is called the **Box–Cox transformation** (Box and Cox, 1964). In this formulation, it is assumed that a linear model of the form

$$T(y, \lambda) = X\beta + \varepsilon$$

holds, for some nonlinear transformation $T(y, \lambda)$ of y depending on an unknown parameter λ, which must be estimated simultaneously with the regression parameters. The original Box–Cox transformation was

$$T(z, \lambda) = \begin{cases} \dfrac{z^\lambda - 1}{\lambda} & \text{if } \lambda \neq 0, \\ \ln z & \text{if } \lambda = 0, \end{cases}$$

which is continuous with respect to λ for all z, and includes, up to constants, the identity transformation ($\lambda = 1$), the logarithmic

($\lambda = 0$) and power transformations. The original proposal has been very successful, and similar methods have been put forward for selecting an appropriate transformation of the explanatory variables, providing a solution more refined than the one described earlier, based on the simple insertion of an x^2 term.

A second approach, **graphical diagnostics**, is based on graphical methods which allow us to explore data only in a qualitative, albeit very useful, way. Where a tentative linear model has been fitted to the data, these methods rely widely on residuals

$$y_i - \hat{\mu}_i \quad (i = 1, \ldots, n),$$

possibly in a scaled form, such as

$$e_i = \frac{y_i - \hat{\mu}_i}{s} \quad (i = 1, \ldots, n)$$

or something similar. It can be easily shown that, if the model has been correctly specified, then the e_i are random variables with mean 0 and variance close to 1. Moreover, the e_i should behave in a completely erratic way, since they are essentially estimates (up to the scale factor σ) of the ε_i, which are mutually independent, or at least uncorrelated, variates. Therefore, when standardized residuals behave differently than the above properties lead us to expect, it can be taken as a possible indication of inadequacy of the fitted model.

One very useful tool for checking the residuals is the scatter plot of the pairs $(\hat{\mu}_i, e_i)$. This plot is a very simple and commonly used technique, whose inspection can often help in detecting abnormal behaviour of the residuals, such as an increase in dispersion of the e_i as the $\hat{\mu}_i$ increase, a feature which suggests heteroscedasticity (non-constant variance) of the errors.

The available techniques are very many, not all based explicitly on residuals. Some of them are very simple and quick, such as the one just mentioned; others are more complex both in the interpretation of the result and in the actual construction, up to the point of requiring a computer, especially for those producing dynamic graphics.

The body of graphical methods represents a powerful tool both for suggesting the form of relationships among variables and for detecting inadequacies of a tentative model, e g nonlinearity, presence of anomalous data (commonly called **outliers**), correlation among error terms ε_i, and more.

For a detailed discussion of these methods, see Cook and Weisberg (1982; 1994) and Cleveland (1995).

The third approach, **nonparametric regression**, takes an entirely new route. For the sake of simplicity, consider the case where there are only two variables to be examined, x and y say, and suppose that a relationship of the form

$$y = r(x) + \varepsilon$$

holds, where $r(\cdot)$ is a function for which we do not specify a mathematical form, i e we do not assume a parametric class of functions to which $r(\cdot)$ must belong. We only assume that $r(\cdot)$ satisfies some regularity condition, such as differentiability.

Since the set of possible functions $r(\cdot)$ of this form cannot be parametrized, we are working within the framework of nonparametric statistics. There exist various techniques for tackling this problem; extensive accounts are given in the books by Silverman (1986) and Härdle (1990).

Undoubtedly, a nonparametric approach to the estimation of $r(x)$ is very appealing, since it eliminates our responsibility of specifying a precise mathematical form for $r(x)$. This fact is especially welcome when our information about $r(x)$ is vague. There are, however, also some disadvantages in this approach.

- Often the person who provided the data, or proposed the data analysis, would appreciate being provided with an estimate for $r(x)$ in a simple mathematical form (while often being unable to suggest to us anything about this form!). Nonparametric methods do not fulfil this requirement. This is a consequence of the fact of choosing as an estimate $\hat{r}(x)$ an element of a set of functions for which 'simple mathematical form' is not a property nor could this property hold if we wanted to include in this set all sufficiently regular functions.

- The space of all possible alternatives has been enlarged by passing from a parametric to a nonparametric formulation of $r(x)$. This implies some loss of statistical efficiency, i e an increase of the variance of the estimator and also the presence of bias. When there is only a single explanatory variable, this loss of efficiency is not dramatic, but it becomes more substantial when we move on to the multiple regression case, namely

$$y = r(x_1, \ldots, x_p) + \varepsilon$$

for p explanatory variables.

- The comparison of results between similar data sets is more difficult in the nonparametric setting than in the parametric, since the latter approach allows us to compare a few summary parameters.

These remarks do not intend to dismiss the importance of the nonparametric approach, just to caution against excessive enthusiasm for an approach which frees us, at least partly, from difficulty of model specification.

A related issue is stability of the conclusions to model specification, especially for those aspects of the model which are less reliably established. It is clearly inappropriate that the conclusions of the analysis depend heavily on some aspects of the model which we are particularly unsure about.

A typical illustration of this point is offered by regression models when normality of the error term is assumed for mathematical convenience rather than out of any conviction. In a number of practical problems, it has been found that the actual distribution of the errors (inspected through the residuals) was quite different from normality. Depending on the nature of this departure from normality, likelihood-based methods can be more or less heavily affected, as for bias, loss of efficiency, etc.

One particularly dangerous situation is the presence of outliers, due for instance to mistakes in data recording. In some cases, the presence of outliers can lead to improper or even incorrect conclusions, if they are not taken care of. Various approaches to this problem are possible; one, already mentioned, uses graphical diagnostics to detect and downweight outlying data. A more formal and, in a sense, more systematic approach is via **robust methods**, which are especially designed to be relatively insensitive to the presence of outliers and, more generally, to failure of assumptions. For an authoritative description of robust methods, see the book of Hampel *et al.* (1986).

One must not get the impression that all these approaches are mutually exclusive. They are logically very different, but in practical data analysis they are used to complement each other, in different stages of the analysis.

Usually, graphical techniques are used in the preliminary stage, often called **exploratory analysis**, to formulate a tentative model, to which formal procedures, such as formal maximum likelihood estimation, are later applied. The outcome of these formal procedures has to be considered quite critically, in the light of the empirical origin of the model; this criticism often leads to reconsideration of the original model formulation, e g because the analysis of the residuals exhibits certain inadequacies The original model then has to be modified accordingly

In extreme cases, the above steps have to be repeated several times, until a satisfactory formulation is reached. Therefore, the model is not an entity specified once and for all, but evolves as the data analysis proceeds.

The strategy described above is clearly very different from what we assumed in the previous chapters. On the other hand, this situation is often imposed by the lack of a reliable subject-matter theory able to provide us with a trustworthy model Moreover, even when such a theory exists, it is good scientific practice critically to examine the model, hence even the theory, in the light of the available data.

There exists the potential risk of going too far in this 'interaction with the data', arriving at formulations chosen only to obtain a good fit of the fitted values $\hat{y}_i$ to the observed values y_i, but over elaborate or meaningless from the subject-matter point of view. The appropriate balancing of all these components is a primary difficulty of real data analysis.

It should also be clear that the above procedure for selecting a model does not follow a precise sequence of steps, according to some rigorous protocol. Instead, it is a very complex procedure where various different ideas, coming both from statistics and from subject-matter knowledge, meet Since no formal guidelines exist, experience plays a fundamental role. Unfortunately, experience is not easily coded and transferred, but reading can be very useful; see, for instance, Cox and Snell (1981), Weisberg (1985), Anderson and Loynes (1987) and Chatfield (1988).

Exercises

5.1 Suppose $T_1, \ldots, T_r$ are unbiased estimates of the same parameter θ, and denote by $v_1, \ldots, v_r$ the corresponding variances, assumed known. Assuming that the estimators are mutually independent, obtain the linear combination of $T_1, \ldots, T_r$ representing the unbiased estimator of θ with minimum variance.

5.2 Under the assumptions of section 5.2.1, show that $X^\top X$ is positive definite (hence of rank p).

5.3 Consider a regression model with $p = 1$, $X = (x_1, \ldots, x_n)^\top$. State under which conditions on the sequence $(x_1, x_2, \ldots)$ the estimate of the regression parameter β is consistent.

5.4 Show that the subset of $C(X)$ satisfying the condition $H\beta = 0$, where H is defined as in section 5.2.5, is a linear subspace of dimension $p - q$ of $\mathbb{R}^n$.

5.5 Check that $H\hat{\beta}_0 = 0$ when $\hat{\beta}_0$ is defined by (5.24).

5.6 Verify the statement given in section 5.2.5 that a vector Xc of $C(X)$, such that $Hc = 0$, is orthogonal to $y - \hat{\mu}_0$.

5.7 Verify the statement of section 5.2.5 that $\hat{\mu} - \hat{\mu}_0 \perp \hat{\mu}_0$.

5.8 Show that the matrices $P_0, P - P_0, I_n - P$ are mutually orthogonal and that their respective ranks are as given in section 5.3.3.

5.9 In section 5.4.2, check that $C(X) = C(Z)$.

5.10 In the framework of section 5.4.3, obtain a confidence interval for the treatment effect $\delta = \beta_2 - \beta_1$.

5.11 In section 5.4 4, obtain $\hat{\gamma}$.

5.12 In section 5.4.4, obtain the LRT statistic for the hypothesis $\beta_1 = \ldots = \beta_r$, where $1 < r < m$.

5.13 State the condition on matrix X such that $\sum_i \hat{\varepsilon}_i = 0$ for all vectors y.

5.14 Consider the simple linear regression model, i e

$$y_i = \beta_1 + \beta_2 x_i + \varepsilon_i \qquad (i = 1, \ldots, n).$$

(a) Prove that $\operatorname{var}\left[\hat{\beta}_1\right]$ is minimized if the values x_i are chosen so that $\sum_i x_i = 0$.

(b) In the case where the x_i can be arbitrarily chosen to lie in the closed interval $[a, b]$ and n is even, show that $\operatorname{var}\left[\hat{\beta}_2\right]$ is minimized if $n/2$ elements x_i are chosen equal to a and the remaining $n/2$ elements equal to b.

5.15 If (5.40) is used for the slope of a simple linear regression, it must coincide with (5.41). Verify that this is true.

5.16 Suppose that, for a simple linear regression model, estimates and standard errors of the parameters have been obtained from a sample of n elements. If a new observation is about to be taken at point x_{n+1}, it is natural to predict Y_{n+1} from

$$\hat{y}_{n+1} = \hat{\beta}_1 + \hat{\beta}_2 \, x_{n+1}$$

using the available estimates. Study the properties of this prediction of Y_{n+1}.

5.17 For three vectors y, x_1, x_2 of dimension n, denote by $\hat{\beta}_1^*$ and $\hat{\beta}_2^*$ the regression coefficients of vector y on x_1 and on x_2, respectively, when the regression lines are computed separately, i e

$$\hat{\beta}_1^* = \frac{y^\top x_1}{x_1^\top x_1}, \quad \hat{\beta}_2^* = \frac{y^\top x_2}{x_2^\top x_2}.$$

Now construct the regression matrix $X = (x_1, \, x_2)$ and denote by $\hat{\beta}$ the corresponding vector of regression coefficients. Determine the relationship between the elements of $\hat{\beta}$ and the previous $\hat{\beta}_1^*$, $\hat{\beta}_2^*$. Check the result by applying it to the formulae of Example 5.2.4.

5.18 Suppose that the assumption $\text{var}[\varepsilon] = \sigma^2 I_n$ is replaced by $\text{var}[\varepsilon] = \sigma^2 W$, where W is a positive definite matrix which is known. Determine how the LSE of β changes when, at the same time, the Euclidean distance between y_1 and y_2 is replaced by the more general Mahalanobis distance

$$\|y_1 - y_2\|_W^2 = (y_1 - y_2)^\top W^{-1} (y_1 - y_2).$$

Obtain also the corresponding variance matrix of the estimates. This case is often referred to as **generalized** least-squares method.

A commonly arising situation which leads to a special case of the situation described above occurs when the generic value of y_i is in fact the arithmetic mean of several observations, m_i say, all obtained with the same row vector $\tilde{x}_i$ of explanatory variables. In this case, we have

$$W = \text{diag}(1/m_1, \ldots, 1/m_i, \, . \, ., 1/m_n)$$

since the generic component has a variance which is inversely proportional to the number of (original) observations of which it is the mean. In principle, the above matrix W should be multiplied by σ^2, but this is both unknown and irrelevant, since it is a scale factor which does not affect the resulting estimate. We then say that m_i is the **weight** of the ith observation and the resulting estimation method is called **weighted least squares**.

5.19 Consider a standard linear model of type (5.7), for which the relevant quantities

$$T_n = \{\hat{\beta}_n,\ \hat{\sigma}_n^2,\ V_n\}$$

have already been computed, where $V_n = (X^\top X)^{-1}$, and the subscript n has been added to emphasize that they are obtained from n observed cases. Suppose that an additional observation y_{n+1} becomes available, together with its vector of explanatory variables $\tilde{x}_{n+1}$, and denote by y_*, X_* the new vector of the response variable and the new regression matrix, respectively. Obtain formulae which allow the estimates to be updated without explicitly inverting $X_*^\top X_*$, since this inversion is computationally burdensome when the number of covariates is high. In other words, we wish to express the new triplet T_{n+1} as a function of the old triplet T_n and of the new data $\{y_{n+1}, \tilde{x}_{n+1}\}$. This technique can then be repeated if a sequence of successive observations is provided, and the triplet T_n updated sequentially; this method is sometimes called **recursive least squares**.

5.20 Derive the LRT statistic for the hypotheses (5.33) where σ^2 is known.

CHAPTER 6

Generalized linear models

6.1 Some limitations of linear models

In the introductory section of Chapter 5, various arguments were put forward to explain the relevance of linear models and to illustrate why their usefulness is broader than might first appear. However, there are problems which cannot be reduced to the form required by linear models. Let us examine what specifically are the difficulties and limitations of linear models.

- It may happen that the relationship between response and explanatory variables is of type (5.1) with $r(\cdot)$ nonlinear in the parameters. If the mathematical form of $r(\cdot)$ is known, as in (5.3), then a linearization such as (5.5) is unlikely to be acceptable.

- Even when the mathematical form of $r(\cdot)$ is not known precisely, we often know enough about the nature of the phenomenon to rule out a linear relationship. The most common case of this situation is related to the range of y. For instance, suppose that y represents the maximum fraction of skin affected in the evolution of a certain skin disease, and that the explanatory variables are prognostic factors for that particular disease; obviously, y lies in $[0, 1]$, but a fitted regression hyperplane cannot satisfy such a constraint for all values of the explanatory variables. In particular, suppose that we make use of the fitted relationship to predict, at the beginning of the disease, the fraction y_{n+1} associated with a new patient for whom the prognostic factors are known; then it may well happen that the predicted value $\hat{y}_{n+1}$ is outside the range $[0, 1]$.

- The variance of the error term, and hence of the response variable, should be constant, while in practical situations one often observes data which clearly do not comply with this requirement. For instance, it is frequently the case that the variance of the response variable increases monotonically with its mean value.

- In linear models we assume that the response variable is normally distributed when testing hypotheses or forming confidence intervals; this assumption is not so crucial when it comes to point estimation. If departure from normality is moderate, we can rely to a certain extent on the robustness of the test statistic. In a large number of cases, however, we have to deal with a y variable having clearly non-normal distribution. A technique often used is to transform the response variable using a nonlinear function so as to obtain approximate normality of the response variable; the Box–Cox transformation, described in section 5.5, is a class of functions often used. However, this approach is not a universal solution; in particular, it cannot be used when y is a discrete variate, a common situation.

In the rest of this chapter, we shall describe a class of models that includes the linear ones of Chapter 5, and copes with the above-mentioned concerns.

One should not think that the new class of models will include *all* possible situations. Indeed, a class of models allowing all possible cases would be so wide as to loose all structure, and would then be of little use. The class of models that we shall consider, called **generalized linear models** (GLMs), is not particularly broad from a mathematical point of view, but it is sufficiently flexible to include a large number of practical important subcases.

Moreover, the GLM class has the advantage of allowing a unified treatment for a set of relevant specific models, which in the past were introduced separately from each other, answering quite distinct practical needs, but which now appear as specific instances of the same approach. In fact, some of these special cases have already been mentioned in the previous chapters, as shall soon become clear.

The GLM class was introduced by Nelder and Wedderburn (1972) who showed that a large number of existing models and techniques were in fact special cases of a larger class. An extensive account on this topic is given by McCullagh and Nelder (1989).

Example 6.1.1 Some of the features mentioned above are illustrated by the data in Table 6.1, originally presented by Bissel (1972). For several pieces of cloth, this table gives the length L and the number of observed defects D in each piece. It is natural to re-

Table 6.1 *Cloth data, from Bissel (1972): for each piece of cloth, L represents the length (in metres) and D the number of defects (Data from Biometrika* **59**, *440, reproduced with the permission of the* Biometrika *trustees.)*

L	D	L	D
551	6	543	8
651	4	842	9
832	17	905	23
375	9	542	9
715	14	522	6
868	8	122	1
271	5	657	9
630	7	170	4
491	7	738	9
372	7	371	14
645	6	735	17
441	8	749	10
895	28	495	7
458	4	716	3
642	10	952	9
492	4	417	2

gard L as an explanatory variable and D as the response variable. The latter variable is discrete, and its discreteness is effectively restricted to very few values when L is small, as exhibited by the scatter plot in Figure 6.1. This plot confirms the obvious conjecture that the mean value of D increases with L, and it shows in addition that the variability of D increases with L, hence with its own mean value.

The discreteness of D and its non-constant variability make it inappropriate to fit a linear model to these data. Instead, considerations similar to those of Example 2.3.15 lead to the formulation

$$D \sim Poisson(\beta L)$$

at least as an initial, tentative model; here β is a positive parameter. This model is a special instance of the GLMs to be presented below.   □

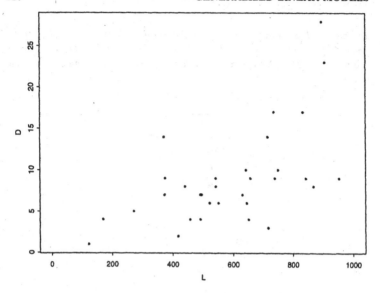

Figure 6 1 *Cloth data: scatter plot of (D, L)*

6.2 Generalized linear models

6.2.1 An interesting class of distributions

Since we shall deal with a class of probability distributions forming a subset of the exponential family, let us introduce a specific notation for this class. For a real random variable Y, we write

$$Y \sim EF(b(\theta), \psi/w)$$

if Y has density function of the form

$$f(y) = \exp\left(\frac{w}{\psi}\{y\,\theta - b(\theta)\} + c(y, \psi)\right) \qquad (6.1)$$

where θ, ψ are scalar parameters, w is a known constant, and $b(\cdot), c(\cdot)$ are known functions, that determine the specific parametric family of distributions.

For any given choice of the **dispersion parameter** ψ, (6.1) forms an exponential family with parameter θ, although (6.1) is not an exponential family when θ, ψ vary simultaneously. However, for a large portion of what we shall say, ψ will be treated as fixed. The variability of Y is actually not regulated by ψ only, but by the ratio

ψ/w; it is, however, convenient to separate this quantity in two components: the unknown parameter ψ and the known constant w, called the **weight**.

For the purposes computing the moments of the distribution, ψ is a constant; therefore, we can proceed analogously to the argument followed to obtain (2.7) and (2.8), and conclude that

$$\mathrm{E}[\,Y\,] = b'(\theta), \qquad \mathrm{var}[\,Y\,] = b''(\theta)\,\psi/w\,.$$

It can be shown that the derivatives $b'(\theta)$, $b''(\theta)$ exist. For use in the algebraic manipulations below, define

$$\mu = b'(\theta), \qquad V(\mu) = b''(\theta)\,.$$

The function $V(\mu)$, called the **variance function**, is regarded as a function of the mean value μ, even if it appears as a function of θ; this is possible by inverting the relationship between θ and μ.

Example 6.2.1 If $Y \sim Poisson(\mu)$, its probability function is

$$
\begin{aligned}
f(y) &= \exp(-\mu)\,\mu^y/y! \\
&= \exp(y \ln \mu - \mu - \ln y!) \\
&= \exp(y\theta - e^\theta - \ln y!)
\end{aligned}
$$

where $y = 0, 1, \ldots$ and $\theta = \ln \mu$. We readily obtain

$$
\begin{aligned}
b(\theta) &= e^\theta\,, \\
\mu = b'(\theta) &= e^\theta\,, \\
b''(\theta) &= e^\theta = \mu\,, \\
\psi &= 1, \quad w = 1\,, \\
c(\psi, y) &= -\ln y!
\end{aligned}
$$

so that $V(\mu) = \mu$. In this case, ψ is effectively not present. □

Example 6.2.2 If $Y \sim N(\mu, \sigma^2)$, the density function is

$$
\begin{aligned}
f(y) &= \exp\left\{ -\frac{1}{2\sigma^2}(y - \mu)^2 - \frac{1}{2}\ln(2\pi\sigma^2) \right\} \\
&= \exp\left\{ \frac{1}{\sigma^2}\left(y\mu - \frac{\mu^2}{2} \right) - \frac{1}{2}\left(\frac{y^2}{\sigma^2} + \ln(2\pi\sigma^2) \right) \right\}
\end{aligned}
$$

for $y \in \mathbb{R}$. This density function is of type (6.1), where

$$
\begin{aligned}
\theta &= \mu\,, \\
b(\theta) &= \frac{\mu^2}{2}\,,
\end{aligned}
$$

$$b'(\theta) = \mu, \qquad b''(\theta) = 1,$$
$$\psi = \sigma^2, \qquad w = 1,$$
$$V(\mu) = b''(\theta)\psi/w = \sigma^2.$$

□

Other important distributions of type (6.1) are listed in Table 6.2, with their characteristics. The meaning of some of these characteristic elements will be explained below.

6.2.2 A new class of models

It is convenient to start by reconsidering the elements which specify a linear model. For the generic ith observation, we define a **linear predictor** $\eta_i = x_i^\top \beta$, where x_i is the ith row* of X $(i = 1, \ldots, n)$; moreover, let us suppose that observation y_i is sampled from $Y_i \sim N(\mu_i, \sigma^2)$, where the relationship between the mean value μ_i and the linear predictor η_i is the identity. In a schematic form, we can write

$$Y_i \sim N(\mu_i, \sigma^2), \qquad \mu_i = \eta_i, \qquad \eta_i = x_i^\top \beta. \qquad (6.2)$$

The GLM class is obtained by extending the above formulation in two directions:

- the distribution of Y_i is not restricted to being normal, but can be any distribution of the form $EF(b(\theta_i), \psi/w_i)$ such that $b'(\theta_i) = \mu_i$ is allowed;

- other forms of relationship between the linear predictor η_i and the mean value μ_i are possible, besides the identity, i e we consider

$$g(\mu_i) = \eta_i$$

where $g(\cdot)$ is a differentiable monotonic function, called the **link function.**

Schematically, we specify a GLM in terms of the following elements:

$$Y_i \sim EF(b(\theta_i), \psi/w_i), \qquad g(\mu_i) = \eta_i, \qquad \eta_i = x_i^\top \beta,$$
$$\text{(error structure)} \qquad \text{(link function)} \qquad \text{(linear predictor)}$$
$$(6.3)$$

---

* In this chapter, x_i denotes the values of the explanatory variables observed on the ith subject, not the ith column of X, as in Chapter 5

Table 6.2 · Characteristics of some distributions of type (6.1)

	Normal	Poisson	Binomial/m	Gamma
	$N(\mu,\sigma^2)$	$P(\mu)$	$Bin(m,\mu)/m$	$Gamma(\omega,\omega/\mu)$
Support	$(-\infty,\infty)$	$\{0,1,2,\dots\}$	$\{0,\frac{1}{m},\frac{2}{m},\dots,1\}$	$(0,\infty)$
ψ	σ^2	1	1	ω^{-1}
w	1	1	m	1
$b(\theta)$	$\theta^2/2$	$\exp(\theta)$	$\log(1+e^\theta)$	$-\log(-\theta)$
$c(y;\psi)$	$-\dfrac{1}{2}\left(\dfrac{y^2}{\psi}+\log(2\pi\psi)\right)$	$-\log y!$	$\log\begin{pmatrix} m \\ m\,y \end{pmatrix}$	$\omega\log(\omega y)-\log y$ $-\log\Gamma(\omega)$
$\mu(\theta)$	θ	$\exp(\theta)$	$e^\theta/(1+e^\theta)$	$-1/\theta$
Canonical link	identity	logarithm	logit	reciprocal
$V(\mu)$	1	μ	$\mu(1-\mu)$	μ^2

where $b'(\theta_i) = \mu_i$. In more detail, a statistical model is a GLM when it satisfies the following requirements:

(a) the observations $y_1, \ldots, y_n$ are sampled from mutually independent random variables $Y_1, \ldots, Y_n$;

(b) each Y_i has a distribution of type $EF(b(\theta_i), \psi/w_i)$ with $E[Y_i] = \mu_i = b'(\theta_i)$, for $i = 1, \ldots, n$;

(c) there exists a function $g(\cdot)$ such that $g(\mu_i) = x_i^\top \beta$, where x_i is a vector of constants and β is a vector of parameters;

(d) the function $b(\cdot)$ and the parameter ψ are common to all Y_i, while the w_i can vary;

(e) the functions $b(\cdot)$, $c(\cdot, \cdot)$, $g(\cdot)$ and the w_i are known.

Notice that, when the link function $g(\ )$ is the identity and the error structure is normal, we are back to a standard linear model.

The phrase 'error term' is maintained because of this connection with linear models, even if in a GLM an explicit 'error term' ε does not necessarily exist as in (5.4) or in (5.7). In fact, in the normal case, we may equivalently write, $Y \sim N(\mu_i, \sigma^2)$, or $Y_i = \mu_i + \varepsilon_i$ with $\varepsilon_i \sim N(0, \sigma^2)$. In this case, there is a complete separation between the systematic component μ_i (which depends on x_i and β, but not on the nature of the random term) and the purely erratic term ε_i which does not depend on x_i and β. In the case of a GLM, this precise separation of the response variable into two components is no longer possible in general, as the following examples will demonstrate.

Example 6.2.3 Suppose that $Y_1, \ldots, Y_n$ are independent Poisson variates with mean value $\mu_1, \ldots, \mu_n$, respectively; suppose in addition that

$$\ln(\mu_i) = x_i^\top \beta$$

where the meaning of x_i and β is as usual. In this situation, called **Poisson regression** with logarithmic link function, the response variable Y_i cannot be decomposed into a form such as

$$Y_i = \mu_i + \varepsilon_i$$

with ε_i representing a purely random term. In fact, to satisfy the above equality, the random variable ε_i should have a distribution depending on μ_i (because its variance is equal to μ_i, to pinpoint just one feature), and it seems inappropriate to call a quantity depending on the linear predictor an 'error term'.   □

Example 6.2.4 In Example 2.4.4, we introduced the logistic regression model. It is easy to check that it falls within the GLM class, with probability distribution of the ith component given by

$$f(y_i) = \exp\{\theta_i\, y_i - \ln(1 + e^{\theta_i})\}$$

where $\theta_i = \text{logit}(\mu_i) = \ln(\mu_i/(1 - \mu_i))$, with the assumption that θ_i is a linear function of an explanatory variable $(i = 1, \ldots, n)$. Therefore, we have

$$b(\theta) = \ln(1 + e^{\theta}),$$

$$b'(\theta) = \frac{e^{\theta}}{1 + e^{\theta}} = \mu,$$

$$b''(\theta) = \frac{e^{\theta}}{(1 + e^{\theta})^2} = \mu(1 - \mu) = V(\mu)$$

We now wish to extend the logistic regression model in two directions. The first extension consists in allowing several explanatory variables, forming the vector x_i, and in assuming that

$$\theta_i = x_i^{\mathsf{T}} \beta.$$

This is an easy extension of the model, providing extra flexibility without formal complications.

The second generalization allows the possibility of replicates of the experiment for any given combination of the experimental factor levels. In the case of a single explanatory variable x, this means that, in general, for a given value x_i of x, there will be several observations, not just one For instance, in Example 2.4.4, this means that, for all or some of the dosage levels of the poison, we shall replicate the experiment with several animals. In the case of two or more explanatory variables, it is intended that replication of the experiment occurs for a given combination of the levels of the variables.

If m_i is the number of replicates at a given combination x_i of the explanatory variables, then the corresponding number of successes is $\tilde{Y}_i \sim Bin(m_i, \mu_i)$. Since our goal is to examine the relationship between the explanatory variables and the probability of success, we do not use $\tilde{Y}_i$ as the response variable, but rather the proportion of successes $Y_i = \tilde{Y}_i/m_i$. In this way, $\mathrm{E}[Y_i] = \mu_i$ is in fact the probability under consideration. The probability function of the generic observation is, omitting the subscript i,

$$
\begin{aligned}
f(y) &= \binom{m}{m\,y}\mu^{m\,y}(1-\mu)^{m(1-y)} \\
&= \exp\left[m\left\{y\theta - \ln(1+e^{\theta})\right\} + \ln\binom{m}{m\,y}\right]
\end{aligned}
$$

for $y = 0, \frac{1}{m}, \frac{2}{m}, \ldots, 1$. The function $b(\theta)$ remains as before, while now $w = m$.

For an extended treatment of binary data and logistic regression, see Cox and Snell (1989) and Agresti (1990). □

6.2.3 Likelihood and Fisher information

Denote by p the dimension of β and by $X = (x_{ij})$ the $n \times p$ matrix having ith row x_i^{T} $(i = 1, \ldots, n)$. Given observations $y_1, \ldots, y_n$, with mean values $(\mu_1, \ldots, \mu_n)$ such that

$$
g(\mu_i) = x_i^{\mathsf{T}}\beta, \tag{6.4}
$$

we wish to make inferences on the parameters β and ψ, with our main interest on β, while ψ is usually a nuisance parameter, when present.

By the hypothesis of independence of the components, we can immediately write the log-likelihood

$$
\begin{aligned}
\ell(\beta) &= \sum_{i=1}^{n}\left(\frac{w_i(y_i\theta_i - b(\theta_i))}{\psi} + c_i(y_i, \psi)\right) \\
&= \sum_{i=1}^{n}\ell_i(\beta).
\end{aligned}
$$

To obtain the likelihood equations, we write

$$
\frac{\partial \ell_i}{\partial \beta_j} = \frac{\partial \ell_i}{\partial \theta_i}\frac{\partial \theta_i}{\partial \mu_i}\frac{\partial \mu_i}{\partial \eta_i}\frac{\partial \eta_i}{\partial \beta_j}
$$

whose terms can be expanded as

$$
\begin{aligned}
\frac{\partial \ell_i}{\partial \theta_i} &= \frac{w_i(y_i - b'(\theta_i))}{\psi} = \frac{w_i(y_i - \mu_i)}{\psi}, \\
\frac{\partial \mu_i}{\partial \theta_i} &= b''(\theta_i) = \frac{w_i\,\mathrm{var}[Y_i]}{\psi}, \\
\frac{\partial \eta_i}{\partial \beta_j} &= x_{ij}.
\end{aligned}
$$

Hence we have

$$\frac{\partial \ell_i}{\partial \beta_j} = \frac{w_i(y_i - \mu_i)}{\psi} \frac{\psi}{w_i \operatorname{var}[Y_i]} \frac{\partial \mu_i}{\partial \eta_i} x_{ij}$$

and the likelihood equations for β are

$$\sum_{i=1}^n \frac{(y_i - \mu_i) x_{ij}}{\operatorname{var}[Y_i]} \frac{\partial \mu_i}{\partial \eta_i} = 0 \qquad (j = 1, \ldots, p). \qquad (6.5)$$

To obtain the Fisher information, consider the second derivatives of $\ell_i(\beta)$, namely

$$-\mathrm{E}\left[\frac{\partial^2 \ell_i}{\partial \beta_j \partial \beta_k}\right] = \mathrm{E}\left[\frac{\partial \ell_i}{\partial \beta_j} \frac{\partial \ell_i}{\partial \beta_k}\right]$$

$$= \mathrm{E}\left[\left(\frac{(Y_i - \mu_i) x_{ij}}{\operatorname{var}[Y_i]} \frac{\partial \mu_i}{\partial \eta_i}\right)\left(\frac{(Y_i - \mu_i) x_{ik}}{\operatorname{var}[Y_i]} \frac{\partial \mu_i}{\partial \eta_i}\right)\right]$$

$$= \frac{x_{ij} x_{ik}}{\operatorname{var}[Y_i]} \left(\frac{\partial \mu_i}{\partial \eta_i}\right)^2.$$

Therefore the (j, k)th entry of the expected information matrix is

$$-\sum_{i=1}^n \mathrm{E}\left[\frac{\partial^2 \ell_i}{\partial \beta_j \partial \beta_k}\right] = \sum_{i=1}^n \frac{x_{ij} x_{ik}}{\operatorname{var}[Y_i]} \left(\frac{\partial \mu_i}{\partial \eta_i}\right)^2.$$

Hence, the information matrix is

$$I(\beta) = X^\top \tilde{W} X \qquad (6.6)$$

where

$$\tilde{W} = \begin{pmatrix} \tilde{w}_1 & 0 & \cdots & 0 \\ 0 & \tilde{w}_2 & \cdots & 0 \\ \vdots & \vdots & \ddots & \vdots \\ 0 & 0 & \cdots & \tilde{w}_n \end{pmatrix},$$

having set

$$\tilde{w}_i = \frac{1}{\operatorname{var}[Y_i]} \left(\frac{\partial \mu_i}{\partial \eta_i}\right)^2 = \frac{w_i}{\psi V(\mu_i)} \left(\frac{\partial \mu_i}{\partial \eta_i}\right)^2. \qquad (6.7)$$

Example 6.2.5 Consider the same situation as in Example 3.1.13 except that we now denote by β_1, β_2 the regression parameters The density function of the generic observation is, omitting the subscript i,

$$\begin{aligned} f(y) &= \exp(-\rho y + \ln \rho) \\ &= \exp(\theta y + \ln(-\theta)) \end{aligned}$$

where

$$\theta = -\rho = -e^{-\eta}$$

so that

$$
\begin{aligned}
b(\theta) &= -\ln(-\theta)\,, \\
b'(\theta) &= -\frac{1}{\theta} = \mu\,, \\
b''(\theta) &= \frac{1}{\theta^2} = \mu^2 = V(\mu)\,, \\
\frac{d\mu}{d\eta} &= -e^{-\eta} = \frac{1}{\theta}\,.
\end{aligned}
$$

By applying (6.5), we obtain

$$\sum_{i=1}^{n} \frac{(y_i + 1/\theta_i)}{1/\theta_i^2} x_{ij} \frac{1}{\theta_i} = 0 \qquad (j = 1, 2)$$

where $x_{i1} = 1, x_{i2} = x_i$. These equations can be rewritten as

$$\sum_i (y_i - 1/\rho_i) x_{ij} \rho_i = 0$$

leading to the equations

$$
\begin{aligned}
\sum y_i \rho_i - n &= 0\,, \\
\sum x_i y_i \rho_i - \sum x_i &= 0\,,
\end{aligned}
$$

which are equivalent to those of Example 3.1.13, after substituting the expression for ρ_i and taking into account that $\sum x_i = 0$

Moreover, (6.7) says that $\tilde{w}_i = 1$, so that from (6.6) we obtain that the Fisher information is

$$I(\beta) = X^\top X,$$

confirming the result of Example 3.2.4.   □

6.2.4 Canonical links and sufficient statistics

The link function $g(\mu)$ does not need to satisfy very stringent conditions, and it can be specified in a number of different ways. However, there exists a specific choice of this function which enjoys special properties, namely

$$g(\mu_i) = \theta_i \qquad \text{so that} \qquad \eta_i = \theta_i\,. \tag{6.8}$$

This choice of $g(\cdot)$ is called the **canonical link**, since the linear predictor η_i coincides with the canonical parameter θ_i of the exponential family. In the case of the canonical link, we have that

$$
\begin{aligned}
\ell(\beta) &= \sum_{i=1}^{n} \{y_i \theta_i - b(\theta_i)\} w_i / \psi + \sum_{i=1}^{n} c_i(y_i, \psi) \\
&= \sum_i \{w_i y_i x_i^\top \beta - w_i b(x_i^\top \beta)\} / \psi + \sum_i c_i(y_i, \psi) \\
&= \left\{ \left(\sum_i w_i y_i x_i\right)^\top \beta - \sum_i w_i b(x_i^\top \beta) \right\} / \psi + \sum_i c_i(y_i, \psi),
\end{aligned}
$$

showing that $\sum_i w_i y_i x_i$ is a sufficient statistic for β when ψ is absent or known. If ψ is unknown but the likelihood is still of exponential form, then $\sum_i w_i y_i x_i$ is still a component of the minimal sufficient statistic.

Example 6.2.6 Suppose Y_i for $Gamma(\omega, \omega/\mu_i)$, with the index ω constant across the observations, and its mean value μ_i satisfies (6 4). The density function of a generic observation is

$$
\begin{aligned}
f(y) &= \exp\left(-\frac{\omega}{\mu} y\right) \left(\frac{\omega}{\mu}\right)^\omega y^{\omega-1} / \Gamma(\omega) \\
&= \exp\left\{ \omega\left(-\frac{y}{\mu} - \ln\mu\right) + (\omega - 1)\ln y - \ln\Gamma(\omega) + \omega\ln\omega \right\} \\
&= \exp\left\{ \omega\left(\theta y + \ln(-\theta)\right) + c(y, \omega) \right\}
\end{aligned}
$$

for $y > 0$, having set $\theta = -1/\mu$ and

$$
c(y, \omega) = (\omega - 1)\ln y - \ln\Gamma(\omega) + \omega\ln\omega.
$$

The canonical link is obtained when

$$
-\frac{1}{\mu_i} = \theta_i = x_i^\top \beta
$$

and in this case

$$
\ell(\beta) = \exp\left\{ \omega\left[\left(\sum_i x_i y_i\right)^\top \beta + \sum_i \ln\left(-x_i^\top \beta\right) \right] + \sum_i c(y_i, \omega) \right\}
$$

confirming that $\sum_i x_i y_i$ is a sufficient statistic if ω is known, as is the case, for example, with the negative exponential distribution ($\omega = 1$). If ω is unknown, then $\sum_i x_i y_i$ is a component of the minimal sufficient statistic ($\sum_i x_i y_i, \sum_i \ln y_i$). $\qquad\square$

Table 6.2 gives canonical link functions for other distributions.

There are other advantages in adopting canonical links. As far as the derivatives of the likelihood function are concerned, we have

$$\frac{\partial \mu_i}{\partial \eta_i} = \frac{\partial \mu_i}{\partial \theta_i} = \frac{\partial b'(\theta_i)}{\partial \theta_i} = b''(\theta_i)$$

taking into account (6.8), so that

$$\frac{\partial \ell_i}{\partial \beta_j} = \frac{(y_i - \mu_i)\, x_{ij}}{\text{var}[\,Y_i\,]}\, b''(\theta_i) = \frac{w_i(y_i - \mu_i)\, x_{ij}}{\psi}\,. \tag{6.9}$$

This fact simplifies (6.5) and, moreover, implies

$$\sum_i y_i\, x_{ij} = \sum_i x_{ij}\, \hat{\mu}_i\,,$$

in the common situation $w_i = 1$, having denoted by $\hat{\mu}_i$ the values of μ_i corresponding to the MLE $\hat{\beta}$ of β. The above relationship can be written in matrix notation as

$$X^{\top} y = X^{\top} \hat{\mu}. \tag{6.10}$$

Suppose that, as is often the case, a column of X is equal to 1_n; then (6.10) is saying that the sum of the observed values y is equal to the sum of the fitted values $\hat{\mu}$, and a similar equality holds for the other columns of X.

Another interesting property of canonical links concerns the Fisher information On differentiating (6.9), we have

$$\frac{\partial^2 \ell_i}{\partial \beta_j\, \partial \beta_k} = -\frac{w_i\, x_{ij}}{\psi}\, \frac{\partial \mu_i}{\partial \beta_k}$$

which does not depend on the observations. Therefore we have that

$$\frac{\partial^2 \ell}{\partial \beta_j\, \partial \beta_k} = \mathrm{E}\!\left[\,\frac{\partial^2 \ell}{\partial \beta_j\, \partial \beta_k}\,\right],$$

so that expected and observed information coincide.

6.2.5 The iteratively reweighted least-squares algorithm

Likelihood equations (6.5) do not, in general, lend themselves to a closed-form solution, and one has to resort to numerical methods.

One of the reasons that contributed to the success of GLMs is the possibility of using a single algorithm for solving (6.5), with little adjustment to be made for the choice of the link function and the type of probability distribution. Moreover, this algorithm operates by solving a sequence of least-squares problems, for which well-tested numerical techniques exist.

The method we will describe works by adjusting the kth approximate solution $\beta^{(k)}$ to obtain the next approximation $\beta^{(k+1)}$, and repeating this process until convergence is obtained. Denote by $u^{(k)}$ the Fisher scores computed at $\beta^{(k)}$, and use a similar notation for the other quantities. To improve the approximate solution $\beta^{(k)}$, we use the Fisher scoring algorithm which can be expressed as

$$\beta^{(k+1)} = \beta^{(k)} + \left(I(\beta^{(k)})\right)^{-1} u^{(k)}$$

which is similar to the Newton–Raphson method (3.5), but with the negative Hessian matrix replaced by the expected Fisher information. The above expression can be rewritten as

$$I(\beta^{(k)}) \beta^{(k+1)} = I(\beta^{(k)}) \beta^{(k)} + u^{(k)} . \tag{6.11}$$

The right-hand side term of this equation is a vector with hth element

$$\sum_j \left[\sum_i \frac{x_{ih} x_{ij}}{\operatorname{var}[Y_i]} \left(\frac{\partial \mu_i}{\partial \eta_i}\right)^2 \right] \beta_j^{(k)} + \sum_i \frac{\left(y_i - \mu_i^{(k)}\right) x_{ih}}{\operatorname{var}[Y_i]} \left(\frac{\partial \mu_i}{\partial \eta_i}\right)$$

so that we can write

$$I(\beta^{(k)})\beta^{(k)} + u^{(k)} = X^\top \tilde{W}^{(k)} z^{(k)},$$

where $z^{(k)}$ is a vector with ith element

$$
\begin{aligned}
z_i^{(k)} &= \sum_j x_{ij} \beta_j^{(k)} + \left(y_i - \mu_i^{(k)}\right) \left(\frac{\partial \eta_i^{(k)}}{\partial \mu_i^{(k)}}\right) \\
&= \eta_i^{(k)} + \left(y_i - \mu_i^{(k)}\right) \left(\frac{\partial \eta_i^{(k)}}{\partial \mu_i^{(k)}}\right) \tag{6.12}
\end{aligned}
$$

and the entries of $\tilde{W}^{(k)}$ are given by (6.7) evaluated at $\beta^{(k)}$. Then, taking into account (6.6), (6.11) becomes

$$(X^\top \tilde{W}^{(k)} X)\beta^{(k+1)} = X^\top \tilde{W}^{(k)} z^{(k)},$$

equivalent to

$$\beta^{(k+1)} = \left(X^\top \tilde{W}^{(k)} X\right)^{-1} X^\top \tilde{W}^{(k)} z^{(k)} . \tag{6.13}$$

This relationship has a form similar to the expression giving the regression parameters of a linear model with *constructed* response variable $z^{(k)}$ using weights $\tilde{w}_i^{(k)}$. Notice that for computing $\tilde{W}^{(k)}$ we can set $\psi = 1$ since the term ψ cancels out. Therefore, knowledge of the dispersion parameter is irrelevant for estimating β.

The algorithm thus has two main steps:

- when $\beta^{(k)}$ is given, $z^{(k)}$ and $\tilde{W}^{(k)}$ are computed using (6.12) and (6.7), respectively;

- then, using (6.13), the next approximation $\beta^{(k+1)}$ is obtained.

These two steps are iterated until the sequence $\beta^{(k)}$ converges. By now, it should be clear why this algorithm is called 'iteratively reweighted least-squares' (IRWLS).

In order to better understand the nature of variable z_i it is useful to consider the following Taylor series expansion of $g(y_i)$ about μ_i,

$$\begin{aligned} g(y_i) &\approx g(\mu_i) + (y_i - \mu_i)\, g'(\mu_i) \\ &= \eta_i + (y_i - \mu_i)\, \frac{\partial \eta_i}{\partial \mu_i} \\ &= z_i\,, \end{aligned}$$

which tells us that z_i is a local approximation to $g(y_i)$.

This remark also provides us with a hint as to how to tackle a problem not addressed so far: starting the algorithm. Initially, we can set $z_i^{(0)} = g(y_i)$ and $\tilde{W}^{(0)}$ equal to the identity matrix; from here, $\beta^{(1)}$ is obtained and the algorithm started. For some models, it will be necessary to modify slightly this choice of $z_i^{(0)}$ to avoid problems such as computing $\log(0)$, as is illustrated by Example 6.2.8 below.

Example 6.2.7 If we apply the IRWLS algorithm to a normal linear regression problem, of the type studied in Chapter 5, we expect a behaviour of the algorithm which is in some sense 'ideal', since (6.13) is a generalization of (5.11). In fact, it turns out that vector $z^{(0)}$ is equal to y and, if $\tilde{W}^{(0)} = I_n$, (6.13) gives directly $\beta^{(1)} = \hat{\beta}$, the usual LSE. Moreover, on inserting this $\beta^{(1)}$ in (6.12), we find that $z^{(1)}$ is still equal to y and $\tilde{W}^{(1)}$ does not change, so that $\beta^{(2)} = \beta^{(1)}$ The algorithm has achieved convergence in exactly one iteration. □

Example 6.2.8 Let us apply the IRWLS algorithm to the case of logistic regression described in Example 6 2.4. To implement the algorithm, we need the following quantities:

$$\text{var}[Y_i] = \frac{\psi}{w_i}\, V(\mu_i) = \frac{1}{m_i}\mu_i(1-\mu_i),$$

$$g(\mu) = \ln\frac{\mu}{1-\mu_i} = \eta, \qquad \mu = \frac{e^\eta}{1+e^\eta},$$

$$\eta_i = x_i^\top \beta = \theta_i,$$

$$\frac{d\mu}{d\eta} = \frac{e^\eta}{(1+e^\eta)^2} = \mu(1-\mu),$$

$$z_i^{(k)} = \eta_i^{(k)} + \frac{y_i - \mu_i^{(k)}}{\mu_i^{(k)}\left(1-\mu_i^{(k)}\right)},$$

$$\tilde{w}_i^{(k)} = m_i\,\mu_i^{(k)}\left(1-\mu_i^{(k)}\right).$$

The equality $\eta_i = \theta_i$ means that the canonical link is the logit function. To start the algorithm, consider setting

$$z_i^{(0)} = g(y_i) = \ln\frac{y_i}{1-y_i} = \ln\frac{\tilde{y}_i}{m_i - \tilde{y}_i}$$

However, if $\tilde{y}_i = 0$ or $\tilde{y}_i = m_i$ for at least one choice of i, this is not computable. To avoid this problem, we make a slight adjustment to our equation for $z_i^{(0)}$:

$$z_i^{(0)} = \ln\frac{\tilde{y}_i + \frac{1}{2}}{m_i - \tilde{y}_i + \frac{1}{2}}$$

called an **empirical logit**. This has been introduced here purely as a numerical adjustment, but there are additional reasons for considering it; see Cox and Snell (1989, pp. 31–33).

Figure 6.2 summarizes schematically the steps to be performed in the practical implementation of the algorithm.

We now demonstrate the behaviour of the algorithm for a specific data set, presented in Table 6.3. These data, from Bliss (1935), summarize the outcome of a biological experiment, giving the number of beetles exposed for 5 hours to gaseous carbon disulphide, and the number of insects killed, at various dose levels of the poison.

Based on past data analyses with similar data, it is reasonable to assume that an adequate description of the relationship between the dosage level and the probability $\pi(x)$ of death is given by

$$\text{logit}(\pi(x)) = \beta_1 + \beta_2 x$$

where x represents the logarithm of the dosage level. If we ap-

initial settings	$z_i \leftarrow \log((\tilde{y}_i + \frac{1}{2})/(m_i - \tilde{y}_i + \frac{1}{2}))$ $\tilde{W} \leftarrow$ identity matrix

$$\downarrow$$

repeat these steps until convergence	$R \leftarrow (X^{\top}\tilde{W}X)^{-1}$ $\beta \leftarrow R\,X^{\top}\tilde{W}z$ $\eta \leftarrow X\beta$ $\mu_i \leftarrow \exp(\eta_i)/(1 + \exp(\eta_i))$ $\Delta_i \leftarrow \mu_i(1 - \mu_i)$ $z_i \leftarrow \eta_i + (\tilde{y}_i - \mu_i)/\Delta_i \qquad (i = 1, \ldots, n)$ $\tilde{W} \leftarrow \text{diag}(\Delta_1\, m_1, \ldots, \Delta_n\, m_n)$

$$\downarrow$$

final settings	$\hat{\beta} \leftarrow$ last values of β $\text{var}\left[\hat{\beta}\right] \leftarrow$ last value of R

Figure 6.2 *Schematic representation of the IRWLS algorithm in the case of logistic regression*

ply the IRWLS algorithm, as specified in Figure 6.2, the sequence of estimates of the parameters and other relevant quantities is as shown in Table 6.4, showing that convergence is achieved in very few iterations. Figure 6.3 displays the observed fractions of events for any given dosage level, as well as the fitted curve

$$\hat{\pi}(x) = \frac{\exp(\hat{\beta}_1 + \hat{\beta}_2 x)}{1 + \exp(\hat{\beta}_1 + \hat{\beta}_2 x)}.$$

□

6.2.6 Estimation of the dispersion parameter

To estimate the dispersion parameters ψ, maximum likelihood estimation could be employed. However, it is more common to use an alternative estimate, which is defined as follows. Once the IRWLS algorithm has been applied to the data and produced the estimate $\hat{\beta}$, the individual mean estimates $(\hat{\mu}_1, \ldots, \hat{\mu}_n)$ are available. Then, we can compute

$$\tilde{\psi} = \frac{1}{n - p} \sum_i \frac{w_i(y_i - \hat{\mu}_i)^2}{V(\hat{\mu}_i)} \qquad (6.14)$$

Table 6.3 *Beetle data: number of* Tribolium confusum *exposed to gaseous carbon disulphide for 5 hours and number killed, at various concentration levels (Data reproduced from Bliss (1935),* The Annals of Applied Biology, **22**, *154.)*

log(dose) $\log (CS_2 \text{ mg/l})$	Number of insects treated	killed
1.6907	59	6
1.7242	60	13
1.7552	62	18
1.7842	56	28
1.8113	63	52
1.8369	59	53
1.8610	62	61
1.8839	60	60

Table 6.4 *Beetle data: successive iterations of IRWLS algorithm*

Cycle	$\hat{\beta}_1$	s.e.$(\hat{\beta}_1)$	$\hat{\beta}_2$	s.e.$(\hat{\beta}_2)$
1	−58.66		34.27	
2	−60.54		34.17	
3	−60.73		34.28	
4	−60.71		34 27	
5	−60.72	5.18	34.27	2.91

which is based on the relationship $\text{var}[Y_i] = V(\mu_i)\,\psi/w_i$. In the case of linear models with identity link function, $\tilde{\psi}$ coincides with s^2 defined by (5.19); this motivates the $n-p$ term in the denominator, instead of just n. This estimate is much simpler to compute and, in some cases, offers greater numerical stability than the MLE; for details, see Nelder and McCullagh (1989, pp. 275–276).

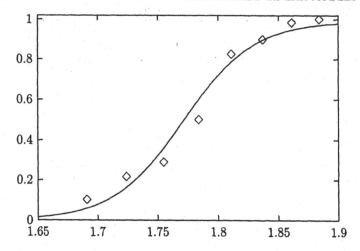

Figure 6.3 *Beetle data: plot of the observed data and of the fitted curve. The diamonds show the observed relative frequencies of events, the continuous curve is the fitted logistic function*

6.3 Examining model adequacy

6.3.1 Deviance

Consider the problem of comparing two GLMs, M_1 and M_2 say, such that $M_2 \subset M_1$, in the sense that that M_1 and M_2 are models of similar type but M_2 imposes additional constraints on the linear predictor. Models satisfying such a condition are called **nested models**.

More specifically, M_1 is a model involving p_1 parameters and M_2 a model with p_2 parameters, in such a way that M_2 is the restriction of M_1 obtained by imposing constraints of the form $g_i(\beta) = 0$ ($i = 1, \ldots, p_2 - p_1$), similarly to (4.19).

Once again, we shall make use of the LRT criterion for comparing M_1 and M_2. However, since GLMs are an extension of linear models, we wish to write the likelihood ratio statistic in such a way as to maintain the connection between GLMs and linear models.

In Chapter 5, we saw that, for linear models, the likelihood ratio statistic is a function of the deviance

$$D = \sum_i (y_i - \hat{\mu}_i)^2 = \|y - \hat{\mu}\|^2$$

associated with each of the two models. In fact, the likelihood itself depends on the data only through D, as is clear on writing

$$\ell(\hat{\beta}) = c - \frac{n}{2}\ln\sigma^2 - \frac{D}{2\sigma^2}\,.$$

Suppose that the error structure of the model is fixed, and consider a variety of models obtained by changing the linear predictor. Define the **saturated model** as the one where the estimates of μ_i coincide with y_i, which is possible in a model containing n parameters. Denote by $\tilde{\mu}_i = y_i$ the value of μ_i for such a model, and define $\tilde{\theta}_i$, $\tilde{\beta}_i$ similarly. Such a model is not of practical use by itself, but it serves as a reference for evalutaing the goodness of fit of a given model M of real interest Technically, it serves to free the log-likelihood from arbitrary constants.

If we compare the log-likelihood of any given model M with that of the saturated model, we obtain that twice the difference between the log-likelihoods of the two models is given by

$$2\{\ell(\tilde{\beta}) - \ell(\hat{\beta})\} = \frac{D}{\sigma^2}$$

and is equal to the deviance itself, up to a constant. The use of symbol $\ell(\cdot)$ for log-likelihoods with different number of parameters is adopted for simplicity of notation.

Moreover, for comparing two nested models, M_1 and M_2 ($M_2 \subset M_1$), the likelihood ratio takes the form

$$W = -2\left\{\ell(\hat{\beta}_2) - \ell(\hat{\beta}_1)\right\} = \frac{D_2 - D_1}{\sigma^2}\,,$$

which is essentially a difference of deviances, up to the constant σ^2, which we suppose to be known, for the moment.

The above scheme can be transferred from linear models to GLMs, by introducing a suitable generalization of the concept of **deviance** If we compare model M with the saturated model via the LRT statistic, we obtain

$$\begin{aligned} W(y) &= -2\left\{\ell(\hat{\beta}) - \ell(\tilde{\beta})\right\} \\ &= -2\sum_i \frac{w_i}{\psi}\left\{\left(y_i\hat{\theta}_i - b(\hat{\theta}_i)\right) - \left(y_i\tilde{\theta}_i - b(\tilde{\theta}_i)\right)\right\} \\ &= \frac{D(y;\hat{\mu})}{\psi} \end{aligned}$$

which is called the **scaled deviance**; its numerator is called the deviance, and can be decomposed as

$$D(y; \hat{\mu}) = \sum_i 2 w_i \left\{ y_i \left(\tilde{\theta}_i - \hat{\theta}_i \right) - b(\tilde{\theta}_i) + b(\hat{\theta}_i) \right\}$$
$$= \sum_i d_i$$

where d_i is the contribution to the deviance from the ith observation.

Where two models are nested $(M_2 \subset M_1)$ and M_2 is correct, then the difference between the corresponding deviances is such that

$$\frac{D(y; \hat{\mu}_2) - D(y; \hat{\mu}_1)}{\psi} \xrightarrow{d} \chi^2_{p_1 - p_2} \qquad (6.15)$$

as $n \to \infty$, from the general results of Chapter 4.

Notice that this asymptotic result holds when both M_1 and M_2 are given models, with respect to n; so p_1 and p_2 are given numbers. In general, (6.15) does not apply when M_1 is the saturated model, which varies with n; it does, however, hold in certain special cases, notably the situations described in section 6.4 below.

Example 6.3.1 For normal variates, the above definition of deviance coincides with that of Chapter 5 in ordinary linear models, namely $D = \sum_i (y_i - \hat{\mu}_i)^2$. $\qquad \square$

Example 6.3.2 In the case of a GLM with a Poisson distribution, $\psi = 1$, $w = 1$, and the scaled deviance coincides with the deviance, both being equal to

$$D = -2 \sum \left\{ (y_i \ln \hat{\mu}_i - \hat{\mu}_i) - (y_i \ln y_i - y_i) \right\}$$
$$= 2 \sum_i \left\{ y_i \ln (y_i / \hat{\mu}_i) - y_i + \hat{\mu}_i \right\}$$

having set $0 \ln 0 = 0$. The above expression simplifies to

$$D = 2 \sum_i y_i \ln \frac{y_i}{\hat{\mu}_i}$$

if $\sum y_i = \sum \hat{\mu}_i$. From (6.10), this condition is satisfied when we use the canonical link and $1_n \in \mathcal{C}(X)$, a frequent situation in practice. $\qquad \square$

6.3.2 Residuals

In Chapter 5, we introduced residuals as a tool for evaluating in an informal way the adequacy of a linear model. We now wish to extend the idea of residuals to GLMs, in order to use the diagnostic and graphical techniques for linear models which we mentioned in section 5.5.

The direct extension of the original definition of the residual is given by

$$\hat{\varepsilon}_i^P = \frac{y_i - \hat{\mu}_i}{\sqrt{V(\hat{\mu}_i)/w_i}}$$

which is called the **Pearson residual**, for reasons to be explained shortly.

The expression for $\hat{\varepsilon}_i^P$ is analogous to that used for linear models, where there exists a component ε_i of which $\hat{\varepsilon}_i$ is somehow an estimate. Since now such an ε_i does not necessarily exist, it makes sense to consider instead other possible definitions of residuals. For example, the contribution to the deviance of the ith observation

$$\hat{\varepsilon}_i^D = \text{sign}\,(y_i - \hat{\mu}_i)\sqrt{d_i}$$

is called the **deviance residual**. Finally, we also mention the **Anscombe residuals**

$$\hat{\varepsilon}_i^A = \frac{A(y_i) - A(\mu_i)}{A'(\mu_i)\sqrt{V(\mu_i)/w_i}}$$

where the transformation

$$A(x) = \int \frac{1}{V^{1/3}(x)}dx$$

is chosen to make the residuals' distribution as close as possible to the normal. For additional details, see McCullagh and Nelder (1989, section 2.4.2).

Example 6.3.3 In the case of normal data, all three type of residuals coincide with the old version, introduced in Chapter 5, namely $\hat{\varepsilon}_i = (y_i - \hat{\mu}_i)$, since $V(\mu) = 1$ and $A(x) = x$. □

Example 6.3.4 For the Poisson distribution, the three kinds of residuals take the following expressions

$$\hat{\varepsilon}_i^P = \frac{y_i - \hat{\mu}_i}{\sqrt{\hat{\mu}_i}}\,,$$

$$\hat{\varepsilon}_i^D = \text{sign}(y_i - \hat{\mu}_i)\sqrt{2\left\{y_i \log(y_i/\hat{\mu}_i) - y_i + \hat{\mu}_i\right\}}\,,$$

$$\hat{\varepsilon}_i^A = \frac{\frac{3}{2}\left(y_i^{2/3} - \hat{\mu}_i^{2/3}\right)}{\hat{\mu}_i^{1/6}}.$$

Since $\sum_i(\hat{\varepsilon}_i^P)^2$ is equal to Pearson's X^2 statistic described in section 4.4 8, it is now clear why they are called Pearson residuals.

□

Example 6.3.5 For a numerical illustration, let us consider again the cloth data introduced in Example 6.1.1. In the discussion following the presentation of the data, we effectively suggested a Poisson regression model with identity link function, using the length L as an explanatory variable for the expected number of defects D. For such a model, the MLE of β allows a closed-form expression, namely $\hat{\beta} = \sum_i D_i / \sum_i L_i$, which is equal to 0 0151 for the present data.

Figures 6 4 and 6.5 present two slightly different ways of examining the adequacy of the model to fit the data. In Figure 6.4, the fitted regression line $\hat{\beta}L$ and two curves having equations $\hat{\beta}L \pm 2\sqrt{\hat{\beta}L}$ have been superimposed on the scatter plot of the observed data. The distance between the regression line $\hat{\beta}L$ and these two curves is regulated by the expected standard deviation under the fitted model, namely $\sqrt{\hat{\beta}L}$, multiplied by a factor of 2 which corresponds to an interval of probability approximately equal to 0.95 if the Poisson distribution is approximated by a normal one.

If the fitted Poisson regression model was correct, we would expect about 5% of the observed data to fall outside the curved bands, that is less than 2 points out of 32 in our case, while in fact there are 4 such points. This discrepancy suggests a partial inadequacy of the model to the data. Notice that this procedure is equivalent to compute the Pearson residuals and comparing them with the reference interval $(-2, 2)$.

Since we are using bands derived from the normal rather than the Poisson distribution, one could object that the above-mentioned lack of fit of the model could be due to inappropriate choice of the reference bands. To improve the approximation to normality, we replace the previous residuals by the Anscombe residuals $\hat{\varepsilon}_i^A$ whose plot is given in Figure 6.5. The new form of plot is slightly different from the previous one, since it represents the pairs $(\hat{\mu}_i, \hat{\varepsilon}_i^A)$ of fitted values and residuals, together with bands having fixed width 2.

Apart from the different type of residuals, the new form of plot

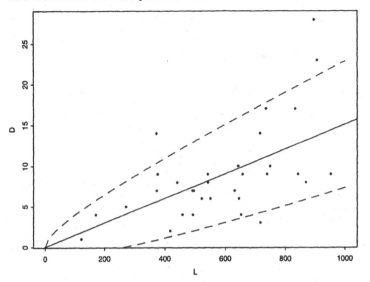

Figure 6.4 *Cloth data: fitted Poisson regression line (continuous line) and expected dispersion bands (dashed curves) with pointwise approximate probability 0.95*

contains essentially the same sort of information as displayed in Figure 6.4; we are simply taking the fitted regression line as the abscissa, measuring distances of residuals from this line, and rescaling the distances so that the bands have fixed width. The type of plot used earlier is more directly interpretable, but it can be used only when there is a single explanatory variable. In general, only the second form of plot is feasible, or some variant of it.

As for the interpretation of the plot itself, Figure 6.5 gives indications analogous to Figure 6.4, in fact even slightly more strongly. The lack of fit of the model is not extreme, but a rediscussion of the choice of model is appropriate. This will be pursued in Example 6.5.1. □

6.4 Applications to frequency tables

6.4.1 Sampling methods

Consider a frequency table such as Table 6.5 which refers to two qualitative variables, *A* and *B*. The term **categorical variable**

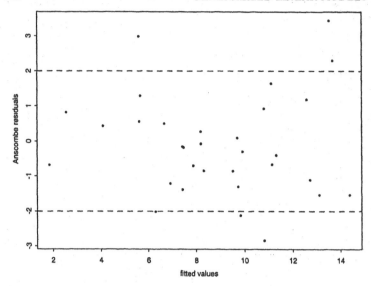

Figure 6 5 *Cloth data: plot of Anscombe residuals versus fitted values and* $(-2, 2)$ *bands*

Table 6.5 *A frequency table.*

	B_1	B_2	$\cdots$	B_c	Total
A_1	y_{11}	y_{12}	$\cdots$	y_{1c}	y_{1+}
A_2	y_{21}	y_{22}	$\cdots$	y_{2c}	y_{2+}
$\vdots$	$\vdots$	$\vdots$	$\ddots$	$\vdots$	$\vdots$
A_r	y_{r1}	y_{r2}	$\cdots$	y_{rc}	y_{r+}
Total	y_{+1}	y_{+2}	$\cdots$	y_{+c}	$y_{++} = N$

is equivalent to 'qualitative variable'; the term **factor** is also often used as a synonym for 'qualitative variable'. We have already seen in section 4.4.9 how to test the hypothesis of independence between factors A and B, but now we wish to reconsider the problem in the light of the concepts just introduced, as these will lead to more general and flexible techniques.

To write down the likelihood, we must specify the random mechanism which generates the data, and this mechanism can be of various different types, depending on the way in which the data have been collected.

Suppose, for instance, that we wish to study the relationship between exposure to some risk factor and occurrence of a given event, such as presence/absence of a certain disease in a subject. There are various ways in which the data can be collected, and correspondingly various probability distributions of the variables.

(a) *Direct observation of the phenomenon.* We observe the phenomenon of interest for a given length of time and classify the observed events. Under fairly general conditions, a generic entry y_{hk} in the frequency table is generated by a Poisson process with mean μ_{hk}, independent of the processes associated with the other entries in the table; hence the joint distribution of all frequency table entries is

$$f(y) = \prod_{h=1}^{r} \prod_{k=1}^{c} \frac{e^{-\mu_{hk}} \mu_{hk}^{y_{hk}}}{y_{hk}!}. \qquad (6.16)$$

Consider, for instance, a study to evaluate the effectiveness of safety belts in alleviating the effects of car accidents. In this case, we have $r = c = 2$, the levels of factor A denoting use (or non-use) of safety belts, the levels of B presence (or absence) of injuries to the people involved. If we consider as the reference 'population' the car accidents recorded in police files, a possible way of collecting data is to examine the files of car accidents for a given period of time. In this case, the use of (6.16) as reference distribution for the data is appropriate.

(b) *Observing a fixed number of events.* If we decide in advance to proceed until data on N car accidents are available, then the distribution of the data changes. Since a priori $N \sim Poisson(\mu_{++})$, where $\mu_{++} = \sum_h \sum_k \mu_{hk}$, then it turns out that, if N is fixed, we have

$$f(y|N) = \frac{\prod_h \prod_k e^{-\mu_{hk}} \mu_{hk}^{y_{hk}}/y_{hk}!}{e^{-\mu_{++}} \mu_{++}^{N}/N!}$$

$$\left(\text{if } \sum_h \sum_k y_{hk} = N\right)$$

$$= \prod_h \prod_k \frac{N!}{y_{hk}!} \pi_{hk}^{y_{hk}}, \qquad (6.17)$$

which is a multinomial distribution with probabilities

$$\pi_{hk} = \mu_{hk}/\mu_{++}$$

for $h = 1, \ldots, r, \ k = 1, \ldots, c$

(c) *Fixing a marginal.* If we do not fix only the total number N of observed events, but we impose fixed values to an entire row or column of marginal frequencies, then the probability distribution changes again. For instance, we could decide in advance to collect data from y_{+1} car accidents without injured people, and y_{+2} accidents with injured people, so effectively constructing two sub-populations, and then we compare the frequencies of people wearing safety belts within each of the two sub-populations. In this case an entire column of marginal frequencies is fixed, and the distribution of the observed frequencies is

$$f(y|y_{+k}, \ k = 1, \ldots, c) = \prod_{k=1}^{c} \left(\prod_{h=1}^{r} \frac{y_{+k}!}{y_{hk}!} \pi_{h|k}^{y_{hk}} \right)$$

where

$$\pi_{h|k} = \frac{\mu_{hk}}{\mu_{+k}}$$

and

$$\sum_{h} y_{hk} = y_{+k} \qquad \text{for all } k.$$

In this case, the distribution is given by the product of c multinomial probability functions.

In Chapter 4, we introduced the problem of analysing frequency tables through the problem of testing independence between two factors A and B, but in fact this is a specific problem. In the rest of this section, we aim to develop a theory for the analysis of frequency data which is applicable to a family of problems, not just to the problem of independence of two factors.

6.4.2 Log-linear models and independence

The term 'log-linear model' applies to a GLM model with logarithmic link function, so that

$$\eta_i = x_i^{\mathsf{T}} \beta = \ln \mathrm{E}[Y_i].$$

Since the logarithmic function is the canonical link for the Poisson distribution, it is natural that log-linear models are most commonly used for the analysis of frequency tables.

In fact, log-linear models are relevant not only for the case of a distribution of type (6.16), but also for the other cases discussed in the same subsection, as we shall explain in the following pages.

Let us adopt the convention that y_i and y_{hk} denote the same observation, but y_{hk} is regarded as the (h, k)th entry of Table 6 2, while y_i is regarded as ith component of the vector y of observations.

For the random variable Y_i with Poisson distribution, denote by μ_i its mean value. Depending on whether the total N of the frequencies is fixed or random, we have

$$\mathrm{E}[Y_i] = \mu_{++}\, \pi_i, \qquad \mathrm{E}[Y_i|N] = N\pi_i, \qquad (6.18)$$

where

$$\pi_i = \frac{\mu_i}{\mu_{++}}$$

represents in both cases the probability that, when an event occurs, it falls in the ith cell.

To start with, consider testing independence between two factors, A and B say; this problem translates directly into the comparison of two log-linear models. The independence hypothesis implies that

$$\pi_{hk} = \pi_{h+}\, \pi_{+k} \qquad (6.19)$$

for all h and k, and then (6.18) becomes

$$\mathrm{E}[Y_i] = \mu_{++}\, \pi_{h+}\, \pi_{+k}, \quad \mathrm{E}[Y_i|N] = N\, \pi_{h+}\, \pi_{+k}$$

or, for the first of these,

$$\eta_{hk} = \ln \mathrm{E}[Y_i] = \ln \mu_{++} + \ln \pi_{h+} + \ln \pi_{+k}$$

The corresponding expression for the case with N fixed is similar, except that μ_{++} is replaced by N but without changing the form of dependence on π_i. On setting

$$\lambda = \ln \mu_{++}, \quad \lambda_h^A = \ln \pi_{h+}, \quad \lambda_k^B = \ln \pi_{+k},$$

we can write the previous expression as

$$\eta_{hk} = \lambda + \lambda_h^A + \lambda_k^B \qquad (6.20)$$

leading to an additive expression for the parameters. The constraints $\sum_h \pi_{h+} = \sum_k \pi_{+k} = 1$ imply a set of nonlinear constraints on the parameters involved in (6.20), leaving $1 + (r - 1) + (c - 1)$ free parameters.

To avoid complications due to the presence of nonlinear constraints, we can choose some other parametrization. For instance, we can keep (6.20) but with

$$\lambda = \ln \mu_{++} + \frac{1}{r} \sum_{h=1}^{r} \ln \pi_{h+} + \frac{1}{c} \sum_{k=1}^{c} \ln \pi_{+k},$$

$$\lambda_h^A = \ln \pi_{h+} - \frac{1}{r} \sum_{h=1}^{r} \ln \pi_{h+},$$

$$\lambda_k^B = \ln \pi_{+k} - \frac{1}{c} \sum_{k=1}^{c} \ln \pi_{+k},$$

so that there are two linear constraints

$$\sum_h \lambda_h^A = \sum_k \lambda_k^B = 0$$

similar to those commonly used in two-way analysis of variance.

Alternatively, in (6.20) we can impose

$$\lambda_1^A = 0, \quad \lambda_1^B = 0,$$

which means we are taking the first level of each factor as the reference level.

As for the estimates $\hat{\eta}_{hk}$ and $\hat{\pi}_{hk}$, it does not matter which parametrization is adopted, provided they are equivalent Neither is the deviance affected. The various parametrizations differ essentially in the interpretation of the individual components of the linear predictor.

To test the independence hypothesis expressed by (6.19), we must specify a general reference hypothesis. The most general one has all the π_{hk} completely free of constraints, except that $\sum_h \sum_k \pi_{hk} = 1$, and corresponds to the saturated model.

To write also the corresponding model in log-linear form, analogously to (6.20), we reparametrize '

$$\eta_{hk} = \ln \mathrm{E}[Y_i] = \ln \mu_{++} + \ln \pi_{h+} + \ln \pi_{+k} + \ln (\pi_{hk}/(\pi_{h+} \pi_{+k}))$$

by

$$\eta_{hk} = \lambda + \lambda_h^A + \lambda_k^B + \lambda_{hk}^{AB} \qquad (6.21)$$

with the additional constraints

$$\sum_h \lambda_{hk}^{AB} = \sum_k \lambda_{hk}^{AB} = 0. \qquad (6.22)$$

The new model for η_{hk} now contains $(r-1)(c-1)$ parameters in addition to the one specified by (6.20).

6.4.3 The likelihood of a log-linear model

It has already been remarked that different likelihood functions can arise, depending on the way in which the data are collected. Consider now the likelihoods associated with the Poisson distribution and the multinomial distribution. In this subsection, we wish to analyse the relationships between these two likelihood functions.

Consider any fixed, but arbitrary, linear predictor and write the two log-likelihood functions for the two alternative probability distributions, ℓ_P and ℓ_M say. Denoting by $\mu_1, \ldots, \mu_n$ the mean values for the given model, we have

$$\ell_P = \sum_i (y_i \ln \mu_i - \mu_i) + c$$

for the Poisson case, and

$$\begin{aligned}
\ell_M &= \sum_i (y_i \ln \pi_i) + c \\
&= \sum_i y_i \ln \mathrm{E}[Y_i|N] + c \\
&= \sum_i (y_i \ln \mu_i - y_i) + c.
\end{aligned}$$

for the multinomial case, taking into account that $\mathrm{E}[Y_i|N] = N\pi_i \propto \mu_i$ The two log-likelihood functions coincide up to an additive constant, provided that the linear predictor includes a constant term, so $\sum \hat{\mu}_i = \sum y_i$ because of (6.10). The above conclusion allows us to fit a log-linear model to data with multinomial distribution using a Poisson likelihood.

An equivalent, but slightly more detailed, argument is as follows. Write

$$\ln \mu_i = \lambda + z_i^{\mathsf{T}} \beta,$$

where we have separated out the parameter λ of the constant explanatory variable, while $z_i^{\mathsf{T}} \beta$ is the component due to the remaining explanatory variable, dropping one element of the original β

vector. We then have

$$
\begin{aligned}
\ell_P(\lambda, \beta) &= \sum_i y_i \ln(\mu_i) - \sum_i \mu_i \\
&= \sum_i y_i(\lambda + z_i^{\mathsf{T}}\beta) - \sum_i \exp(\lambda + z_i^{\mathsf{T}}\beta) \\
&= N\lambda + \left(\sum y_i z_i^{\mathsf{T}}\right)\beta - \mu_{++} \\
&= \left\{ \left(\sum_i y_i z_i^{\mathsf{T}}\right)\beta - N\ln\sum_i \exp(z_i^{\mathsf{T}}\beta) \right\} \\
&\quad + [N\ln(\mu_{++}) - \mu_{++}]
\end{aligned}
\tag{6.23}
$$

since

$$
\mu_{++} = \sum_i \mu_i = \sum_i \exp(\lambda + z_i^{\mathsf{T}}\beta)
$$

so that

$$
\ln\mu_{++} = \lambda + \log\left[\sum_i \exp(z_i^{\mathsf{T}}\beta)\right].
$$

Changing parametrization, consider (6.23) as a function of β and μ_{++}. Notice that the term inside square brackets in (6.23) is the log-likelihood of a Poisson(μ_{++}) random variable and it corresponds to the unconditional log-likelihood for a total of N events. The term inside braces depends on β only and, on setting

$$
\pi_i = \frac{\mu_i}{\sum_s \mu_s} = \frac{\exp(z_i^{\mathsf{T}}\beta)}{\sum_s \exp(z_s^{\mathsf{T}}\beta)},
\tag{6.24}
$$

we notice that the term in curly brackets can be written as $\sum_i y_i \ln\pi_i$, which is the log-likelihood of a multinomial random variable with observed frequencies $(y_1, \ldots, y_{n-1})$, and y_n fixed by the constraint $\sum_{i=1}^n y_i = N$.

To summarise, (6.23) can be decomposed as

$$
\ell_P(\mu_{++}, \beta) = \ell_M(\beta) + [N\ln(\mu_{++}) - \mu_{++}].
$$

On the right-hand side, the second term is the Poisson log-likelihood corresponding to the total number of events, while the first term gives the log-likelihood of the observed frequencies, conditional on the total number of observed events. Since the second term does not depend on β, the value $\hat\beta$ maximizing ℓ_P is the same $\hat\beta$ maximizing ℓ_M. Moreover, even the derivatives with respect to β are the same for the two functions, so the standard errors of the estimates coincide too.

In (6.24), we make use of the values $\hat{\beta}$ obtained from ℓ_P to get estimates of the probabilities π_i of ℓ_M, while λ, or equivalently μ_{++}, is not present in ℓ_M.

Remark 6.4.1 In the above discussion, we compared the log-likelihood for the case of fixed N with the case of random N, showing that they are equivalent for estimating the regression parameters of the non-constant covariates. For the case that row or column frequencies are fixed, it is possible to show an equivalence property between likelihoods, analogous to that just proved.   □

6.4.4 More complex problems

By an appropriate use of indicator variables, we can specify a log-linear model that corresponds to a given pattern of the probabilities π_{hk} involved. In particular, by using indicator variables as in the two-way analysis of variance, we can specify the independence hypothesis expressed by (6.20) or define a saturated model. This gives rise to two deviances whose difference coincides with the test statistic G^2 of section 4.4.9, obviously with the same asymptotic distribution $\chi^2_{(r-1)(c-1)}$ under the null model.

We have already remarked that the independence hypothesis between two factors in a two-way frequency table is a very special case of the problems which one can encounter in frequency table analysis. The theoretical framework developed in the previous subsections allows us to deal with these more complex situations.

First of all, it is not necessary for the frequency table to be a two-way table; three-way and higher-dimensional tables are currently used, especially in social science applications.

Secondly, it is not necessary that both factors involved, A and B, are purely qualitative with unordered levels. In fact, it is even more frequent to have **ordered levels**, such as 'poor', 'fair', 'good', 'excellent'; the term 'ordered qualitative' or 'categorical' variable is then used In these cases, it is desirable that the data analysis reflects the ordering of the levels.

If the factor under consideration has k levels, one will prefer not to make use of $k-1$ indicator variables and parameters, attempting instead to reduce the model complexity, in terms of the number of parameters involved. There are various ways of pursuing such a task; the simplest is to introduce a quantitative variable z with scores $1, \ldots, k$, or possibly some other scoring scheme, associated

Table 6.6 *A* 2 × 2 *probability table*

Exposure to risk	Occurrence of event		Total
	$\bar{E}$	E	
$\bar{X}$	π_{00}	π_{01}	π_{0+}
X	π_{10}	π_{11}	π_{1+}
Total	π_{+0}	π_{+1}	1

with the levels of the ordinal qualitative variable, and then estimate a single parameter. If appropriate, one could also introduce some function of z, for instance z^2, besides z. There is a clear similarity with the strategy adopted in standard linear models.

A full treatment of log-linear models would lead us far beyond the scope of this book. For an extended account of this topic, see the books by Agresti (1990, Chapters 5–8) and McCullagh and Nelder (1989, Chapters 5–6).

6.4.5 Odds ratio

If an event E has probability p of success, the ratio $p/(1-p)$ is called the **odds** for that event. The logarithm of the odds is the **logit** of p

Let us consider a 2 × 2 table, in which one of the factors represents an exposure to some risk factor X (e.g. presence of a certain substance in the subject's diet) and the other factor represents the occurrence of some relevant event E (e.g. occurrence of problems in the blood circulation). The situation is summarized in Table 6.6, which shows the probabilities for each of the four possible outcomes of an observation.

In a **prospective study**, we choose in advance the number of observations of type X and of type $\bar{X}$ to consider, and then classify each observation as being of type E or $\bar{E}$. In this case, the difference in the logits is

$$\log \frac{\pi_{01}}{\pi_{00}} - \log \frac{\pi_{11}}{\pi_{10}} = \log \frac{\pi_{01}\,\pi_{10}}{\pi_{00}\,\pi_{11}}.$$

In a **retrospective study**, we choose in advance the number of observations of type $\bar{E}$ and of type E to analyse, and then classify the observations according to the other factor. For a retrospective study, the difference in the logits is

$$\log \frac{\pi_{10}}{\pi_{00}} - \log \frac{\pi_{11}}{\pi_{01}} = \log \frac{\pi_{01} \, \pi_{10}}{\pi_{00} \, \pi_{11}}.$$

Therefore, the differences on the logit scale coincide; hence, after exponentiation, the **odds ratio** is in both cases equal to

$$\omega = \frac{\pi_{01} \, \pi_{10}}{\pi_{00} \, \pi_{11}}$$

which can then be studied independently of the sampling mechanism. An equivalent name for ω is **cross-product ratio**.

This invariance property with respect to the sampling mechanism becomes essential when one is interested in studying a phenomenon in a prospective manner, but there are practical reasons forcing a retrospective analysis. In the above-mentioned example on the effect of a certain diet on blood circulation, the ideal way of conducting the study would be to select two homogeneous groups of people without blood circulation problems, to treat them with two different diets for a length of time, and in the final stage to classify the subjects according to the state of their blood circulation; this is exactly what we mean by a prospective study. If, however, there are reasons to believe that the potentially dangerous substance used in one of the two diets displays its effect only after several years of treatment, then it becomes impractical to proceed with the experiment just described. We would then be forced to use instead a retrospective study in the following way: two groups of people are selected, one with blood circulation problems, the other one free of such problems, and for each subject we try to establish whether his or her diet included over a given period of time the substance of interest. For the reasons explained, the estimated odds ratio derived from the retrospective study can be transferred to the odds ratio in the prospective framework.

In this kind of situation, $\omega = 1$ is the reference value since it corresponds to independence between factors. Values greater than 1 denote positive **association**, meaning a higher probability for the pairs of events (X, E) and $(\bar{X}, \bar{E})$ than for the pairs $(\bar{X}, E)$ and $(X, \bar{E})$. Similarly, $\omega < 1$ denotes negative association, which means the risk factor and the event of interest are associated in the other direction.

It is worth noting that the invariance property with respect to the sampling mechanism holds for the odds ratio, but not for all other measures of association between categorical variables.

The odds ratio is directly related to the parameters of a log-linear model In fact, on using parametrization (6.21) with constraints (6.22), a simple computation shows that

$$\ln \omega = 4 \lambda_{00}^{XE}.$$

If one adopts a different parametrization or different constraints, the above relationship changes, but in any case $\ln \omega$ remains a function of the λ_{hk}'s

6.5 Quasi-likelihood

Throughout this book, our discussion has relied almost completely on the likelihood function, except for the odd digression, the most relevant of which was the introduction of the least-squares criterion. The latter method provides us with estimates of the regression parameters without specifying a full stochastic model; it only requires us

(a) to define the relationship between the explanatory variables and the mean value of the response variable;

(b) to assume functional independence of the above mean value and the scale parameter of the error term (which is also the variance of the response variable at a given design point); this means that σ^2 is a quantity not related to the mean value of the response variable

In the present section, we wish to explore further in this direction, specifically in the sense of introducing a relationship between mean and variance. We therefore assume that

$$E[Y_i] = \mu_i, \qquad \text{var}[Y_i] = \psi \, V(\mu_i).$$

It turns out that

$$u = u(\mu, Y) = \frac{Y - \mu}{\psi \, V(\mu)},$$

(omitting the subscript i) behaves like a score function, in the sense that

$$E[u] = 0, \quad \text{var}[u] = \frac{1}{\psi \, V(\mu)}, \quad -E\left[\frac{\partial u}{\partial \mu}\right] = \frac{1}{\psi \, V(\mu)}.$$

Since the integral of u should behave like a log-likelihood, we define

$$Q(\mu; y) = \int_y^\mu u(t; y)dt = \int_y^\mu \frac{y - t}{\psi V(t)} dt$$

to be a **quasi-likelihood** (although it should really be called a quasi-log-likelihood). If n observations from independent variables are available, let

$$Q(\mu; y) = \sum_{i=1}^n Q(\mu_i; y_i).$$

Besides possessing many of the formal properties of a log-likelihood, there are cases when Q is in fact a proper log-likelihood, although this is not true in general. Where this property holds, i e there exists a log-likelihood function ℓ such that

$$\frac{\partial \ell}{\partial \mu} = \frac{y - \mu}{\psi V(\mu)}$$

with $\mathrm{E}[Y] = \mu$, $\mathrm{var}[Y] = \psi V(\mu)$, then it can be shown that ℓ has exponential family structure.

So far μ_i has been kept free of constraints. In a regression model, μ_i is a function of the parameter β and a simple computation shows that the quasi-likelihood equations are

$$\sum_{i=1}^n \frac{(y_i - \mu_i)}{V(\mu_i)} \frac{\partial \mu_i}{\partial \beta_j} = 0 \qquad (j = 1, \ldots, p)$$

which do not depend on the dispersion parameter ψ.

If the relations between β and μ_i are of type (6.4), then the above equations reduce to standard likelihood equations (6.5).

Notice that the quasi-likelihood function is defined essentially by specifying the mean value function (as a function of the covariates) and the variance function as a function of the mean value. When the variance function is equal to 1, the method reduces to least squares.

It is also possible to introduce a concept analogous to the concept of deviance. For a single observation y, define the **quasi-deviance** as

$$D(y; \hat{\mu}) = -2\psi Q(\hat{\mu}; y) = 2 \int_{\hat{\mu}}^y \frac{y - t}{V(t)} dt$$

which is non-negative, and in fact positive if we rule out the case $y = \hat{\mu}$. For a set of n independent observations, we sum the n quasi-deviance components.

Table 6.7 summarises the most important types of quasi-likelihood functions.

Example 6.5.1 We continue the discussion of the cloth data considered in Examples 6.1.1 and 6.3.5, and address the question of the lack of fit in the initially proposed Poisson regression model. The observed inadequacy of the model appears to be due to the presence of a variability larger than the expected variability in a Poisson random variable. A phenomenon of this kind, called **overdispersion**, can be interpreted in various ways. One approach is to assume

$$D_i \sim Poisson(\beta_i L_i) \qquad (i = 1, \ldots, n)$$

where now β_i is a random variable. A possible physical interpretation of this new hypothesis is that the various pieces of cloth are produced under different conditions (e.g. different textile machines), each one having its own rate of defects. Bissel (1972) assumed that β_i is sampled from gamma random variable, and derived the corresponding likelihood analysis.

If we do not want to commit ourselves to a specific distributional assumption, such as the gamma distribution, an alternative strategy is possible using the quasi-likelihood approach, which requires us to specify the form of the first two moments only. In our case, the assumption $E[D] = \beta L$ does not need to be changed. To allow for overdispersion, we say that

$$\mathrm{var}[D] = \psi E[D] = \psi \beta L$$

where ψ varies in $(0, \infty)$. This expression says that the variance of D increases linearly with L, as is apparent from Figure 6.1, but we no longer require $\mathrm{var}[D] = E[D]$.

To estimate ψ from the data, (6.14) takes the form

$$\tilde{\psi} = \sum_i \frac{(D_i - \hat{\beta} L_i)^2}{(n-1)\hat{\beta} L_i}.$$

For our data, we obtain $\tilde{\psi} = 2.194$, a value substantially different from 1, which was intrinsic to the earlier Poisson regression model. We can now produce a plot similar to that of Figure 6.4 but with bands wider by a factor of $\sqrt{\tilde{\psi}}$, as shown in Figure 6.6. The wider bands accommodate some of the overdispersed points, leaving out only two, a number in agreement with the expected 5%. □

Table 6.7 Quasi-likelihoods associated with various variance functions

Variance function $V(\mu)$	Quasi-likelihood $Q(\mu;y)$	Canonical parameter θ	Distribution	Constraints
1	$-(y-\mu)^2/2$	μ	Normal	—
μ	$y\log\mu - \mu$	$\log\mu$	Poisson	$\mu > 0,\ y \geq 0$
μ^2	$-y/\mu - \log\mu$	$-1/\mu$	Gamma	$\mu > 0,\ y \geq 0$
μ^3	$-y/(2\mu^2) + 1/\mu$	$-1/(2\mu^2)$	Inverse Gaussian	$\mu > 0,\ y \geq 0$
μ^ω	$\mu^{-\omega}\left(\dfrac{\mu y}{1-\omega} - \dfrac{\mu^2}{2-\omega} \right)$	$\dfrac{1}{(1-\omega)\mu^{\omega-1}}$	—	$\mu > 0,\ \omega \neq 0,1,2$
$\mu(1-\mu)$	$y\log\left(\dfrac{\mu}{1-\mu} \right) + \log(1-\mu)$	$\log\left(\dfrac{\mu}{1-\mu} \right)$	Binomial/m	$0 < \mu < 1,$ $0 \leq y \leq 1$
$\mu + \mu^2/\omega$	$y\log\left(\dfrac{\mu}{\mu+\omega} \right) + \omega\log\left(\dfrac{\omega}{\mu+\omega} \right)$	$\log\left(\dfrac{\mu}{\omega+\mu} \right)$	Negative binomial	$\omega > 0,\ \mu > 0,$ $y \geq 0$

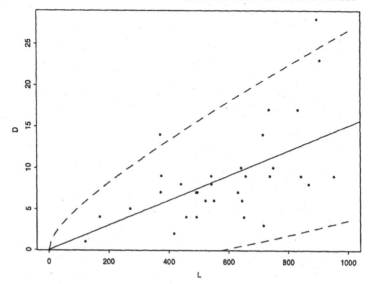

Figure 6.6 *Cloth data: fitted regression line (continuous line) and expected dispersion bands (dashed curves) of approximate pointwise probability 0 95 under overdispersion model*

Exercises

6.1 Complete Tables 6.2 and 6.7 by developing the entries which are not derived in the text; in particular, obtain $w, b(\cdot), c(\cdot)$, the canonical link and deviance for

   (a) the gamma distribution with known index,
   (b) the inverse Gaussian distribution,
   (c) the negative binomial distribution with known index ω.

6.2 In section 6.2.2, it was stated that, in general, a precise separation between the error term and the response variable is not possible for a GLM. Find another example, besides standard linear models with normal error distribution, where this separation is in fact possible.

6.3 The class of GLMs includes some of the models of type (5.1) with normal error term ε, provided the function $r(\cdot)$ is of a special form. Specify the form which $r(\cdot)$ takes. What is the appropriate GLM?

6.4 In Exercise 5.18, we introduced weighted least squares. Show how they can be embedded in the framework of the GLMs.

6.5 For the data in Table 6.3, obtain estimates of β_1, β_2 when the logit link function is replaced by the **probit** link

$$\Phi^{-1}(\pi(x)) = \beta_1 + \beta_2 x$$

and compare the new estimated curve with the old one displayed in Figure 6.3.

6.6 Verify the statement in the text that the difference of deviances for testing independence of two factors in a two-way frequency table coincides with test function G^2 of section 4.4.9.

6.7 Check that, after fitting model (6.20) or one equivalent to it, the fitted frequency table has the same marginal frequencies as the observed table.

Complements of probability theory

The aim of this appendix is to recall some standard concepts of probability theory, and to develop some additional topics which are not usually presented in probability textbooks.

A.1 Inequalities

A.1.1 The Chebyshev inequality

Denote by X a random variable with finite second moment, and let $\sigma = \sqrt{\text{var}[X]}$. The Chebyshev inequality states that, for any positive real h,

$$\Pr\{|X - E[X]| \geq h\sigma\} \leq \frac{1}{h^2}.$$

This proposition is a corollary of the following result

Theorem A.1.1 *If Y is a non-negative random variable with finite expected value, then*

$$\Pr\{Y \geq t\} \leq \frac{E[Y]}{t}$$

for any positive real t.

Proof. Denoting by $F(\cdot)$ the distribution function of Y, we have

$$
\begin{aligned}
E[Y] &= \int_{[0,\infty)} u\, dF(u) \\
&\geq \int_{[t,\infty)} u\, dF(u) \\
&\geq t \int_{[t,\infty)} dF(u).
\end{aligned}
$$

□

The Chebyshev inequality follows from Theorem A.1.1 by setting

$$Y = \left(\frac{X - \mathrm{E}[X]}{\sigma}\right)^2$$

and $h^2 = t$.

A.1.2 The Schwarz inequality

Theorem A.1.2 *If X and Y are random variables with finite second moment, then*

$$(\mathrm{E}[XY])^2 \le \mathrm{E}[X^2]\,\mathrm{E}[Y^2]. \qquad (A.1)$$

Proof. The statement does not change if X and Y are multiplied by constant factors; hence, we can restrict ourselves to the case $\mathrm{E}[X^2] = \mathrm{E}[Y^2] = 1$. Since $(X \pm Y)^2 \ge 0$, then

$$2\,|XY| \le X^2 + Y^2$$

and

$$(\mathrm{E}[XY])^2 \le \{\mathrm{E}[|XY|]\}^2 \le 1.$$

$\square$

The Schwarz inequality also holds for sequences of real numbers. If we assign a constant probability $1/n$ to each pair of numbers (x_i, y_i) for $i = 1, \ldots, n$, from the above general results we obtain

$$\left(\sum_{i=1}^{n} x_i y_i\right)^2 \le \left(\sum_{i=1}^{n} x_i^2\right)\left(\sum_{i=1}^{n} y_i^2\right). \qquad (A.2)$$

A.1.3 Convexity and the Jensen inequality

Definition A.1.3 *A real-valued function $f(\cdot)$ is said to be convex in the interval I (finite or infinite) of the real line if, for any interior point c of I, there exists a number b such that*

$$f(x) \ge f(c) + b(c - x) \quad \text{for all } x \in I; \qquad (A.3)$$

if the strict inequality holds for all $x \ne c$, we say that $f(\ )$ is strictly convex.

A different definition is often introduced: it is required that, for any two interior points x_1 and x_2 ($x_1 < x_2$) in I, the chord joining

$(x_1, f(x_1))$ and $(x_2, f(x_2))$ is entirely above the function $f(x)$ for all $x \in (x_1, x_2)$, i e it is required that

$$p f(x_1) + (1 - p) f(x_2) \geq f(p x_1 + (1 - p) x_2), \quad \text{for all } p \in (0, 1). \tag{A.4}$$

To verify that condition (A.3) implies (A.4), choose $c = p x_1 + (1 - p) x_2$; on setting first $x = x_1$ and then $x = x_2$, we obtain

$$f(x_1) \geq f(c) + b(x_1 - c), \quad f(x_2) \geq f(c) + b(x_2 - c).$$

By multiplying the first inequality by p, and the second by $1 - p$, and then summing them, we obtain

$$\begin{aligned} p f(x_1) + (1 - p) f(x_2) &\geq f(c) + b\{p(x_1 - c) + (1 - p)(x_2 - c)\} \\ &= f(c) + b\{p x_1 + (1 - p) x_2 - c\} \\ &= f(c) \\ &= f(p x_1 + (1 - p) x_2). \end{aligned}$$

It can even be shown that the opposite implication holds, namely (A.4) implies (A.3); the proof is left to the reader.

If f is twice differentiable, it can be shown that condition (A.3) is equivalent to $f''(x) \geq 0$ for all interior points x of I

It can immediately be deduced that the sum of n functions all convex in the interval I is a convex function. In fact, let $f_1(x), \ldots, f_n(x)$ denote the n convex functions, and let $g(x) = f_1(x) + \cdots + f_n(x)$ be their sum. Then, for any given interior point c of I, there exists a constant b_i, for each $f_i(\cdot)$, such that

$$f_i(x) \geq f_i(c) + b_i(x - c) \quad \text{for all } x \in I \quad (i = 1, \ldots, n).$$

By summing all terms, and setting $B = b_1 + \cdots + b_n$, we have

$$g(x) \geq g(c) + B(x - c) \quad \text{for all } x \in I$$

We now wish to show that a local minimum of a convex function is essentially a global minimum. Suppose there exist two local minima, x_1 and x_3 say (where $x_1 < x_3$). Then a neighbourhood I_1 of x_1 exists such that $f(x_1) \leq f(x)$ for all $x \in I_1$, with strict inequality for at least one point in I_1; a similar fact holds for a neighbourhood I_3 of x_3. If $f(x)$ is not constant in $[x_1, x_3]$, then there must exist a point x_2 between x_1 and x_3 such that $f(x_1) < f(x_2) > f(x_3)$. Try now to choose a number b_2 such that

$$f(x) \geq f(x_2) + b_2(x - x_2) \quad \text{for all } x \in I.$$

If $b_2 \geq 0$, then $f(x_3) < f(x_2)$ cannot hold; if $b_2 \leq 0$, then $f(x_1) < f(x_2)$ cannot hold. Then $f(x)$ must be constant in $[x_1, x_3]$. To summarize, if a point is a local minimum, then it is a global minimum, except for the possibility that there exists an interval where $f(x)$ is constantly equal to the minimum.

If X is a random variable taking values in I, with $E[X] = \mu$, and $f(\ )$ is a convex function in I, then the inequality

$$f(X) \geq f(\mu) + b(X - \mu)$$

holds with probability 1, for some suitable b. Notice that, if X is not a constant, μ is interior to I; otherwise, the above inequality is effectively an equality. By integrating both sides of the inequality with respect to the distribution of X, obtain

$$E[f(X)] \geq f(E[X]) \qquad (A.5)$$

provided the left-hand side term exists; (A.5) is called the **Jensen inequality** for convex functions If $f(\cdot)$ is strictly convex, and X is not a degenerate random variable, then

$$\Pr\{f(X) > f(\mu) + b(X - \mu)\} > 0$$

and then, in (A.5), the strict inequality holds.

A function $f(\cdot)$ is said to be **concave** in I if $-f(\cdot)$ is convex in I. Using an argument similar to that for convex functions, we obtain

$$E[f(X)] \leq f(E[X])$$

for a concave function f, provided that $E[f(X)]$ exists.

Example A.1.4 If X is a positive non-degenerate random variable, and its mean value is defined, then

$$E\left[\frac{1}{X}\right] > \frac{1}{E[X]},$$

since $f(x) = 1/x$ is strictly convex on the positive half-line.   □

Example A.1.5 If X is a positive non-degenerate random variable, then

$$E[\log_b X] < \log_b\{E[X]\}$$

for any positive base b (provided the expectations exist) since the logarithm is a strictly concave function.   □

A.2 Some univariate continuous distributions

A.2.1 The normal distribution

A continuous random variable Y is said to have normal (or Gaussian) distribution with parameters μ and σ^2, if its density function at t is

$$\frac{1}{\sqrt{2\pi}\sigma}\exp\left\{-\frac{1}{2}\left(\frac{t-\mu}{\sigma}\right)^2\right\} \qquad (-\infty < t < \infty); \qquad (A.6)$$

for the sake of brevity, we shall write $Y \sim N(\mu,\sigma^2)$. The density function (A.6) is unimodal and symmetric around the mode $t = \mu$.

If $Y \sim N(0,1)$, we say that Y is a standardized normal random variable, and its density function and distribution function are denoted by

$$\phi(t) = \frac{1}{\sqrt{2\pi}}e^{-t^2/2}, \quad \Phi(t) = \int_{-\infty}^{t} \phi(u)\,du, \qquad (A.7)$$

respectively. Since this distribution function does not have an explicit representation, either numerical tables or numerical approximations are used for its computation; see for instance Abramowitz and Stegun (1965, chap. 26). A simple approximate formula, sufficiently accurate for many practical purposes, is

$$\Phi(t) \approx \frac{1}{1 + \exp\{-at(1 + bt^2)\}} \qquad (A.8)$$

where $a = 1.5976$ and $b = 0.044715$; the maximum absolute deviation from the correct value is less than 0.0002 (Page, 1977).

If $Y \sim N(\mu,\sigma^2)$, its characteristic function is

$$\mathrm{E}\big[e^{itY}\big] = \exp(it\mu - \tfrac{1}{2}\sigma^2 t^2)$$

from which one can obtain several important properties of the normal distribution.

- The moments of Y exists up to any finite order; in particular we have

$$\mathrm{E}[Y] = \mu, \quad \mathrm{var}[Y] = \sigma^2,$$

$$\mathrm{E}\big[(Y-\mu)^k\big] = \begin{cases} 0 & \text{for } k \text{ odd,} \\ 1 \times 3 \times \cdots \times (k-1)\sigma^k & \text{for } k \text{ even.} \end{cases}$$

- If $Y \sim N(\mu,\sigma^2)$ and a, b are constants, then

$$a + bY \sim N(a + b\mu, b^2\sigma^2).$$

This means that the entire family of distributions can be generated by linear transformations, starting from any member of the family. We then say that the class of normal distributions forms a **location and scale family** of distributions.

- If $Y_1 \sim N(\mu_1, \sigma_1^2)$ and $Y_2 \sim N(\mu_2, \sigma_2^2)$ with Y_1 and Y_2 mutually independent, then $Y_1 + Y_2 \sim N(\mu_1 + \mu_2, \sigma_1^2 + \sigma_2^2)$. This result extends readily to linear combinations of normal random variables; specifically, if $(Y_1, \ldots, Y_n)$ are independent normal random variables with distribution $N(\mu_i, \sigma_i^2)$ respectively, and a_i, b_i are constants $(i = 1, \ldots, n)$, then

$$\sum_{i=1}^{n} a_i Y_i + b_i \sim N\left(\sum_{i=1}^{n} a_i \mu_i + b_i, \sum_{i=1}^{n} a_i^2 \sigma_i^2\right).$$

Further important properties of the normal distribution will be presented in section A.5.

A.2.2 *The uniform distribution*

A continuous random variable Y with density function at t

$$f(t; a, b) = \begin{cases} \dfrac{1}{b-a} & \text{for } t \in (a, b), \\ 0 & \text{otherwise,} \end{cases}$$

is said to be uniformly distributed in (a, b), or simply uniform in (a, b); the notation $Y \sim U(a, b)$ will be used. The expected value and the variance are

$$\mathrm{E}[Y] = \frac{a+b}{2}, \quad \mathrm{var}[Y] = \frac{(b-a)^2}{12}.$$

The class of uniform distributions forms a location and scale family.

An important result concerning this distribution is the following: if Z is a continuous random variable with distribution function $F(\cdot)$, then $W = F(Z) \sim U(0, 1)$. In fact, for any $t \in (0, 1)$, we can write

$$
\begin{aligned}
\Pr\{W \leq t\} &=, \ \Pr\{F(Z) \leq t\} \\
&= \ \Pr\{Z \leq F^{-1}(t)\} \\
&= \ F(F^{-1}(t)) \\
&= \ t
\end{aligned}
$$

which is the distribution function of $U(0, 1)$ at t for $0 < t < 1$. If

$t \leq 0$, then clearly $\Pr\{W \leq t\} = 0$; furthermore, if $t \geq 1, \Pr\{W \leq t\} = 1$. Notice that there may be reveral choices for the value of $F^{-1}(t)$ in the second equality above; however, any of these values can be chosen without affecting the result. The transformation $W = F(Z)$ is called **integral transformation** of Z.

A.2.3 The gamma distribution

Before introducing the next distribution, define the gamma function as

$$\Gamma(x) = \int_0^\infty t^{x-1} e^{-t} dt \qquad (A.9)$$

if $x > 0$. On integrating (A.9) by parts, we obtain

$$\Gamma(x+1) = x\,\Gamma(x)$$

and then, if x is an integer,

$$\begin{aligned}
\Gamma(x) &= (x-1)\,\Gamma(x-1) \\
&= (x-1)\,(x-2)\cdots 1\,\Gamma(1) \\
&= (x-1)!
\end{aligned}$$

since $\Gamma(1) = 1$ Another important relationship, not proved here, is

$$\Gamma(\tfrac{1}{2}) = \sqrt{\pi}.$$

A simple approximate expression for computing the gamma function is given by Stirling's formula, that is

$$\Gamma(x) \sim \sqrt{2\pi}\, x^{x-1/2}\, e^{-x} \qquad (x \to \infty) \qquad (A.10)$$

where the symbol $\sim$ means that the ratio of the two sides converges to 1 as $x \to \infty$ If the argument is an integer, (A.10) becomes

$$n! \sim \sqrt{2\pi}\, n^{n+1/2}\, e^{-n} \qquad (n \to \infty) \qquad (A.11)$$

after suitable arrangement of the terms.

We then say that a continuous random variable Y has gamma distribution with index (or shape parameter) ω and scale parameter λ if its density function at t is

$$f(t; \omega, \lambda) = \begin{cases} \lambda^\omega\, t^{\omega-1}\, e^{-\lambda t} / \Gamma(\omega) & \text{for } t > 0, \\ 0 & \text{for } t \leq 0, \end{cases} \qquad (A.12)$$

for ω and λ positive; in this case, we write $Y \sim Gamma(\omega, \lambda)$.

The gamma distribution is often used in practice because of its flexible shape, controlled by ω. In fact, if $0 < \omega \leq 1$, then $f(t; \omega, \lambda)$

decreases over the whole positive half-line; if $\omega > 1$ the density function first increases and then decreases. Moreover, a closer look shows that

$$f'(0; \omega, \lambda) = \begin{cases} -\infty & \text{for } 0 < \omega < 1, \\ -\lambda^2 & \text{for } \omega = 1, \\ \infty & \text{for } 1 < \omega < 2, \\ \lambda^2 & \text{for } \omega = 2, \\ 0 & \text{for } \omega > 2. \end{cases}$$

For a generic ω, (A.12) cannot be integrated in closed form, and numerical methods must be used. If ω is an integer, repeated integration by parts of (A.12) gives the explicit expression

$$\Pr\{Y > t\} = \sum_{k=0}^{\omega-1} \frac{(\lambda t)^k e^{-\lambda t}}{k!}. \tag{A.13}$$

The characteristic function of Y is

$$\mathrm{E}\big[e^{itY}\big] = (1 - it/\lambda)^{-\omega}. \tag{A.14}$$

From this expression, or by direct integration, we obtain

$$\mathrm{E}\big[Y^k\big] = \lambda^{-k}\omega(\omega + 1) \cdots (\omega + k - 1), \quad \text{for } k = 0, 1, 2, \ldots$$

and, in particular,

$$\mathrm{E}[Y] = \omega/\lambda, \quad \mathrm{var}[Y] = \omega/\lambda^2.$$

From (A.14), we can immediately conclude that, if $Y_1 \sim Gamma(\omega_1, \lambda)$ and $Y_2 \sim Gamma(\omega_2, \lambda)$, and they are independent random variables, then $Y_1 + Y_2 \sim Gamma(\omega_1 + \omega_2, \lambda)$.

A.2.4 The exponential distribution

Setting $\omega = 1$ in (A.12), we obtain the density function of the (negative) exponential distribution. Although it is a special case of the gamma distribution, it deserves special mention because of its frequent usage.

An important property of the exponential distribution is the following. If $Y \sim Gamma(1, \lambda)$, then

$$\begin{aligned} \Pr\{Y > t + s \mid Y > s\} &= \frac{\exp\{-\lambda(t + s)\}}{\exp\{-\lambda s\}} \\ &= \exp(-\lambda t) \\ &= \Pr\{Y > t\}. \end{aligned}$$

If Y represents the waiting time to a certain event, the above relationships can be expressed as follows: 'if I have already been waiting s time units for an event to occur, the probability that I have to wait t more time units would be the same as if I started waiting now'. This fact is commonly referred to as 'the memoryless property' of the exponential distribution.

A.2.5 The beta distribution

The beta function is defined as

$$B(p,q) = \int_0^1 x^{p-1}(1-x)^{q-1}\,dx$$

for p and q real positive arguments. It can be shown that the following connection with the gamma function holds:

$$B(p,q) = \frac{\Gamma(p)\,\Gamma(q)}{\Gamma(p+q)}.$$

We then say that the continuous random variable Y has beta distribution with parameters (p,q) if its density function at t is

$$f(t;p,q) = \begin{cases} \dfrac{t^{p-1}(1-t)^{q-1}}{B(p,q)} & \text{for } t \in (0,1), \\ 0 & \text{otherwise.} \end{cases} \qquad (A.15)$$

Depending on p and q, the shape of (A.15) can vary considerably in the interval $(0,1)$:

- if $p>1, q>1$, the density is bell-shaped with a single mode at $(p-1)/(p+q-2)$;
- if $p<1, q<1$, it is U-shaped with antimode at $(p-1)/(p+q-2)$;
- if $p>1, q \leq 1$, the density is monotonically increasing;
- if $p \leq 1, q>1$, it is monotonically decreasing;
- if $p = q$, the density function is symmetric around the abscissa $\frac{1}{2}$;
- if $p = q = 1$, we obtain the $U(0,1)$ density.

If p and q are interchanged in (A.15), the function is reflected about the abscissa $t = \frac{1}{2}$.

By direct integration of (A.15), we obtain

$$\mathrm{E}[Y^s] = \frac{\Gamma(p+q)\,\Gamma(p+s)}{\Gamma(p)\,\Gamma(p+q+s)} \quad \text{for } s>0$$

and then, in particular,

$$\mathrm{E}[Y] = \frac{p}{p+q}, \quad \mathrm{var}[Y] = \frac{pq}{(p+q)^2(p+q+1)}.$$

A.2.6 The inverse Gaussian distribution

A continuous random variable Y is said to be inverse Gaussian, written $Y \sim IG(\mu, \lambda)$, if its density function is

$$f(t; \mu, \lambda) = \begin{cases} 0 & \text{if } t \leq 0, \\ \left(\dfrac{\lambda}{2\pi t^3}\right)^{\frac{1}{2}} \exp\left(-\dfrac{\lambda}{2\mu^2}\dfrac{(t-\mu)^2}{t}\right) & \text{if } t > 0, \end{cases}$$

for positive reals μ and λ.

The name 'inverse Gaussian' is due to a number of properties which characterize this distribution as the 'dual' of the normal distribution. In particular, its cumulant generating function is the inverse function of the analogous function of the normal distribution; in fact, the cumulant generating function of Y is

$$K(s) = \frac{\lambda}{\mu}\left\{1 - \left(1 - \frac{2\mu^2 s}{\lambda}\right)^{1/2}\right\},$$

and, by solving the equation $K(s) = u$ with respect to s, one obtains indeed the cumulant generating function of a normal distribution. For a detailed discussion of these and additional aspects, see the monograph of Seshadri (1993). Notice that

$$\mathrm{E}[Y] = \mu, \qquad \mathrm{var}[Y] = \frac{\mu^3}{\lambda}.$$

A.2.7 The Cauchy distribution

A continuous random variable Y with density function at t

$$\frac{\lambda}{\pi\{\lambda^2 + (t-\theta)^2\}} \quad \text{for } -\infty < t < \infty$$

is said to be a Cauchy random variable with location parameter θ and scale parameter λ, where $\lambda > 0$. Its distribution function is

$$\frac{1}{2} + \frac{1}{\pi}\arctan\left(\frac{t-\theta}{\lambda}\right),$$

showing that θ is the median of the distribution. The corresponding characteristic function is

$$\exp(it\theta - \lambda|t|)$$

which cannot be differentiated at $t = 0$, so that this distribution does not have moments of any order.

There is a connection between this distribution and the normal one: if Z_1 and Z_2 are independent $N(0,1)$ random variables, then Z_1/Z_2 is a Cauchy random variable with location parameter 0 and scale parameter 1.

A.3 Some univariate discrete distributions

A.3.1 The binomial distribution

A discrete random variable Y, taking values $0, 1, \ldots, n$ with

$$\Pr\{Y = t\} = \binom{n}{t} p^t (1-p)^{n-t} \quad (t = 0, 1, \ldots, n), \qquad (A.16)$$

for $0 < p < 1$ and n a positive integer, is said to be binomial with index n and parameter p; we write $Y \sim Bin(n,p)$. Formula (A.16) provides the probability of obtaining t 'successes' in n independent replicates of an experiment with two possible outcomes (Bernoulli trials), such that the probability of 'success' in a single trial is constantly equal to p.

The characteristic function of Y is

$$E[e^{itY}] = \{1 + p(e^{it} - 1)\}^n \qquad (A.17)$$

from which we obtain

$$E[Y] = np, \quad var[Y] = np(1-p). \qquad (A.18)$$

Moreover, from (A.17), we can immediately deduce the following additive property: if $Y_1 \sim Bin(n_1, p)$ and $Y_2 \sim Bin(n_2, p)$ with Y_1 and Y_2 mutually independent, then $Y_1 + Y_2 \sim Bin(n_1 + n_2, p)$.

A.3.2 The hypergeometric distribution

Let Y be a discrete random variable such that

$$\Pr\{Y = t\} = \frac{\binom{M}{t}\binom{N-M}{n-t}}{\binom{N}{n}} \tag{A.19}$$

where t is an integer such that $\max(0, n - N + M) \leq t \leq \min(n, M)$ and N, n, M are positive integers ($N \geq M, N \geq n$). We then say that Y is an hypergeometric random variable with parameters (N, n, M). Formula (A.19) provides the probability of drawing t 'success' balls, when n balls are drawn without replacement from an urn containing M 'success' balls and $N - M$ 'failure' balls.

To establish some of the properties of Y, it is convenient to write $Y = \sum_{i=1}^{n} I_i$, where I_i is an indicator variable taking values 1 or 0, depending on whether the ith draw is a 'success' or a 'failure'. It then turns out that

$$\Pr\{I_i = 1\} = \frac{M}{N}, \quad \Pr\{I_i = I_j = 1\} = \frac{M(M-1)}{N(N-1)},$$

$$\mathrm{cov}[\,I_i, I_j\,] = -\frac{M(N-M)}{N^2(N-1)}$$

for $i, j = 1, \ldots, n$ ($i \neq j$), leading to

$$\mathrm{E}[Y] = n\frac{M}{N}, \quad \mathrm{var}[Y] = n\frac{M}{N}\left(1 - \frac{M}{N}\right)\left(\frac{N-n}{N-1}\right). \tag{A.20}$$

Replacing the ratio M/N of the number of favourable cases to the total number of cases by p in (A.20), we obtain expressions similar to (A.18), except for the final factor of the variance, a factor close to 1 if $N \gg n$.

A.3.3 The Poisson distribution

A discrete random variable Y taking values $0, 1, 2, \ldots$ with

$$\Pr\{Y = t\} = \frac{e^{-\lambda}\lambda^t}{t!} \quad (t = 0, 1, \ldots) \tag{A.21}$$

for some $\lambda > 0$ is said to be a Poisson random variable with parameter λ. The characteristic function is readily computed to be

$$\mathrm{E}\left[e^{itY}\right] = \exp\{\lambda(e^{it} - 1)\}, \tag{A.22}$$

which gives

$$E[Y] = \lambda, \quad \text{var}[Y] = \lambda.$$

From (A.22), we can immediately conclude that, if Y_1 and Y_2 are independent Poisson variables with parameters λ_1 and λ_2, respectively, then $Y_1 + Y_2$ is a Poisson variable with parameter $\lambda_1 + \lambda_2$.

Expression (A.21) can be obtained as a limit form of (A.16), if in the latter we set $p = \lambda/n$, for some positive constant λ, and allow n to tend to ∞.

Formula (A.13) establishes a connection between the Poisson and the gamma distribution functions.

A.3.4 The negative binomial distribution

A discrete random variable Y which can take non-negative values with

$$\Pr\{Y = t\} = \binom{t + r - 1}{t} p^r (1-p)^t \qquad (t = 0, 1, \ldots), \quad \text{(A.23)}$$

for some positive real r and $0 < p < 1$, is said to be negative binomial with index r and parameter p. Notice that the binomial coefficient of (A.23) is meaningful even if r is not an integer, since by definition

$$\binom{v}{n} = \frac{v(v-1) \times \cdots \times (v - n + 1)}{n!}$$

for a positive real v and a positive integer n, and

$$\binom{v}{0} = 1.$$

If r is an integer, (A.23) is sometimes called the **Pascal distribution**, and it can be generated as follows. Consider an infinite sequence of trials whose outcome is 'success' with probability p, and 'failure' with probability $1 - p$, and assume that the trials are mutually independent. Formula (A.23) gives the probability of experiencing t 'failures' before achieving r 'successes' Thus, in a sense, Y represents a 'waiting time'.

For any value of r, the characteristic function of Y is

$$E\left[e^{itY}\right] = \left(\frac{p}{1 - (1-p)e^{it}}\right)^r \qquad \text{(A.24)}$$

which gives

$$E[Y] = r\frac{1-p}{p}, \quad \text{var}[Y] = r\frac{1-p}{p^2}.$$

From (A.24), we can immediately conclude that the sum of two independent negative binomial random variables with common parameter p and indices r and s is also a negative binomial random variables with parameter p and index $r + s$.

The negative binomial distribution can be considered a discrete analogue of the gamma distribution in the sense which is now explained. Divide Y by n and, at the same time, modify p in such a way that the mean value remains unchanged, μ say $(\mu > 0)$, i e

$$\frac{r(1-p)}{np} = \mu$$

which gives

$$p = \frac{r}{r + n\mu}.$$

Substituting this expression in (A.24), and allowing n to tend to ∞, we obtain an expression of type (A.14). Dividing Y by n means that, in a time unit, n trials instead of 1 take place, but p is modified so that the mean waiting time is unchanged. In other words, although time is made to approach continuity, the mean waiting time remains unchanged.

If $r = 1$, (A.23) is called the **geometric distribution**. For the reasons already given, the geometric distribution is the discrete analogue of the exponential distribution. Notice that, if r is an integer, we can think of a negative binomial random variable as the sum of r independent geometric random variables.

A.4 Random vectors: general concepts

A.4.1 Basic matrix algebra

Matrices will usually be denoted by capital Roman letters; often a notation of the form $A = (a_{ij})$ is adopted, where $a_{ij} \in \mathbb{R}$ is the generic entry of A. If A has m rows and n columns, we say that A is an $m \times n$ matrix. The transpose of A is denoted by $A^\top$.

A matrix v of size $n \times 1$ is said to be a (column) vector of size n or, equivalently, an $n \times 1$ vector, and in this case we write $v \in \mathbb{R}^n$; similarly, a $1 \times n$ matrix is called a row vector. Vectors are commonly denoted by lower-case Roman letters.

Denote by I_k the identity matrix of order k, by 1_n the $n \times 1$ vector with all elements equal to 1, and by 0 a matrix with all elements equal to zero (the dimensions of this matrix will be clear from the context).

For a $k \times k$ square matrix A, we recall the following facts:

1. A is symmetric if $A^\top = A$.

2. $|A|$ denotes the determinant of A; for two conformable matrices A and B, i.e. where the matrix product AB is defined, we have that $|AB| = |A| \times |B|$.

3. If $|A| \neq 0$, then A is non-singular, and there exists an inverse matrix A^{-1} such that $AA^{-1} = A^{-1}A = I_k$; moreover, $(A^\top)^{-1} = (A^{-1})^\top$ and $(AB)^{-1} = B^{-1}A^{-1}$ if both inverses exist.

4. A is positive semidefinite, and we write $A \geq 0$, if A is symmetric and $u^\top Au \geq 0$ for all non-null vectors $u \in \mathbb{R}^k$; moreover, we use the notation $A \geq B$ to express the inequality $A - B \geq 0$.

5. A is positive definite, and we write $A > 0$, if A is symmetric such that $u^\top Au > 0$ for all non-null vectors $u \in \mathbb{R}^k$; moreover, we use the notation $A > B$ to express the strict inequality $A - B > 0$.

6. A is orthogonal if its inverse coincides with its transpose, i e $A^\top = A^{-1}$; in this case $|A| = \pm 1$.

7. $\mathrm{tr}(A)$ denotes the trace of A, i e the sum of the main diagonal elements; for any two matrices, A and B, the property $\mathrm{tr}(AB) = \mathrm{tr}(BA)$ holds, provided both products exist.

8. A is idempotent if $A = A^2$; for an idempotent matrix, the rank is equal to the trace, i.e $\mathrm{rk}(A) = \mathrm{tr}(A)$.

9. A is a diagonal matrix if all elements outside the main diagonal are 0; in this case, we write $A = \mathrm{diag}(a_1, \ldots, a_k)$ where $(a_1, \ldots, a_k)$ are the diagonal elements.

10. The following equality holds:

$$(A + BCD)^{-1} = A^{-1} - A^{-1}B(C^{-1} + DA^{-1}B)^{-1}DA^{-1}$$
$$\text{(A.25)}$$

when the matrices are conformable and the inverses exist; in particular, if b and d are vectors, (A.25) becomes

$$(A + bd^\top)^{-1} = A^{-1} - \frac{1}{1 + d^\top A^{-1}b} A^{-1}bd^\top A^{-1}. \quad \text{(A.26)}$$

Theorem A.4.1 (Spectral decomposition) *If A is a symmetric $k \times k$ matrix, there exist real numbers $\lambda_1, \ldots, \lambda_k$ and an orthogonal matrix Q such that*

$$A = Q\Lambda Q^\mathsf{T}, \quad \Lambda = \mathrm{diag}(\lambda_1, \ldots, \lambda_k).$$

The numbers $\lambda_1, \ldots, \lambda_k$ are called **eigenvalues** of A and the jth column of Q is called the **eigenvector** corresponding to λ_j. It is immediate that, if $A \geq 0$, then $\lambda_j \geq 0$ for all j; similarly, $A > 0$ implies $\lambda_j > 0$ for all j. From the above theorem and from the properties of the determinant of an orthogonal matrix, it follows that

$$|A| = |\Lambda| = \prod_{j=1}^{k} \lambda_j.$$

We now wish to introduce the notion of the **square root** of a matrix; more specifically, given a semidefinite positive matrix A, we wish to find a matrix B such that $A = BB^\mathsf{T}$. In the basic case of a diagonal matrix, $A = \mathrm{diag}(a_1, \ldots, a_k)$ say, it is natural to set

$$B = \mathrm{diag}(a_1^{1/2}, \ldots, a_k^{1/2}).$$

In the general case of a positive semidefinite matrix A, put

$$B = Q\,\Lambda^{1/2} = Q\,\mathrm{diag}(\lambda_1^{1/2}, \ldots, \lambda_k^{1/2})$$

such that

$$BB^\mathsf{T} = (Q\,\Lambda^{1/2})\,(\Lambda^{1/2}Q^\mathsf{T}) = Q\Lambda Q^\mathsf{T} = A.$$

We then say that B is *a* square root of A. Notice that, if $A > 0$,

$$B^{-1} = (Q\,\Lambda^{1/2})^{-1} = \Lambda^{-1/2}Q^\mathsf{T}.$$

A.4.2 Random vectors

If $X_1, \ldots, X_k$ are random variables defined on the same probability space, we define the random vector, or **multivariate** random variable, X, as

$$X = \begin{pmatrix} X_1 \\ X_2 \\ \vdots \\ X_k \end{pmatrix}.$$

The **mean value** of X is obtained by forming the vector of the mean values of the components, namely

$$E[X] = \begin{pmatrix} E[X_1] \\ E[X_2] \\ \vdots \\ E[X_k] \end{pmatrix};$$

similarly, we define the **variance matrix** (also called the dispersion matrix) as

$$\text{var}[X] = \begin{pmatrix} \text{var}[X_1] & \text{cov}[X_1, X_2] & \cdots & \text{cov}[X_1, X_k] \\ \text{cov}[X_2, X_1] & \text{var}[X_2] & \cdots & \text{cov}[X_2, X_k] \\ \vdots & \vdots & \ddots & \vdots \\ \text{cov}[X_k, X_1] & \text{cov}[X_k, X_2] & \cdots & \text{var}[X_k] \end{pmatrix},$$

assuming that all elements exist. In fact, the existence of the diagonal elements is sufficient to ensure the existence of the remaining elements. Notice that $\text{var}[X]$ is a symmetric matrix.

Closely related to the variance matrix is the concept of the **correlation matrix** whose generic element is

$$\text{corr}[X_i, X_j] = \frac{\text{cov}[X_i, X_j]}{\sqrt{\text{var}[X_i]\,\text{var}[X_j]}} \qquad (i, j = 1, \ldots, k)$$

which lies in $[-1, 1]$. This matrix can be viewed as the variance matrix of the standardized variables, defined as

$$(X_j - E[X_j])/\sqrt{\text{var}[X_j]} \quad \text{for } j = 1, \ldots, k.$$

If $\text{var}[X]$ is diagonal, then so is the correlation matrix, and X is said to have uncorrelated components.

A.4.3 Some general properties

We develop here the basic properties of the mean value and of variance matrices of multivariate random variables. Throughout this subsection, let $X = (X_1, \ldots, X_k)^\top$, with $E[X] = \mu$, $\text{var}[X] = V$.

Lemma A.4.2 *If $A = (a_{ij})$ is an $n \times k$ matrix and $b = (b_1, \ldots, b_n)^\top$ is an $n \times 1$ vector, define*

$$Y = AX + b.$$

Then

(i) $\mathrm{E}[Y] = A\mu + b$,

(ii) $\mathrm{var}[Y] = AVA^{\mathsf{T}}$.

Proof. First of all, we show that $\mathrm{E}[Y_i]$ exists, if Y_i is the ith component of Y. In fact

$$
\begin{aligned}
\mathrm{E}[|Y_i|] &= \mathrm{E}\left[\left|\sum_{j=1}^{k} a_{ij}X_j + b_i\right|\right] \\
&\leq \mathrm{E}\left[\sum_{j} |a_{ij}X_j| + |b_i|\right] = \sum_{j} |a_{ij}|\,\mathrm{E}[|X_j|] + |b_i| \\
&< \infty
\end{aligned}
$$

$(i = 1, \dots, k)$. Then $\mathrm{E}[Y_i]$ exists such that

$$
\mathrm{E}[Y_i] = \mathrm{E}\left[\sum_{j=1}^{k} a_{ij}X_j + b_i\right] = \sum_{j} a_{ij}\mathrm{E}[X_j] + b_i
$$

$(i = 1, \dots, k)$. In matrix notation, these relationships take the form

$$
\mathrm{E}[Y] = A\,\mathrm{E}[X] + b = A\mu + b.
$$

This shows part (i) of the theorem. For part (ii), the proof is analogous, taking into account that

$$
\mathrm{cov}[a_{ij}X_j, a_{rs}X_s] = a_{ij}\,a_{rs}\,\mathrm{cov}[X_j, X_s].
$$

$\square$

Lemma A.4.3 *The variance matrix V of the random vector X is positive semidefinite, and it is positive definite if there exists no vector b (other than $b = 0$) such $b^{\mathsf{T}}X$ is a degenerate random variable.*

Proof. If $Y = b^{\mathsf{T}}X$, then $0 \leq \mathrm{var}[b^{\mathsf{T}}X] = b^{\mathsf{T}}Vb$, and then $V \geq 0$. If $b^{\mathsf{T}}Vb = 0$, then $\mathrm{var}[b^{\mathsf{T}}X] = 0$, i e $b^{\mathsf{T}}X$ is a constant with probability 1. $\square$

Lemma A.4.4 *If $\mathrm{var}[X] = V > 0$, there exists a square matrix C such that $Y = CX$ has uncorrelated components with unit variance, i.e. $\mathrm{var}[Y] = I_k$.*

Proof. By using the results of section A.4.1, we can write $V = BB^{\mathsf{T}}$, where B is the non-singular square root of A. If we define

$Y = B^{-1}X$, it follows that

$$
\begin{aligned}
\text{var}[Y] &= \text{var}\left[B^{-1}X\right] = B^{-1}\,\text{var}[X]\,(B^{-1})^\mathsf{T} \\
&= B^{-1}V(B^\mathsf{T})^{-1} = B^{-1}BB^\mathsf{T}(B^\mathsf{T})^{-1} \\
&= I_k.
\end{aligned}
$$

$\square$

Lemma A.4.5 *If $A = (a_{ij})$ is a $k \times k$ matrix, then*

$$
\mathrm{E}\left[X^\mathsf{T}AX\right] = \mu^\mathsf{T}A\mu + \text{tr}(AV).
$$

Proof.

$$
\begin{aligned}
\mathrm{E}\left[X^\mathsf{T}AX\right] &= \mathrm{E}\left[\sum_{i=1}^{k}\sum_{j=1}^{k}X_i a_{ij} X_j\right] \\
&= \sum_i \sum_j a_{ij}\mathrm{E}[X_i X_j] \\
&= \sum_i \sum_j a_{ij}(\mu_i\mu_j + v_{ij}) \\
&= \sum_i \sum_j a_{ij}\mu_i\mu_j + \sum_i \sum_j a_{ij}v_{ji} \\
&= \mu^\mathsf{T}A\mu + \sum_i (AV)_{ii} \\
&= \mu^\mathsf{T}A\mu + \text{tr}(AV)
\end{aligned}
$$

where μ_i and v_{ij} denote the generic elements of μ and V, and $(AV)_{ii}$ is the generic diagonal element of AV. This completes the proof.

The same computation could be written in more compact notation as follows:

$$
\begin{aligned}
\mathrm{E}\left[X^\mathsf{T}AX\right] &= \mathrm{E}\left[\text{tr}(X^\mathsf{T}AX)\right] \\
&= \mathrm{E}\left[\text{tr}(AXX^\mathsf{T})\right] \\
&= \text{tr}(A\,\mathrm{E}[XX^\mathsf{T}]) \\
&= \text{tr}(A(\mu\mu^\mathsf{T} + V)) \\
&= \mu^\mathsf{T}A\mu + \text{tr}(AV),
\end{aligned}
$$

taking into account that the trace operator is linear, so that it can be interchanged with the expectation operator. $\square$

A.5 The multivariate normal distribution

A.5.1 The density function

Before introducing the definition of the multivariate normal distribution, we present some preliminaries which help to motivate the rationale of the definition. Consider a random vector $Z = (Z_1, \ldots, Z_k)^\top$ where $Z_1, \ldots, Z_k$ are independent and identically distributed random variables $N(0, 1)$, and define

$$Y = AZ + \mu$$

for some non-singular $k \times k$ matrix A, and a $k \times 1$ vector μ. The components of Y are linear combinations of independent univariate normal random variables, and we have already seen in section A.2.1 that each one of these components has a normal distribution. It is then reasonable to think of the whole vector Y as a k-dimensional generalization of the normal distribution.

To obtain the density function of Y, start from the density function of Z at $t \in \mathbb{R}^k$, namely

$$f_Z(t) = \frac{1}{(2\pi)^{k/2}} \exp(-\tfrac{1}{2} t^\top t).$$

Since

$$Z = A^{-1}(Y - \mu),$$

the Jacobian of the transformation is

$$\left| \left(\frac{\partial Z_i}{\partial Y_j} \right) \right| = |A|^{-1} = |V|^{-1/2}$$

taking into account that

$$|V| = |A\,A^\top| = |A|^2.$$

Moreover, on setting $y = At + \mu$, we obtain

$$
\begin{aligned}
t^\top t &= \{A^{-1}(y - \mu)\}^\top \{A^{-1}(y - \mu)\} \\
&= (y - \mu)^\top V^{-1}(y - \mu).
\end{aligned}
$$

Therefore, the density function of Y at y is

$$f_Y(y) = \frac{1}{(2\pi)^{k/2}|V|^{1/2}} \exp\{-\tfrac{1}{2}(y - \mu)^\top V^{-1}(y - \mu)\}. \quad \text{(A.27)}$$

Now, we define that a multivariate random variable $Y = (Y_1, \ldots, Y_k)^\top$ having density function of type (A.27) as a **multivariate normal** random variable with parameters μ and V (however we obtain Y, which need not necessarily be expressed as $Y = AZ + \mu$), and write

$$Y \sim N_k(\mu, V).$$

If $k = 1$, the parameters μ and V reduce to scalars and $N_1(\mu, v)$ is simply $N(\mu, v)$.

Notice that, although we said that any random variable with density function (A.27) is multivariate normal, it is still true that (A.27) is the density function of the transformation $AZ + \mu$, so that any property belonging to the distribution of $AZ + \mu$ belongs to distribution (A.27). This remark allows a straightforward proof of certain results which would otherwise be quite tedious to obtain directly from (A.27). For instance, using Lemma A.4.2, we can immediately write

$$\mathrm{E}[Y] = \mathrm{E}[AZ + \mu] = \mu, \quad \mathrm{var}[Y] = \mathrm{var}[AZ + \mu] = V$$

since $\mathrm{E}[Z] = 0, \mathrm{var}[Z] = I_k$. Moreover, if $X = BY + b$, where B is a non-singular $k \times k$ matrix and b is a $k \times 1$ vector, then

$$X = BAZ + (B\mu + b), \quad (BA)(BA)^{\mathsf{T}} = BVB^{\mathsf{T}}$$

and

$$X \sim N_k(B\mu + b, BVB^{\mathsf{T}}).$$

The set of points in $\mathbb{R}^k$ with equal density (A.27) are those satisfying the equation $(y - \mu)^{\mathsf{T}} V^{-1}(y - \mu) = $ constant, which is the equation of an ellipsoid centred at μ. In particular, if $k = 2$, this ellipsoid becomes an ellipse whose main axis forms an angle

$$\alpha = \tfrac{1}{2} \arctan\left(\frac{2\,v_{12}}{v_{11} - v_{22}}\right)$$

with the coordinate axis y_1

Figure A.1 provides graphical representations of (A.27) in the bivariate case with standardized marginals, i e with $k = 2$ and

$$\mu = \begin{pmatrix} 0 \\ 0 \end{pmatrix}, \quad V = \begin{pmatrix} 1 & \rho \\ \rho & 1 \end{pmatrix},$$

where the density function at (y_1, y_2) is

$$\frac{1}{2\pi\sqrt{1 - \rho^2}} \exp\left(-\frac{1}{2(1 - \rho^2)}(y_1^2 - 2\rho y_1 y_2 + y_2^2)\right). \qquad \text{(A.28)}$$

In this case, since the variances are equal, $\alpha = \mathrm{sgn}(\rho)\pi/4$. Figure A.1 provides three perspective views of (A.28) as well as the

(a)

(b)

(c)

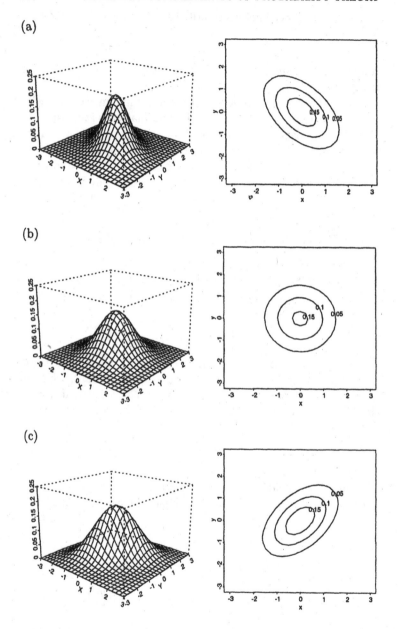

Figure A.1 *The bivariate normal density function with standardized marginals, when ρ is equal to (a) $-\frac{1}{2}$, (b) 0, (c) $\frac{1}{2}$.*

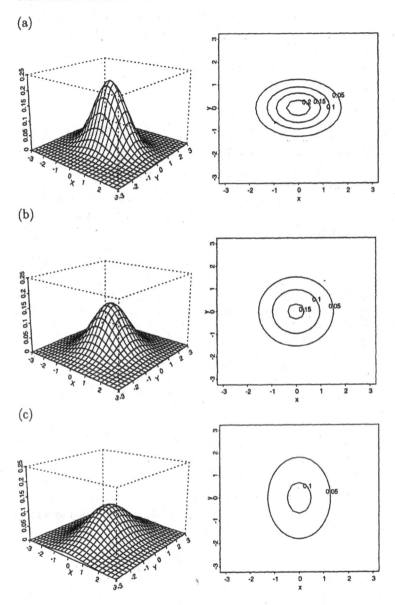

Figure A.2 *The bivariate normal density functions with independent components, 0 marginal means, $v_{11} = 1$, and v_{22} equal to (a) $\frac{1}{2}$, (b) 1, (c) 2.*

correponding contour-level plots, whose curves are ellipses, when ρ is equal to $-\frac{1}{2}, 0, \frac{1}{2}$, respectively. Figure A.2 contains similar plots for the case

$$\mu = \begin{pmatrix} 0 \\ 0 \end{pmatrix}, \quad V = \begin{pmatrix} 1 & 0 \\ 0 & v_{22} \end{pmatrix}$$

for $v_{22} = \frac{1}{2}, 1, 2$.

Remark A.5.1 We have effectively restricted ourselves to the case $V > 0$. This is not really necessary, and the general case $V \geq 0$ could be considered. However, this extension is outside the scope of this book, and it is of little practical relevance. □

Remark A.5.2 From (A.27), it follows immediately that, if V is a diagonal matrix, $Y_1, \ldots, Y_k$ are independent random variables. In other words, if the components of a multivariate normal variable are uncorrelated, then they are independent. This conclusion is sometimes rephrased as 'if the random variables $Y_1, \ldots, Y_k$ are normally distributed and uncorrelated, then they are independent', but this statement is not correct, since the fact that $Y_1, \ldots, Y_k$ are normal variables does not imply that $(Y_1, \ldots, Y_k)^\top$ is a multivariate normal variable (see Exercise A.7). □

A.5.2 The characteristic function

Lemma A.5.3 *If A is a $k \times k$ positive definite matrix and b is a $k \times 1$ vector, then*

$$\int_{\mathbb{R}^k} \frac{1}{(2\pi)^{k/2}} \exp\{-\tfrac{1}{2}(y^\top A y - 2b^\top y)\}\, dy = \frac{\exp(\tfrac{1}{2} b^\top A^{-1} b)}{|A|^{1/2}}$$

where dy stands for $dy_1 \cdots dy_k$.

Proof. Denote by I the integral on the left-hand side of the equality. Let $\mu = A^{-1} b$ and, within the integral, expand $\exp(\cdot)$ by adding and subtracting $\frac{1}{2} \mu^\top A \mu$, so that

$$\begin{aligned}
I &= \cdot \; |A^{-1}|^{1/2} \exp(\tfrac{1}{2} \mu^\top A \mu) \int_{\mathbb{R}^k} g(y)\, dy \\
&= \frac{\exp(\tfrac{1}{2} b^\top A^{-1} b)}{|A|^{1/2}}
\end{aligned}$$

where $g(y)$ is the density function (A.27) with V replaced by A^{-1}. □

Computation of the moment generating function of Y is now immediate:

$$
\begin{aligned}
\mathrm{E}\!\left[\exp(t^\mathsf{T} Y)\right] &= \int_{\mathbb{R}^k} \exp(t^\mathsf{T} y) f_Y(y)\,dy \\
&= \exp(-\tfrac{1}{2}\mu^\mathsf{T} V^{-1}\mu) \\
&\quad \int_{\mathbb{R}^k} \frac{\exp\left\{-\tfrac{1}{2}\left(y^\mathsf{T} V^{-1} y - 2(V^{-1}\mu + t)^\mathsf{T} y\right)\right\}}{(2\pi)^{k/2}\,|V|^{1/2}}\,dy \\
&= \exp(\tfrac{1}{2}t^\mathsf{T} V t + t^\mathsf{T}\mu),
\end{aligned}
$$

using the previous lemma with $A = V^{-1}$ and $b = V^{-1}\mu + t$. The characteristic function of Y is

$$
\mathrm{E}\!\left[\exp(it^\mathsf{T} Y)\right] = \exp(it^\mathsf{T}\mu - \tfrac{1}{2}t^\mathsf{T} V t).
$$

A.5.3 Marginal distributions

Let us partition Y as

$$
Y = \begin{pmatrix} Y_1 \\ Y_2 \end{pmatrix}
$$

where $Y_1 \in \mathbb{R}^r$, $Y_2 \in \mathbb{R}^{k-r}$ $(1 < r < k)$, and correspondingly let

$$
\mu = \begin{pmatrix} \mu_1 \\ \mu_2 \end{pmatrix}, \quad t = \begin{pmatrix} t_1 \\ t_2 \end{pmatrix}, \quad V = \begin{pmatrix} V_{11} & V_{12} \\ V_{21} & V_{22} \end{pmatrix}
$$

where $V_{21} = V_{12}^\mathsf{T}$. The marginal distribution of Y_1 is obtained by setting $t_2 = 0$ in the expression of the characteristic function of Y, obtaining

$$
\exp\left(it_1^\mathsf{T}\mu_1 - \tfrac{1}{2}t_1^\mathsf{T} V_{11} t_1\right)
$$

which is the characteristic function of $N_r(\mu_1, V_{11})$.

This means that all marginal distributions of a multivariate normal are multivariate normal themselves. In general, the opposite implication does not hold; there are examples of non-normal multivariate distributions whose marginals are normals; see Exercise A.7.

A.5.4 Conditional distributions

We first obtain some expressions useful for manipulating the inverse of a variance matrix, but in fact valid more generally for any non-singular symmetric matrix.

With the same block partition of the previous subsection, write

$$V^{-1} = W = \begin{pmatrix} W_{11} & W_{12} \\ W_{21} & W_{22} \end{pmatrix}$$

where $W_{21} = W_{12}^{\mathsf{T}}$. From the identity

$$\begin{pmatrix} W_{11} & W_{12} \\ W_{21} & W_{22} \end{pmatrix} \begin{pmatrix} V_{11} & V_{12} \\ V_{21} & V_{22} \end{pmatrix} = \begin{pmatrix} I_r & 0 \\ 0 & I_{k-r} \end{pmatrix},$$

we deduce

$$W_{11}V_{12} + W_{12}V_{22} = 0$$

from which we obtain

$$W_{12} = -W_{11}V_{12}V_{22}^{-1}.$$

Replacing this equality in

$$W_{11}V_{11} + W_{12}V_{21} = I_r,$$

we deduce

$$W_{11}(V_{11} - V_{12}V_{22}^{-1}V_{21}) = I_r.$$

Proceeding in a similar way for the other relationships of the above identity, we obtain

(a) $W_{11} = (V_{11} - V_{12}V_{22}^{-1}V_{21})^{-1}$,

(b) $W_{22} = (V_{22} - V_{21}V_{11}^{-1}V_{12})^{-1}$,

(c) $W_{12} = -W_{11}V_{12}V_{22}^{-1}$,

(d) $W_{21} = -W_{22}V_{21}V_{11}^{-1}$.

We now wish to obtain the density function of Y_1 conditional on $Y_2 = y_2$, where $y_2 \in \mathbb{R}^{k-r}$. This density is proportional to

$$\exp(-\tfrac{1}{2}Q) = \exp\left\{ -\tfrac{1}{2}\left(z^{\mathsf{T}}V^{-1}z - z_2^{\mathsf{T}}V_{22}^{-1}z_2 \right) \right\},$$

where $z = y - \mu$, $z_2 = y_2 - \mu_2$. Using expression (b) with the symbols W and V interchanged, and setting

$$V_{11\,2} = V_{11} - V_{12}V_{22}^{-1}V_{21}$$

we obtain

$$Q = (z_1^{\mathsf{T}}, z_2^{\mathsf{T}}) \begin{pmatrix} W_{11} & W_{12} \\ W_{21} & W_{22} \end{pmatrix} \begin{pmatrix} z_1 \\ z_2 \end{pmatrix} - z_2^{\mathsf{T}}(W_{22} - W_{21}W_{11}^{-1}W_{12})z_2$$

$$= (z_1 + W_{11}^{-1}W_{12}z_2)^{\mathsf{T}} W_{11} (z_1 + W_{11}^{-1}W_{12}z_2)$$

$$= (z_1 - V_{12}V_{22}^{-1}z_2)^{\mathsf{T}} V_{11\,2}^{-1} (z_1 - V_{12}V_{22}^{-1}z_2)$$

which, for fixed y_2, is a quadratic form of y_1. Therefore, the distribution of Y_1 conditional on $Y_2 = y_2$ is

$$N_r \left(\mu_1 + V_{12} V_{22}^{-1} (y_2 - \mu_2), V_{11\,2} \right). \tag{A.29}$$

If y_2 is regarded as variable, the expression for the conditional mean value is the equation of a hyperplane passing through the point $(\mu_1^\top, \mu_2^\top)^\top$, called a **regression hyperplane**. The variance matrix of the conditional distribution does not depend on y_2, and it is always 'smaller' than that of the unconditional distribution, in the sense that the matrix $V_{11} - V_{11\,2}$ is positive definite.

A.5.5 Mahalanobis distance

In connection with a random variable $Y \sim N_k(\mu, V)$, the Mahalanobis distance is often useful. This distance, associated with the norm

$$\|a\|_V = \left(a^\top V^{-1} a \right)^{1/2}$$

(for $a \in \mathbb{R}^k$), is a generalization of the usual Euclidean distance, and it reduces to the latter when $V = I_k$.

The Mahalanobis distance enjoys an important invariance property: if the coordinates are transformed by a non-singular matrix C, we have

$$
\begin{aligned}
a &\rightarrow Ca \\
V &\rightarrow C V C^\top \\
\|a\|_V^2 &\rightarrow (Ca)^\top (C V C^\top)^{-1} (Ca) = a^\top V^{-1} a = \|a\|_V^2
\end{aligned}
$$

so that the distance of a from the origin is unchanged after the coordinate transformation.

This distance is closely related to the normal density, since (A.27) can be written, up to a normalizing constant, as

$$\exp(-\tfrac{1}{2} \|y - \mu\|_V^2).$$

This means that points with equal density are those with equal Mahalanobis distance from μ.

A.5.6 The chi-squared distribution

If $Z = (Z_1, \ldots, Z_k)^\top \sim N_k(0, I_k)$, we say that

$$U_k = Z^\top Z = \sum_{i=1}^k Z_i^2$$

is a (central) chi-squared random variable with k **degrees of freedom**, and write

$$U_k \sim \chi_k^2.$$

We shall also often use a notation such as $W \sim c\,\chi_k^2$, for some positive constant c, to indicate that $W/c \sim \chi_k^2$.

To compute the density function of U_k, it is convenient to start from the case $k = 1$. For $t > 0$, write

$$\Pr\{U_1 \le t\} = \Pr\{Z_1^2 \le t\} = \Pr\{-\sqrt{t} \le Z_1 \le \sqrt{t}\} = 2\{\Phi(\sqrt{t}) - \tfrac{1}{2}\}$$

whose derivative is

$$f_1(t) = \frac{d}{dt}\{2\Phi(\sqrt{t}) - 1\} = \frac{t^{-1/2}e^{-t/2}}{\sqrt{2\pi}}.$$

This is the density function of a gamma random variable with index $\frac{1}{2}$ and scale parameter $\frac{1}{2}$. Therefore, by the additive property of the gamma distribution, $U_k \sim Gamma(\frac{1}{2}k, \frac{1}{2})$, with density function

$$f_k(t) = \frac{1}{2^{k/2}\Gamma(k/2)}t^{(k/2)-1}e^{-t/2} \qquad (t > 0).$$

Moreover, from the expression of the moments of the gamma distribution, we can immediately write

$$\mathrm{E}[U_k] = k, \qquad \mathrm{var}[U_k] = 2k.$$

It is also clear that, if $W_r \sim \chi_r^2$, and U_k and W_r are independent, then $W_r + U_k \sim \chi_{r+k}^2$ by the analogous property of the gamma distribution.

A.5.7 Distribution of the sample mean and variance

Suppose that $(Y_1, \ldots, Y_n)$ are independent and identically distributed random variables $N(\mu, \sigma^2)$ or, equivalently,

$$Y = (Y_1, \ldots, Y_n)^\top \sim N_n(\mu 1_n, \sigma^2 I_n).$$

Denote by

$$\overline{Y} = n^{-1}Y^\top 1_n = \frac{1}{n}\sum_{i=1}^{n} Y_i,$$

$$S^2 = (n-1)^{-1}(Y - 1_n\overline{Y})^\top(Y - 1_n\overline{Y}) = \frac{1}{n-1}\sum_{i=1}^{n}(Y_i - \overline{Y})^2$$

the sample mean and the corrected sample variance. We wish to show that $\overline{Y}$ and S^2 are independent random variables such that

$$\overline{Y} \sim N(\mu, \sigma^2/n), \qquad S^2 \sim \frac{\sigma^2}{n-1}\chi^2_{n-1}.$$

Consider any orthogonal matrix A with the entries of the first row all equal to $1/\sqrt{n}$. For instance, we could choose the Helmert matrix A, given by

$$\begin{pmatrix} \frac{1}{\sqrt{n}} & \frac{1}{\sqrt{n}} & \frac{1}{\sqrt{n}} & \frac{1}{\sqrt{n}} & \cdots & \frac{1}{\sqrt{n}} \\ \frac{1}{\sqrt{1\times2}} & -\frac{1}{\sqrt{1\times2}} & 0 & 0 & \cdots & 0 \\ \frac{1}{\sqrt{2\times3}} & \frac{1}{\sqrt{2\times3}} & -\frac{2}{\sqrt{2\times3}} & 0 & \cdots & 0 \\ \frac{1}{\sqrt{3\times4}} & \frac{1}{\sqrt{3\times4}} & \frac{1}{\sqrt{3\times4}} & -\frac{3}{\sqrt{3\times4}} & \cdots & 0 \\ \vdots & \vdots & \vdots & \vdots & \ddots & \vdots \\ \frac{1}{\sqrt{n(n-1)}} & \frac{1}{\sqrt{n(n-1)}} & \frac{1}{\sqrt{n(n-1)}} & \frac{1}{\sqrt{n(n-1)}} & \cdots & -\frac{n-1}{\sqrt{n(n-1)}} \end{pmatrix}$$

Then

$$Z = AY \sim N_n(\mu_z, \sigma^2 I_n)$$

where $\mu_z = (\mu\sqrt{n}, 0, \ldots, 0)^\top$. In addition

$$Z_2^2 + \cdots + Z_n^2 = Z^\top Z - Z_1^2 = \sum_{j=1}^{n} Y_j^2 - n\overline{Y}^2 = (n-1)S^2$$

and then

$$\frac{1}{\sigma^2}\left(\sum_{j=2}^{n} Z_j^2\right) = \frac{n-1}{\sigma^2}S^2 \sim \chi^2_{n-1}.$$

Since the Z_j components are independent, then S^2 is independent of $Z_1 = \sqrt{n}\,\overline{Y}$.

A.5.8 Non-central chi-squared distribution

If $Z = (Z_1, \ldots, Z_k)^\top \sim N_k(\mu, I_k)$, we say that

$$U_k = Z^\top Z = \sum_{i=1}^{k} Z_i^2$$

is a non-central chi-squared random variable with k degrees of freedom and non-centrality parameter $\delta = \mu^\top \mu$, and write

$$U_k \sim \chi_k^2(\delta),$$

since it can be shown that the distribution of U_k depends on μ only through δ. If $\delta = 0$, we return to the central chi-squared case.

Theorem A.5.4 *If* $Y \sim N_k(\mu, V)$ *where* $V > 0$, *then*

$$Q = Y^\top V^{-1} Y = \|Y\|_V^2 \sim \chi_k^2(\mu^\top V^{-1} \mu).$$

Proof. Decomposing V as $V = B B^\top$, we can write

$$Q = Y^\top (BB^\top)^{-1} Y = (B^{-1}Y)^\top B^{-1} Y = Z^\top Z,$$

where $Z = B^{-1} Y \sim N_k(B^{-1}\mu, I_k)$. Then $Z^\top Z \sim \chi_k^2(\delta)$, where

$$\delta = (B^{-1}\mu)^\top B^{-1}\mu = \mu^\top V^{-1}\mu.$$

$\square$

This theorem says that the squared Mahalanobis distance of a normal random variable from the origin is a non-central chi-squared variate. It is clear that

$$\|Y - \mu\|_V^2 \sim \chi_k^2,$$

i.e the squared distance from the mean value is a central chi-square.

Theorem A.5.5 (Fisher–Cochran) *Suppose that* $Y \sim N_k$ (μ, I_k), *and that the positive semidefinite matrices* $A_1, \ldots, A_m$ *with respective ranks* $r_1, \ldots, r_m$ *are such that*

$$A_1 + A_2 + \cdots + A_m = I_k.$$

Then each of the following statements implies the other:

(i) *The quadratic forms* $Q_j = Y^\top A_j Y$ $(j = 1, \ldots, m)$ *are independent* $\chi_{r_j}^2(\mu^\top A_j \mu)$ *random variables.*

(ii) $r_1 + r_2 + \cdots + r_m = k$.

Proof. We show first that (i) implies (ii). In fact

$$\begin{aligned} Y^\top Y &= Y^\top (A_1 + A_2 + \cdots + A_m) Y \\ &= Q_1 + \cdots + Q_m \\ &\sim \chi_r^2(\delta) \end{aligned}$$

where

$$r = \sum_j r_j, \quad \delta = \mu^\top \left(\sum_{j=1}^m A_j \right) \mu = \mu^\top \mu.$$

On the other hand, $Y^\top Y \sim \chi_k^2(\mu^\top \mu)$, and then $r = k$.

Next, we show that (ii) implies (i). Write $A_j = B_j B_j^\top$, where B_j is a $k \times r_j$ matrix; this is possible by taking the square root of A_j, as explained at the end of section A.4.1, and then deleting the columns associated with null eigenvalues. Then $B = (B_1, \ldots, B_m)$ is a square matrix, because of assumption (ii), and

$$BB^\top = B_1 B_1^\top + \cdots + B_m B_m^\top = A_1 + \cdots + A_m = I_k$$

showing that B is orthogonal. Then $Z = B^\top Y \sim N_k(B^\top \mu, I_k)$, and

$$Q_j = Y^\top A_j Y = Y^\top B_j B_j^\top Y = Z_j^\top Z_j$$

where $Z^\top = (Z_1^\top, \ldots, Z_m^\top)$. The Z_j random variables are independent $N_{r_j}(B_j^\top \mu, I_{r_j})$, and the Q_j are mutually independent $\chi_{r_j}^2(\mu^\top A_j \mu)$ for $j = 1, \ldots, m$. $\qquad\square$

Corollary A.5.6 *If* $Y \sim N_k(\mu, I_k)$ *and* A *is a symmetric idempotent matrix of order* k, *then* $Y^\top A Y \sim \chi_r^2(\mu^\top A \mu)$, *where* $r = \text{rk}(A) = \text{tr}(A)$.

Proof. Apply the Fisher–Cochran theorem to the case $m = 2$ with $A_1 = A$, $A_2 = I_k - A$, which satisfy the required condition since

$$\text{rk}(A_1) + \text{rk}(A_2) = \text{tr}(A) + \text{tr}(I_k - A) = \text{tr}(I_k) = k.$$

$\qquad\square$

Remark A.5.7 We shall more often use the Fisher–Cochran theorem in the case $\text{var}[Y] = \sigma^2 I_k$ than for $\text{var}[Y] = I_k$. The theorem is then applied to the transformed variable $X = \sigma^{-1} Y$, and, if the required conditions are fulfilled, $X^\top A_j X \sim \chi^2(\mu^\top A_j \mu / \sigma^2)$, which means $Y^\top A_j Y \sim \sigma^2 \chi^2(\mu^\top A_j \mu / \sigma^2)$. $\qquad\square$

Example A.5.8 Suppose $Y \sim N_n(\mu 1_n, \sigma^2 I_n)$, as in section A.5.7, and

$$
\begin{aligned}
A_1 &= \left(I_n - \frac{1}{n} 1_n 1_n^\top\right)^\top \left(I_n - \frac{1}{n} 1_n 1_n^\top\right) \\
&= I_n - \frac{1}{n} 1_n 1_n^\top, \\
A_2 &= I_n - A_1 = \frac{1}{n} 1_n 1_n^\top.
\end{aligned}
$$

The equalities

$$Y^\top A_1 Y = \sum_{j=1}^{n} (Y_j - \overline{Y})^2,$$

$$Y^\top A_2 Y = \frac{\left(\sum_{j=1}^n Y_j\right)^2}{n} = n\overline{Y}^2$$

hold, where

$$\overline{Y} = \frac{1}{n}\sum_{j=1}^n Y_j.$$

Notice that $A_1 Y$ represents the vectors of deviations between Y_j $(j = 1, \ldots, n)$ and $\overline{Y}$. The matrices A_1, A_2 are idempotent, with respective ranks $n - 1$ and 1. Since $A_1 + A_2 = I_n$ and $\mathrm{rk}(A_1) + \mathrm{rk}(A_2) = n$, then

$$Y^\top A_1 Y \sim \sigma^2 \chi^2_{n-1}, \quad Y^\top A_2 Y \sim \sigma^2 \chi^2_1(n\mu^2/\sigma^2)$$

in agreement with the results of section A.5.7. The non-centrality parameter of $Y^\top A_1 Y$ is 0 since $A_1 1_n = 0$.   □

A.5.9 The t and F distributions

In sections A.5.6 and A.5.8, we introduced two probability distributions related to the normal. Here are two more distributions of particular relevance to statistics.

If $Z \sim N(0,1)$ and $U \sim \chi^2_k$, and Z and U are independent, we say that

$$T = \frac{Z}{\sqrt{U/k}} \tag{A.30}$$

has Student's t distribution with k degrees of freedom, and write $T \sim t_k$. The following properties of the t distribution are stated without proof.

1. Its density function is positive over the real axis and symmetric about 0.

2. Its moments are defined up to order $k - 1$.

3. If $k = 1$, the density function of t_k is equal to the Cauchy density function with location parameter 0 and scale parameter 1.

4. As k tends to ∞, the distribution function of T converges to the $N(0,1)$ distribution function; this statement follows easily from the theorems stated in section A.8.2.

If the assumption $Z \sim N(0,1)$ is replaced by the more general assumption $Z \sim N(\delta, 1)$, then we say that T is non-central Student's t, with non-centrality parameter δ, and write $T \sim t_k(\delta)$.

Consider again the quantities $\overline{Y}$ and S^2 introduced in section A.5.7. Then

$$T^* = \frac{\overline{Y}\sqrt{n}}{S}$$

is a Student's t random variable with $n-1$ degrees of freedom, and is a central or non-central t, depending on whether or not μ is zero. In fact, we can write

$$T^* = \frac{\overline{Y}/(\sigma/\sqrt{n})}{\sqrt{S^2/\sigma^2}},$$

where the numerator is $N(\mu\sqrt{n}/\sigma, 1)$ and the denominator is an independent χ^2_{n-1} divided by $n-1$, taking into account the results of section A.5.7. The non-centrality parameter is $\sqrt{n}\mu/\sigma$.

To define the F distribution, let us introduce an additional random variable $V \sim \chi^2_m$, independent of U. Then

$$F^* = \frac{V/m}{U/k}$$

is said to be distributed as an F random variable with m degrees of freedom in the numerator and k degrees of freedom in the denominator; we write $F^* \sim F_{m,k}$. Tables of the percentage points of the F distribution function are available for a large number of pairs (m, k). It can immediately be verified that $T^2 \sim F_{1,k}$.

If V is a non-central χ^2, then the F distribution is also said to be non-central; however, U must always be a central χ^2.

In statistics, the F distribution arises in the following manner. Consider the Fisher–Cochran theorem setting, with the additional assumption

$$A_m \mu = 0.$$

Then $Y^\mathsf{T} A_m Y \sim \chi^2_{r_m}$ with non-centrality parameter $\mu^\mathsf{T} A_m \mu = 0$. Therefore

$$\frac{Y^\mathsf{T} A_j Y/r_j}{Y^\mathsf{T} A_m Y/r_m} \sim F_{r_j, r_m}.$$

A.6 The multinomial distribution

A.6.1 Definition and probability function·

The outcome of a trial may be one of $k+1$ mutually exclusive events $E_0, \ldots, E_k$ with probabilities

$$p_j = \Pr\{E_j\}, \qquad (j = 0, 1, \ldots, k)$$

such that $p_0 + p_1 + \cdots + p_k = 1$. If n independent replicates of the trial take place, denote by Y_j the number of times that E_j occurs $(j = 0, \ldots, k)$. We then say that $Y = (Y_1, \ldots, Y_k)^\top$ is a multinomial random variable with index n and parameter $p = (p_1, \ldots, p_k)^\top$, and write $Y \sim Bin_k(n, p)$.

The notation $Bin_k(n, p)$ emphasizes that Y is a multivariate extension of the binomial distribution to which it reduces if $k = 1$. The frequency Y_0 is not included as a component of Y, since it is determined by the remaining frequencies and the overall total n. Obviously, the choice of Y_0 as 'complement frequency' is arbitrary; any other Y_j could be used with the same role.

The probability distribution of Y is obtained in a similar way to the binomial. Since $Y_0 = n - Y_1 - \cdots - Y_k$, we can write

$$\Pr\{Y = t\} = \Pr\{Y_0 = t_0, \ldots, Y_k = t_k\}$$
$$= \begin{pmatrix} n \\ t_1 \cdots t_k \end{pmatrix} p_0^{t_0} p_1^{t_1} \cdots p_k^{t_k}$$

for a vector $t = (t_1, \ldots, t_k)^\top$ of non-negative integers such that $t_0 = n - t_1 - \cdots - t_k \geq 0$; the symbol

$$\begin{pmatrix} n \\ t_1 \cdots t_k \end{pmatrix} = \frac{n!}{t_0! \, t_1! \cdots t_k!}$$

denotes the **multinomial coefficient**.

A.6.2 Some simple properties

From the very definition of the multinomial random variable, some interesting implications readily follow. Let Y be defined as in the previous subsection.

(a) If $1 \leq r \leq k$, then $(Y_1, \ldots, Y_r)^\top \sim Bin_r(n, (p_1, \ldots, p_r)^\top)$, on setting $Y_0 = n - Y_1 - \cdots - Y_r$, i.e. the marginals of a multinomial variable are themselves also multinomial random variables.

(b) If $X \sim Bin_k(m, p)$ and is independent of Y, then $X + Y \sim Bin_k(n + m, p)$.

(c) If $1 \leq r \leq k$, and Y^* is defined as

$$Y^* = \sum_{j=r}^{k} Y_j, \quad p^* = \sum_{j=r}^{k} p_j$$

then

$$(Y_1,\ldots,Y_{r-1},Y^*)^\top \sim Bin_r(n,(p_1,\ldots,p_{r-1},p^*)^\top).$$

(d) If $1 \le r < k$, then the distribution of $(Y_1,\ldots,Y_r)^\top$ conditional on $Y_{r+1} = t_{r+1},\ldots,Y_k = t_k$ is $Bin_r(n-T,(p_1/\pi,\ldots,p_r/\pi)^\top)$ where

$$T = \sum_{j=r+1}^{k} t_j, \quad \pi = \sum_{j=0}^{r} p_j.$$

Statement (d) is obtained from the computation

$$\Pr\{Y_1 = t_1,\ldots,Y_r = t_r | Y_{r+1} = t_{r+1},\ldots,Y_k = t_k\}$$

$$= \frac{\Pr\{Y_1 = t_1,\ldots,Y_r = t_r, Y_{r+1} = t_{r+1},\ldots,Y_k = t_k\}}{\Pr\{Y_{r+1} = t_{r+1},\ldots,Y_k = t_k\}}$$

$$= \frac{\binom{n}{t_1 \cdots t_k} p_0^{t_0} \cdots p_r^{t_r} p_{r+1}^{t_{r+1}} \cdots p_k^{t_k}}{\binom{n}{t_{r+1} \cdots t_k} \pi^{n-T} p_{r+1}^{t_{r+1}} \cdots p_k^{t_k}}$$

$$= \binom{n-T}{t_1 \cdots t_r} \left(\frac{p_0}{\pi}\right)^{t_0} \left(\frac{p_1}{\pi}\right)^{t_1} \cdots \left(\frac{p_r}{\pi}\right)^{t_r}$$

taking into account that $t_0 + t_1 + \cdots + t_r = n - T$.

The above statements refer to the first r components of Y just for notational simplicity; any other subset of the k components could be considered.

A.6.3 Characteristic function and moments

If $n = 1$, the characteristic function of Y is

$$E\left[\exp\left(i\sum_{j=1}^{k} t_j Y_j\right)\right] = p_0 + \sum_{j=1}^{k} p_j e^{it_j}$$

since Y_j can only take the value 0 or 1, and $\sum_j Y_j \le 1$. For generic n, we obtain

$$E\left[\exp\left(i\sum_{j=1}^{k} t_j Y_j\right)\right] = \left(p_0 + \sum_{j=1}^{k} p_j e^{it_j}\right)^n,$$

taking into account property (b) given above. This expression leads to

$$E[Y_j] = n\,p_j, \qquad \text{var}[Y_j] = n\,p_j(1 - p_j),$$
$$\text{cov}[Y_i, Y_j] = -n\,p_i\,p_j \quad \text{for } i \neq j.$$

A.7 Order statistics

A.7.1 Definition

Let $(Y_1, \ldots, Y_n)$ be independent and identically distributed real-valued random variables whose common distribution function is $F(\cdot)$. A permutation $(Y_{(1)}, \ldots, Y_{(n)})$ of $(Y_1, \ldots, Y_n)$ such that

$$Y_{(1)} \leq \cdots \leq Y_{(n)} \tag{A.31}$$

gives the **order statistics** of $(Y_1, \ldots, Y_n)$, and the jth component $Y_{(j)}$ is called the jth order statistic $(1 \leq j \leq n)$.

In general, there are several permutations fulfilling (A.31), unless all the Y_j take different values, a condition satisfied with probability 1 if the Y_j are continuous random variables.

The **sample median** is a real number which has to its left as many Y_j as there are to its right. In particular, if n is odd, say $n = 2k + 1$ where k is an integer number, the median is $Y_{(k+1)}$. If n is even, $n = 2k$ say, any number between $Y_{(k)}$ and $Y_{(k+1)}$ satisfies the requirement. In this case, it is common usage to set the median equal to $\frac{1}{2}(Y_{(k)} + Y_{(k+1)})$, but there is no definitive reason for this choice. Jackson (1921) proposed a rationally motivated criterion for selecting a single value of the median regardless of whether n is even or odd.

A.7.2 Distribution of the smallest and largest values

To compute the distribution of $Y_{(n)}$ when the random variables $(Y_1, \ldots, Y_n)$ are continuous with common density function $f(\cdot) = F'(\cdot)$, write

$$\begin{aligned}
\Pr\{Y_{(n)} \leq t\} &= \Pr\{Y_{(1)} \leq t, \ldots, Y_{(n)} \leq t\} \\
&= \Pr\{Y_1 \leq t, \ldots, Y_n \leq t\} \\
&= \prod_{j=1}^{n} \Pr\{Y_j \leq t\} \\
&= \{F(t)\}^n
\end{aligned}$$

for any real t; the corresponding density function is

$$n\{F(t)\}^{n-1}f(t).$$

Similarly, for $Y_{(1)}$, write

$$\Pr\{Y_{(1)} > t\} \;=\; \Pr\{Y_1 > t, \ldots, Y_n > t\}$$
$$=\; \{1 - F(t)\}^n$$

and the density function is

$$n\{1 - F(t)\}^{n-1}f(t).$$

A.7.3 Joint distribution of several order statistics

Maintaining the assumption of the previous subsection that the variables are continuous, we wish to obtain the joint density function of $(Y_{(k_1)}, Y_{(k_2)}, \ldots, Y_{(k_m)})$ where $1 \leq k_1 < k_2 < \cdots < k_m \leq n$. For any set of values $t_1, \ldots, t_m$ for which $f(\cdot)$ is continuous, such that $t_1 < \cdots < t_m$, divide the real axis into intervals

$$(-\infty, t_1], (t_1, t_1 + dt_1], (t_1 + dt_1, t_2], \ldots, (t_m, t_m + dt_m], (t_m + dt_m, \infty)$$

with corresponding probabilities

$$F(t_1),\; f(t_1)dt_1 + o(dt_1),\; F(t_2) - F(t_1) + O(dt_1), \ldots,$$
$$f(t_m)dt_m + o(dt_m),\; 1 - F(t_m) + O(dt_m),$$

and compute the probability of having $k_1 - 1$ elements of $(Y_1, \ldots, Y_n)$ in the first interval, 1 element in the second interval, $k_2 - k_1 - 1$ elements in the third interval, $\ldots$, $n - k_m$ elements in the last. Using the multinomial distribution, the required probability is

$$\frac{n!}{(k_1 - 1)!\, 1!\, (k_2 - k_1 - 1)! \cdots 1!\, (n - k_m)!}$$
$$\times F(t_1)^{k_1 - 1}\, f(t_1)\, \{F(t_2) - F(t_1)\}^{k_2 - k_1 - 1} \cdots f(t_m)$$
$$\times \{1 - F(t_m)\}^{n - k_m}\, dt_1 \cdots dt_m$$

$$(\mathrm{A}.32)$$

up to terms of order smaller than $dt_1 dt_2 \cdots dt_m$. Therefore, after removing the differentials, (A.32) gives the density function, computed at $(t_1, t_2, \ldots, t_m)^{\top}$.

An important special case is given by $m = n$, $k_j = j$ for $j = 1, \ldots, n$, i e the joint distribution of the whole set of order statistics is required. In this case, (A.32) becomes

$$n!\, f(t_1)\, f(t_2)\, \cdots\, f(t_n) \qquad (t_1 < t_2 < \cdots < t_n).$$

Setting instead $m = 1$ and either $k_m = 1$ or $k_m = n$ in (A.32), the density functions of $Y_{(1)}$ and of $Y_{(n)}$ are reobtained.

A.7.4 The sample distribution function

Closely connected to the idea of order statistics is the notion of the sample distribution function (or empirical distribution function), defined as

$$F_n(t) = \frac{1}{n} \sum_{j=1}^{n} I_{(-\infty,\,t]}(Y_j),$$

where $I_A(\cdot)$ is the indicator function of the set A, i e.

$$I_A(x) = \begin{cases} 1 & \text{if } x \in A, \\ 0 & \text{otherwise.} \end{cases} \tag{A.33}$$

In other words, $F_n(t)$ is the proportion of Y_i lying between $-\infty$ and t. The connection with order statistics is given by

$$Y_{(j)} = \inf_{t} \{t : F_n(t) \geq j/n\} \quad \text{for } j = 1,\ldots,n.$$

Moreover, the evocative equality

$$\frac{1}{n} \sum_{j=1}^{n} Y_j^k = \int_{\mathbb{R}} y^k \, dF_n(y)$$

holds, i.e the kth sample moment of $(Y_1,\ldots,Y_n)^\top$ can be regarded as the kth moment of the probability distribution $F_n(\cdot)$. Notice that $nF_n(t) \sim Bin(n, F(t))$, for any fixed value of t, and this implies that

$$E[\,F_n(t)\,] = F(t), \qquad \text{var}[\,F_n(t)\,] = \frac{1}{n} F(t)(1 - F(t)).$$

The final result of this section is stated without proof.

Theorem A.7.1 (Glivenko–Cantelli) *If*

$$D_n = \sup_{-\infty < t < \infty} |F_n(t) - F(t)|$$

then

$$\Pr\left\{ \lim_{n \to \infty} D_n = 0 \right\} = 1.$$

A.8 Sequences of random variables

The next two subsections simply recall basic definitions and the most commonly used results on sequences of random variables, while the remaining part of this section develops in greater detail some additional material which is not commonly included in standard probability textbooks, but which is useful in asymptotic statistical theory.

A.8.1 Definitions

Denote by $\{X_n\} = (X_1, X_2, \ldots, X_n, \ldots)$ an infinite sequence of random variables and let X denote an additional random variable. All these random variables are defined on the same probability space.

(i) The sequence $\{X_n\}$ converges **almost surely** (or with probability 1) to X if

$$\Pr\left\{\lim_{n\to\infty} X_n = X\right\} = 1,$$

and we write $X_n \longrightarrow X$ (a.s.).

(ii) The sequence $\{X_n\}$ converges **in probability** to X if, for all $\varepsilon > 0$,

$$\lim_{n\to\infty} \Pr\left\{|X_n - X| < \varepsilon\right\} = 1,$$

and we write $X_n \xrightarrow{p} X$.

(iii) If $F_n(\cdot)$ is the distribution function of X_n $(n = 1, 2, \ldots)$ and $F(\cdot)$ is the distribution function of X, and the equality

$$\lim_{n\to\infty} F_n(t) = F(t)$$

holds for all real t where $F(\cdot)$ is continuous, then $\{X_n\}$ converges **in distribution** to X, and we write $X_n \xrightarrow{d} X$.

This definition also holds when not all random variables are defined on the same probability space. With slight abuse of terminology, we often indicate only the distribution of X, for instance by writing $X_n \xrightarrow{d} N(0,1)$.

A.8.2 Theorems

The following results are stated without proof; proofs which can be found in many probability textbooks.

(a) If $X_n \longrightarrow X$ (a.s.), then $X_n \xrightarrow{p} X$.

(b) If $\lim\limits_{n\to\infty} \mathrm{E}\big[(X_n - X)^2\big] = 0$, then $X_n \xrightarrow{p} X$.

(c) If $X_n \xrightarrow{p} X$, then $X_n \xrightarrow{d} X$.

(d) If $X_n \xrightarrow{d} c$, where c is a constant, then $X_n \xrightarrow{p} c$.

(e) If $X_n \xrightarrow{d} X$ and $g(\cdot)$ is continuous, then $g(X_n) \xrightarrow{d} g(X)$. Combining this statement with the previous two, we conclude that $X_n \xrightarrow{p} c$ implies $g(X_n) \xrightarrow{p} g(c)$, if g is continuous.

(f) If $X_n \xrightarrow{d} X$ and $Y_n \xrightarrow{d} c$, where c is a constant, then

$$
\begin{aligned}
X_n + Y_n &\xrightarrow{d} X + c, \\
X_n - Y_n &\xrightarrow{d} X - c, \\
X_n Y_n &\xrightarrow{d} Xc, \\
X_n / Y_n &\xrightarrow{d} X/c \quad \text{if } c \neq 0.
\end{aligned}
$$

This theorem still holds if we replace all $\xrightarrow{d}$ symbols by $\xrightarrow{p}$.

(g) If $X_n \xrightarrow{d} X$, then

$$
\lim_{n\to\infty} \mathrm{E}[\exp(itX_n)] = \mathrm{E}[\exp(itX)]
$$

for all real t.

(h) If the sequence of characteristic functions of X_n $(n = 1, 2, \ldots)$ converges (as $n \to \infty$) to the function $c(\cdot)$ continuous at 0, then $c(\cdot)$ is the characteristic function of a random variable X such that $X_n \xrightarrow{d} X$.

(i) **Strong law of large numbers.** If $X_1, X_2, \ldots$ are independent and identically distributed random variables such that $\mathrm{E}[X_1]$ exists, then

$$
\frac{1}{n} \sum_{i=1}^{n} X_i \longrightarrow \mathrm{E}[X_1] \quad \text{(a.s.)}.
$$

(j) **Central limit theorem.** If $X_1, X_2, \ldots$ are independent and identically distributed random variables such that $\mathrm{E}[X_1^2]$ exists, then

$$
\frac{\sum_{i=1}^{n} X_i - n\mathrm{E}[X_1]}{\sqrt{n \operatorname{var}[X_1]}} \xrightarrow{d} N(0, 1).
$$

Example A.8.1 Consider t_k as defined in (A.30). If $k \to \infty$, then $\text{var}[U/k] \to 0$ and, taking into account $E[U/k] = 1$, we conclude $U/k \xrightarrow{p} 1$ using result (b) above. From results (c) and (e), we deduce that $\sqrt{U/k} \xrightarrow{p} 1$, and so we finally have $t_k \xrightarrow{d} N(0,1)$, using the fourth statement of (f). □

A.8.3 Order in probability

If $\{r_n; n = 1, 2, \ldots\}$ and $\{s_n; n = 1, 2, \ldots\}$ are sequences of real numbers, then r_n is said to be of smaller order than s_n, written $r_n = o(s_n)$, if
$$\lim_{n \to \infty} \frac{r_n}{s_n} = 0.$$
If instead
$$\lim_{n \to \infty} \left| \frac{r_n}{s_n} \right| < \infty,$$
then r_n is said to be of order $O(s_n)$, written $r_n = O(s_n)$.

We wish to extend these concepts from sequences of real numbers to sequences of random variables, following Mann and Wald (1943).

Definition A.8.2 *If $\{X_n\}$ is a sequence of random variables and $\{r_n\}$ a sequence of positive real numbers, then we say that*

(i) X_n is in probability of order $o(r_n)$, written $X_n = o_p(r_n)$, if
$$\frac{X_n}{r_n} \xrightarrow{p} 0;$$

(ii) X_n is in probability of order $O(r_n)$, written $X_n = O_p(r_n)$, if for all $\varepsilon > 0$ there exists a real number M_ε such that
$$\Pr\left\{ \left| \frac{X_n}{r_n} \right| > M_\varepsilon \right\} < \varepsilon$$
for all n greater than N_ε, which depends on ε.

Remark A.8.3 If $\{r_n\}$ is a constant sequence, so that we can set $r_n \equiv 1$ without loss of generality, then $X_n = o_p(1)$ is equivalent to $X_n \xrightarrow{p} 0$, and $X_n = O_p(1)$ means that $\{X_n\}$ does not 'explode' as n diverges, in the sense of tending to $\pm\infty$ or increasing its variance indefinitely. □

Example A.8.4 If $X_n \sim U(0, (n \log_2 n)^{-1})$ for $n > 1$, then $\{X_n\} \xrightarrow{p} 0$. We wish to find whether $X_n = o_p(n^{-1})$. Since
$$(X_n/n^{-1}) \sim U(0, (\log_2 n)^{-1}),$$

then
$$\Pr\left\{\left|\frac{X_n}{n^{-1}}\right| > \varepsilon\right\} = \max\{0, 1 - \varepsilon\log_2 n\} \to 0,$$

for all $\varepsilon > 0$. Therefore, $X_n/n^{-1} \xrightarrow{p} 0$ and $X_n = o_p(n^{-1})$. It is clear that $X_n = O_p((n\log_2 n)^{-1})$ because $X_n/(n\log_2 n)^{-1} \sim U(0,1)$, and the condition required by the definition is satisfied by some number $M_\varepsilon > 1 - \varepsilon$. □

Example A.8.5 We wish to show that, if Y_n is an exponential random variable with mean value $3/n + 1/n^2$ $(n = 1, 2, \ldots)$, then $Y_n = O_p(n^{-1})$. In fact, for $\delta > 0$,

$$\Pr\left\{\left|\frac{Y_n}{n^{-1}}\right| > \delta\right\} = \Pr\left\{Y_n > \delta/n\right\} = \exp\{-\delta/(3 + n^{-1})\}$$

and, denoting by ε the rightmost term, we obtain

$$\delta = -(3 + n^{-1})\ln\varepsilon.$$

Therefore, the required condition is satisfied by $M_\varepsilon = -4\ln\varepsilon$. □

Theorem A.8.6 *Let $\{r_n\}$ and $\{s_n\}$ be sequences of positive real numbers and $\{X_n\}$, $\{Y_n\}$ be two sequences of random variables.*

(i) If $X_n = o_p(r_n)$, $Y_n = o_p(s_n)$, then

$$X_n Y_n = o_p(r_n s_n), \quad X_n + Y_n = o_p(\max\{r_n, s_n\}).$$

(ii) If $X_n = O_p(r_n)$, $Y_n = O_p(s_n)$, then

$$X_n Y_n = O_p(r_n s_n), \quad X_n + Y_n = O_p(\max\{r_n, s_n\}).$$

(iii) If $X_n = O_p(r_n)$, $Y_n = o_p(s_n)$, then

$$X_n Y_n = o_p(r_n s_n).$$

Proof. Only part (i) is shown. If $|(X_n Y_n)/(r_n s_n)| > 1$, then at least one of the inequalities $|X_n/r_n| > 1$, $|Y_n/s_n| > 1$ is satisfied. The assumptions imply that, given $\varepsilon > 0$, $\delta > 0$, there exists N_ε such that

$$\Pr\left\{\left|\frac{X_n}{r_n}\right| > \varepsilon\right\} < \delta, \qquad \Pr\left\{\left|\frac{Y_n}{s_n}\right| > \varepsilon\right\} < \delta,$$

for all $n > N_\varepsilon$. Therefore

$$\Pr\left\{\left|\frac{X_n Y_n}{r_n s_n}\right| > \varepsilon^2\right\} \leq \Pr\left\{\left\{\left|\frac{X_n}{r_n}\right| > \varepsilon\right\} \cup \left\{\left|\frac{Y_n}{s_n}\right| > \varepsilon\right\}\right\}$$

$$\leq \quad \Pr\left\{\left|\frac{X_n}{r_n}\right| > \varepsilon\right\} + \Pr\left\{\left|\frac{Y_n}{s_n}\right| > \varepsilon\right\}$$

$$< \quad 2\delta$$

for all $n > N_\varepsilon$, hence showing the first statement of (i).

For the second statement, put $q_n = \max\{r_n, s_n\}$. Then, with the same $\varepsilon, \delta, N_\varepsilon$ as used earlier, we have

$$\Pr\left\{\left|\frac{X_n}{q_n}\right| > \varepsilon\right\} < \delta, \qquad \Pr\left\{\left|\frac{Y_n}{q_n}\right| > \varepsilon\right\} < \delta,$$

so that

$$\Pr\left\{\left|\frac{X_n + Y_n}{q_n}\right| > 2\varepsilon\right\} \quad \leq \quad \Pr\left\{\left|\frac{X_n}{q_n}\right| + \left|\frac{Y_n}{q_n}\right| > 2\varepsilon\right\}$$

$$\leq \quad \Pr\left\{\left\{\left|\frac{X_n}{q_n}\right| > \varepsilon\right\} \cup \left\{\left|\frac{Y_n}{q_n}\right| > \varepsilon\right\}\right\}$$

$$\leq \quad \Pr\left\{\left|\frac{X_n}{q_n}\right| > \varepsilon\right\} + \Pr\left\{\left|\frac{Y_n}{q_n}\right| > \varepsilon\right\}$$

$$< \quad 2\delta$$

for all $n > N_\varepsilon$. This completes the proof of (i); the proof of (ii) and (iii) are similar. $\square$

In practice, this theorem says that the rules for sums and products of sequences of random variables are the same rules used for ordinary sequences of reals.

Since it is not generally easy to establish the order in probability of a sequence of random variables, the next theorem is useful, because it involves only the sequence of second moments.

Theorem A.8.7 *Let $\{X_n\}$ be a sequence of random variables with $\mathrm{E}\left[X_n^2\right] = r_n^2 < \infty$, and $\{s_n\}$ a sequence of positive reals. Then*

$$(i) \quad \text{if } r_n^2 = O(s_n^2), \quad X_n = O_p(s_n);$$
$$(ii) \quad \text{if } r_n^2 = o(s_n^2), \quad X_n = o_p(s_n).$$

Proof. Consider first case (i). The assumptions imply that there exists a positive real a such that $\mathrm{E}\left[X_n^2\right] < as_n^2$ for all n. By Theorem A.1.1, we can write

$$\Pr\left\{\left|\frac{X_n}{s_n}\right| \geq t\right\} \quad = \quad \Pr\left\{X_n^2 \geq t^2 s_n^2\right\}$$

$$\leq \quad \frac{r_n^2}{t^2 s_n^2} < \frac{a}{t^2}$$

for positive t. On choosing $\varepsilon = a/t^2$, $M_\varepsilon = \sqrt{a/\varepsilon}$, the conclusion follows.

For part (ii), the argument is the same except that the ratio r_n^2/s_n^2 is now $o(1)$; hence $X_n/s_n \xrightarrow{p} 0$. $\qquad\square$

Theorem A.8.8 *Let $\{X_n\}$ be a sequence of random variables and $\{r_n\}$ a sequence of reals such that*

$$r_n(X_n - c) \xrightarrow{d} Z,$$

where c is a constant and Z a non-degenerate random variable. Then

$$X_n = c + O_p(r_n^{-1}).$$

Proof. Denoting by $F(\cdot)$ the distribution function of Z and by t any real number where F is continuous, there exists a positive sequence $\{s_n\}$ converging to 0, such that

$$G(t) - s_n < \Pr\{|r_n(X_n - c)| > t\} < G(t) + \varepsilon \qquad (\text{A.34})$$

where $G(t) = 1 - F(t) + F(t^-) = \Pr\{|Z| > t\}$. Since Z is not degenerate, we can find a suitable positive value t_ε such that $G(t_\varepsilon) < \varepsilon$, for any given positive ε. Moreover, since s_n tends to 0, there exists n_0 such that, for $n > n_0$,

$$\Pr\{|r_n(X_n - c)| > t_\varepsilon\} < G(t_\varepsilon) + s_n < \varepsilon$$

which is the required condition for writing $X_n = c + O_p(r_n^{-1})$. Notice that, on applying the same argument to $G(t) - s_n$ in (A.34), it follows that $X_n - c$ is not $o_p(r_n^{-1})$. $\qquad\square$

A.8.4 Stochastic series expansions

Theorem A.8.9 *Let $\{X_n\}$ be a sequence of random variables such that*

$$X_n = c + O_p(r_n),$$

where c is a constant and $\{r_n\}$ is a sequence of reals converging to 0. If $f(\cdot)$ has k continuous derivatives, then

$$f(X_n) = f(c) + f'(c)(X_n - c) + \cdots + \frac{f^{(k)}(c)}{k!}(X_n - c)^k + o_p(r_n^k).$$

Proof. Expand $f(x)$ as a Taylor series at $x = c$ up to the kth term, with remainder term

$$\frac{(X_n - c)^k}{k!}\left(f^{(k)}(Z_n) - f^{(k)}(c)\right),$$

where Z_n lies between c and X_n. Since $Z_n \xrightarrow{p} c$ and $f^{(k)}(\cdot)$ is continuous, then $f^{(k)}(Z_n) - f^{(k)}(c) = o_p(1)$, while $(X_n - c)^k = O_p(r_n^k)$, and by Theorem A.8.6 their product is $o_p(r_n^k)$. $\qquad \square$

Corollary A.8.10 *Let $\{X_n\}$ be a sequence of random variables and c a constant such that $\sqrt{n}(X_n - c) \xrightarrow{d} Z$, where Z is a non-degenerate random variable. If $f(\cdot)$ is a function with continuous derivative, then*

$$\sqrt{n}\{f(X_n) - f(c)\} \xrightarrow{d} f'(c)\, Z.$$

This result is sometimes referred to as the **delta method** for the computation of the asymptotic distribution of $f(X_n)$.

Example A.8.11 Let $\{X_n\}$ be a sequence of independent and identically distributed Poisson random variables with common mean value λ. Denote by $\{\overline{X}_n\}$ the corresponding sequence of successive arithmetic means, i e

$$\overline{X}_n = \frac{1}{n}\sum_{i=1}^{n} X_i.$$

The central limit theorem implies that

$$\sqrt{n}(\overline{X}_n - \lambda) = \frac{\sum_{i=1}^{n} X_i - n\lambda}{\sqrt{n}}$$

converges in distribution to $N(0, \lambda)$. By applying the above corollary to $f(x) = \sqrt{x}$, with $c = \lambda$, we obtain that

$$\sqrt{n}\left(\sqrt{\overline{X}_n} - \sqrt{\lambda}\right)(2\sqrt{\lambda}) \xrightarrow{d} N(0, \lambda)$$

so that

$$\sqrt{n}\left(\sqrt{\overline{X}_n} - \sqrt{\lambda}\right) \xrightarrow{d} N(0, 1/4).$$

In this case, the square-root transformation is such that, to the first order of approximation, the distribution of $f(\overline{X}_n) - f(c)$ does not depend on λ, while $\text{var}[\overline{X}_n]$ does depend on λ. We say that the transformation has 'stabilized the variance'. Notice that the above relationship can be written as

$$\sqrt{Y} - \sqrt{n\lambda} \xrightarrow{d} N(0, 1/4)$$

where $Y = \sum X_i$ has mean value $n\lambda$. This implies that we can transform a Poisson random variable Y with large mean value to a random variable $\sqrt{Y}$ with approximate variance $1/4$, independent of the mean value.   □

Remark A.8.12 Corollary A.8.10 deals with the specific sequence $r_n = 1/\sqrt{n}$. This restriction is by no means essential, and the results could be reformulated in more general terms. However, the sequence $r_n = 1/\sqrt{n}$ is by far the most relevant in statistics, as the statement of the central limit theorems suggests, and as demonstrated by various asymptotic results in Chapters 3 and 4.   □

Remark A.8.13 Under the assumption of Theorem A.8.9, suppose that $E[X_n]$ exists and is constant, $E[X_n] = \mu$ say. Restricting ourselves to the case $r_n = 1/\sqrt{n}$ (just for the sake of simplicity), we write

$$f(X_n) = f(\mu) + f'(\mu)(X_n - \mu) + \tfrac{1}{2}f''(\mu)(X_n - \mu)^2 + o_p(1/n^2).$$

Application of the expectation and the variance operator to both sides of this equality suggests the following approximate expressions of the mean value and of the variance:

$$\begin{aligned} E[f(X_n)] &= f(\mu) + \tfrac{1}{2}f''(\mu)\,\mathrm{var}[X_n] + o(1/n) \\ \mathrm{var}[f(X_n)] &= f'(\mu)^2\,\mathrm{var}[X_n] + o(1/n) \end{aligned}$$

where $\mathrm{var}[X_n] = O(1/n)$. A rigorous proof of these relationships would require some additional regularity conditions and a more elaborate argument, which we shall not pursue here.   □

Exercises

A.1 Show that (A.4) implies (A.3).

A.2 By applying the definition of the mean value, check that the mean value of the Cauchy distribution does not exist.

A.3 For the geometric distribution, formulate a memoryless property similar to that of the exponential distribution.

A.4 Consider a random variable Y with density function at t

$$f_r(t; \theta) = \exp\left(-\frac{1}{r}|t|^r\right) / c_r \qquad (-\infty < t < \infty)$$

where r is a positive real and c_r a suitable normalizing constant.

  (a) Determine c_r.

(b) Which type of density function do we obtain when $r = 1, r = 2, r \to \infty$?

(c) Show that $E[Y^m]$ exists for any positive integer m, and compute this expectation.

A.5 Let $(Y_1, \ldots, Y_n)$ be independent and identically distributed negative exponential random variables with scale parameter λ.

(a) Obtain the joint density function of the first r components of the order statistics $(Y_{(1)}, \ldots, Y_{(r)})$ for $1 < r \le n$.

(b) Show that the random variables

$$
\begin{aligned}
V_1 &= n Y_{(1)} \\
V_2 &= (n-1)(Y_{(2)} - Y_{(1)}) \\
&\vdots \\
V_r &= (n-r+1)(Y_{(r)} - Y_{(r-1)})
\end{aligned}
$$

are independent and identically distributed $Gamma$ $(1, \lambda)$.

A.6 Prove the statement of section A.4.2 that the existence of the diagonal elements of the variance matrix implies the existence of the off-diagonal elements.

A.7 Suppose $Z \sim N(0,1)$ and define

$$
Y = \begin{cases} -Z & \text{if } |Z| < c, \\ Z & \text{if } |Z| \ge c, \end{cases}
$$

where c is a positive constant.

(a) Show that $Y \sim N(0,1)$ for all choices of c.

(b) What is the support of (Y, Z)?

(c) Calculate the correlation between Z and Y.

(d) Show that, for a suitably chosen c, this correlation is 0, and comment on this result.

A.8 Consider a bivariate normal variable $Y = (Y_1, Y_2)^\top$ with standardized marginal and correlation coefficient ρ. Determine the ellipse in the (y_1, y_2) plane having probability $1 - \alpha$ and minimum area.

A.9 If $X \sim N_k(\mu, \Omega_Y)$, $Y = X + \varepsilon$, where $\varepsilon \sim N_k(0, \Omega_\varepsilon)$, obtain the distribution of X given $Y = y$, for $y \in \mathbb{R}^k$.

A.10 If $(Y_0, Y_1, \ldots, Y_r)$ are independent Poisson variables with respective parameters $(\mu_0, \mu_1, \ldots, \mu_r)$, show that the distribution of $(Y_1, \ldots, Y_r)$ conditional on $\sum_{j=0}^{r} Y_j = t$ is multinomial with index t and parameter $(\pi_1, \ldots, \pi_r)$ where $\pi_k = \mu_k / \sum_{j=0}^{r} \mu_j$ for $k = 1, \ldots, r$.

A.11 Denote by $(Y_1, \ldots, Y_n)$ continuous independent and identically distributed random variables with common distribution function $F(\cdot)$ and let $(Y_{(1)}, \ldots, Y_{(n)})$ be the corresponding order statistics. Determine the distribution of the **range** $R = Y_{(n)} - Y_{(1)}$.

A.12 Complete the proof of Theorem A.8.6, parts (ii) and (iii).

A.13 If $Y \sim Bin(n, p)$, where $0 < p < 1$, compute the asymptotic distribution up to order $O_p(n^{-1/2})$ of:

  (a) $\log\{(Y + \frac{1}{2})/(n - Y + \frac{1}{2})\}$, called the **empirical logit**;
  (b) $\arcsin \sqrt{Y/n}$

as $n \to \infty$.

Main abbreviations and symbols

a.s.	almost surely
IRWLS	iteratively reweighted least squares
GLM	generalized linear model
LRT	likelihood ratio test (function)
LSE	least-squares estimate
MLE	maximum likelihood estimate
MS	mean of squares
s r s	simple random sample
SS	sum of squares

$\mathbb{R}$	the set of real numbers
I_n	the identity matrix of order n
$I_A(\cdot)$	the indicator function of set A
E[ ]	expected value, mean value
var[ ]	variance (matrix)
cov[ ]	covariance
corr[ ]	correlation
$\Pr\{A\}$, $\Pr\{A;\theta\}$	probability of event A (depending on θ)
$\sim$	is distributed as
$\rightsquigarrow$	is approximately distributed as
$\xrightarrow{p}$	converges in probability (see section A.8.1)
$\xrightarrow{d}$	converges in distribution (see section A.8.1)
O_p, o_p	order in probability (see section A.8.3)
$N(\mu,\sigma^2)$	the normal distribution with mean value μ and variance σ^2 (see section A.2.1)
$N_k(\mu,\Omega)$	the k-dimensional normal distribution with mean value μ and variance matrix Ω (see section A.5)
$\phi(\cdot)$	the $N(0,1)$ density function (see (A.7))
$\Phi(\cdot)$	the $N(0,1)$ distribution function (see (A.7))
$U(a,b)$	the uniform distribution on the interval (a,b) (see section A.2.1)

$Gamma(\omega, \lambda)$	the gamma distribution with index ω and scale parameter λ (see section A.2.3)
$IG(\mu, \lambda)$	the inverse Gaussian distribution with mean parameter μ and scale parameter λ (see section A.2.6)
$\chi_k^2(\delta)$	the chi-squared distribution with k degrees of freedom and non-centrality parameter δ (see sections A.5.6 and A.5.8)
$Bin(n, p)$	the binomial distribution with index n and parameter p (see section A.3.1)
$Bin_k(n, p)$	the k-dimensional multinomial distribution with index n and parameter vector p (see section A.6)
$Poisson(\lambda)$	the Poisson distribution with mean parameter λ (see section A.3.3)
$\mathcal{Y}$	sample space (see section 2.1)
Θ	parameter space (see section 2.1)
$\mathcal{F}$	statistical model (see section 2.1)
$L(\theta)$, $L(\theta; y)$	likelihood function for the parameter θ
$\ell(\theta)$	log-likelihood function
$\hat{\theta}, \hat{\psi}, \ldots$	maximum likelihood estimate of $\theta, \psi, \ldots$
$I(\theta)$	expected information (matrix)
$\mathcal{I}(\hat{\theta})$	observed information (matrix)
$\lambda(y)$	likelihood ratio test statistic
W, W_e, W_u	test statistics related to the likelihood (see section 4.2.2)

Answers to selected exercises

2.2 $\mathcal{Y} = \mathbb{R}^+ \times \mathbb{R}^+$, $A_t = \{(x,t) : x \in (0,t)\} \cup \{(t,y) : y \in (0,t)\}$
for $t \in \mathbb{R}^+$.

2.3 (a) $\mathbb{R}^n$. (b) $L(\theta) = ce^{n\theta} I_{(-\infty, y_{(1)})}(\theta)$.

2.4 If we write $f(y; \theta) = f(y; \theta_0)g(u(y), \theta)$, this distribution is of type (2.4) since $f(y; \theta_0)$ does not depend on θ.

2.7 (a) $c_\theta = -\ln(1 - \theta)$.

2.10 The exponential family is not regular since the minimal sufficient statistic $(\sum y_i, \sum y_i^2)$ has dimension 2, while θ is a scalar.

2.11 $E[T] = \text{var}[T] = \theta \sum_i x_i$, with canonical parameter $\ln \theta$.

2.13 (b) Since $\cos(t - \alpha) = \cos t \cos \alpha + \sin t \sin \alpha$, we can write

$$f(t; \kappa, \alpha) = \frac{1}{2\pi} \exp\{\psi_1(\kappa, \alpha) \cos t + \psi_2(\kappa, \alpha) \sin t - \ln I_0(\kappa)\}$$

where $\psi_1(\kappa, \alpha) = \kappa \cos \alpha$, $\psi_2(\kappa, \alpha) = \kappa \sin \alpha$, showing the structure of an exponential family of order 2.

2.14 The order is $k + k(k+1)/2$. The minimal sufficient statistic is given by $\sum_i y_i$ and by the upper (or lower) triangle of $\sum_i y_i y_i^\mathsf{T}$, including the diagonal elements.

3.1 If m_1, m_2 denote the first two sample moments, then $\tilde{\omega} = m_1^2/v$, $\tilde{\lambda} = m_1/v$ where $v = m_2 - m_1^2$.

3.6 (a) No. (b) and (c) See Feller (1968, pp. 45–47).

3.7

$$\begin{pmatrix} \dfrac{n - 1 - (n-3)\rho^2}{(1 - \rho^2)^2} & \dfrac{\rho}{\sigma^2(1 - \rho^2)} \\ \dfrac{\rho}{\sigma^2(1 - \rho^2)} & \dfrac{n}{2\sigma^2} \end{pmatrix}$$

3.8 If $T_x = \sum_i x_i$ and $T_y = \sum_i y_i$, then

$$L(\theta) = c \frac{\theta^{T_x + T_y} e^{-m\theta}}{(1 + \theta)^{T_y + n}},$$

$$\ell'(\theta) = -m + \frac{T_x + T_y}{\theta} - \frac{T_y + n}{1 + \theta}.$$

The likelihood equation is

$$m\theta^2 - (T_x - n - m)\theta - (T_x + T_y) = 0$$

whose left-hand side is non-positive at $\theta = 0$, and goes to ∞ as $\theta \to \pm\infty$. Therefore, the equation has one and only one solution in $(0, \infty)$, except for the case of all sample elements equal to 0.

3.9 (c) Differentiability of $f(\theta)$ equivalent to differentiability of $f(\theta) = \sum_r \theta^r a_r$, which is a power series whose derivative exists on the set where the series itself converges.

   (d) The likelihood is

$$L(\theta) = c\, \frac{\theta^T}{f(\theta)^n}$$

where $T = \sum_i y_i$, hence

$$\ell(\theta) = c + T \ln \theta - n \ln f(\theta),$$

and the likelihood equation is

$$0 = T/\theta - n f'(\theta)/f(\theta)$$

which is equivalent to the equation in the text.

   (e) The mean value and variance of Y can be computed by using (2.7) and (2.8), or by direct computation, i e

$$\mathrm{E}[Y] = \frac{\theta}{f(\theta)} \sum_{r=1}^{\infty} r\theta^{r-1} a_r = \frac{\theta f'(\theta)}{f(\theta)},$$

$$\mathrm{E}[Y(Y-1)] = \frac{\theta^2}{f(\theta)} \sum_{r=2}^{\infty} r(r-1)\theta^{r-2} a_r = \frac{\theta^2 f''(\theta)}{f(\theta)},$$

$$\mathrm{var}[Y] = \frac{\theta^2 f''(\theta)}{f(\theta)} + \frac{\theta f'(\theta)}{f(\theta)} - \left(\frac{\theta f'(\theta)}{f(\theta)}\right).$$

To obtain the expected information, write

$$-\ell''(\theta) = \frac{T}{\theta^2} + n\, \frac{f''(\theta) f(\theta) - (f'(\theta))^2}{f(\theta)^2},$$

and compute its expectation, with $\mathrm{E}[T]$ replaced by $n\,\mathrm{E}[Y]$, obtaining $I(\theta) = n\,\mathrm{var}[Y]/\theta^2$.

3.12 (a) $y_{(1)}$

(b) Since the integral of the density function is 1, it follows that

$$k(\theta) = \frac{1}{\int_\theta^\infty g(y)\,dy}$$

whose denominator is not increasing with respect to θ.

(c) $y_{(1)}$

3.15 The estimates of the mean values and of the variances are given by the corresponding sample means and variances, while the MLE of ρ is given by the *sample correlation*

$$\hat{\rho} = r = \frac{\sum(x_i - \bar{x})(y_i - \bar{y})}{\sqrt{\sum(x_i - \bar{x})^2 \sum(y_i - \bar{y})^2}}.$$

The asymptotic distribution of the estimates has

$$\begin{aligned}
\text{var}\big[\hat{\sigma}_i^2\big] &\sim 2\sigma^4/n \quad (i = 1, 2),\\
\text{var}[\hat{\rho}] &\sim (1 - \rho^2)^2/n,\\
\text{corr}\big[\hat{\sigma}_1^2, \hat{\sigma}_2^2\big] &\sim \rho^2,\\
\text{corr}\big[\hat{\rho}, \hat{\sigma}_i^2\big] &\sim \rho/\sqrt{2} \quad (i = 1, 2).
\end{aligned}$$

3.16 The MLE $\hat{\sigma}$ of σ is biased, but its bias can be removed by multiplying the estimate by a suitable scale factor, obtaining

$$\tilde{\sigma} = \sqrt{\sum_i (y_i - \bar{y})^2} \, \frac{\Gamma(\frac{n-1}{2})}{\sqrt{2}\,\Gamma(\frac{n}{2})} \approx \sqrt{\frac{\sum_i (y_i - \bar{y})^2}{n - 1.45}}$$

where $\bar{y} = \sum_i y_i/n$.

3.18 Assuming, without loss of generality that the elements of x are in increasing order, the required condition is that y contains at least a 0 with a 1 on each side, and correspondingly at least a 1 with a 0 on each side; see Silvapulle (1981).

3.19 (a) $(n_1 + n_2, n_3)$

(b) The MLE is the positive root of the equation

$$n\theta^2 - \{n_0 - 2(n_1 + n_2) - n_3\}\theta - 2n_3 = 0.$$

For additional aspects and a discussion of the biological background, see Fisher (1958, pp. 299ff.).

4.4 The MLE is $\hat{\theta} = n/Q$, and the LRT statistic is

$$W = -2\ln\lambda = 2(c - n\ln Q + Q)$$

where c is a constant. As a function of Q, W is first decreasing and then increasing; therefore, the rejection region at level α is given by the interval (q_1, q_2), where q_1 and q_2 are such that

$$\Pr\{Q \in (q_1, q_2)\} = 1 - \alpha$$

under H_0, and in addition

$$-n\ln q_1 + q_1 = -n\ln q_2 + q_2.$$

To avoid solving these nonlinear equations, it is customary to replace them by

$$\Pr\{Q < q_1\} = \Pr\{Q > q_2\} = \alpha/2.$$

For computing the distribution of the test statistic, notice that $-\ln Y_i \sim Gamma(1, \theta)$, implying $Q \sim Gamma(n, \theta)$ without approximation; hence $2Q \sim \theta\chi^2_{2n}$. This distribution can be used to compute the power of the test procedure.

4.5 Log-transform the data, and then use Student's t test. The test procedure has an exact significance level since the transformed data are normally distributed

4.6 The MLEs of θ_1, θ_2 are $\hat{\theta}_1 = S_1/n$, $\hat{\theta}_2 = S_2/T$, respectively, where $S_1 = \sum_i y_i$, $S_2 = \sum t_i y_i$ and $T = \sum t_i^2$. The LRT criterion gives

$$W = -2\ln\lambda = \frac{S_1^2}{n} + \frac{S_2^2}{T}.$$

To obtain the distribution of W, notice that (S_1, S_2) has a bivariate normal distribution with

$$\mathrm{E}[S_1] = n\theta_1, \quad \mathrm{E}[S_2] = T\theta_2,$$
$$\mathrm{var}[S_1] = n, \quad \mathrm{var}[S_2] = T,$$
$$\mathrm{cov}[S_1, S_2] = 0.$$

Hence $W \sim \chi^2_2(n\theta_1 + \theta_2 T)$ without approximation.

4.7 The test statistic is t as defined in section 4.4.3, but the rejection region is one-sided, as in section 4.4.2.

4.8 Denote by s_x^2 and s_z^2 the sample variances of the x_i and of the z_i, respectively, and by $s^2 = (n\,s_x^2 + m\,s_z^2)/(n+m)$ the pooled estimate. Then

$$
\begin{aligned}
\lambda(y)^2 &= \frac{(s_x^2)^n\,(s_z^2)^m}{(s^2)^{n+m}} \\[2mm]
&= \left(\frac{s_x^2(n+m)}{ns_x^2 + ms_z^2}\right)^n \left(\frac{s_z^2(n+m)}{ns_x^2 + ms_z^2}\right)^m \\[2mm]
&= \frac{(n+m)^{n+m}}{nm(1 + R\,m/n)(1 + R^{-1}\,n/m)}
\end{aligned}
$$

where $R = s_z^2/s_x^2$. Then $W = -2\ln\lambda(y)$ is a function of R, or equivalently of $F = R(n-1)/(m-1)$, and this function is first decreasing and then increasing. The corresponding test procedure rejects the null hypothesis when F is outside an interval, (c_1, c_2) say. For computing c_1 and c_2, notice that $F \sim F_{m-1,n-1}$ under the null hypothesis. Similarly to Example 4.5.3, c_1 and c_2 are chosen in such a way that each interval $(0, c_1)$ and (c_2, ∞) has probability $\alpha/2$. Under H_1, the F distribution is multiplied by the scale factor σ_z^2/σ_x^2, hence allowing computation of the power of the test procedure.

4.11 [First part.] If $z = \sum_i y_i/(n\,\mu_0)$, then $W = -2n(\ln z + z - 1)$. Using Theorem A.8.9, deduce that $W = n(z-1)^2 + o_p(1)$ and the conclusion follows from the central limit theorem and Theorem A.8.2(e).

4.17 If N denotes the number of sample elements falling in the interval $(-\infty, 1]$, then $N \sim Bin(n, \theta)$, where $\theta = F(1)$. Apply the method described in section 4.4.7 to obtain a confidence interval for θ.

5.1 Using the Lagrange multipliers method, obtain

$$
T = \sum_i \left(\frac{1/v_i}{\sum_j 1/v_j}\right) T_i.
$$

5.2 If $a \in \mathbb{R}^p$, $a \neq 0$, then $a^\top X^\top X a = (Xa)^\top (Xa) = \|Xa\|^2 > 0$, taking into account that $Xa \neq 0$ since the columns of X are linearly independent by the assumption on the rank of X.

5.3 $\sum_{i=1}^n x_i^2 \to \infty$ as $n \to \infty$.

5.6 $(y - \hat{\mu}_0)^\top Xc = y^\top (I - P + P_H)^\top Xc = 0$ taking into account

$P_H X c = 0$ by the assumptions, and $PXc = Xc$ since $Xc \in \mathcal{C}(X)$.

5.7 It is sufficient to show $(P - P_0)^\top P_0 = 0$, i e. $P_H(P - P_H) = 0$. The statement is true since P_H is idempotent and $P_H P = P_H$, as can be checked by direct multiplication.

5.10 A confidence interval of level $1 - \alpha$ is

$$(\bar{z} - \bar{x} \pm t' \, s \, \sqrt{1/n + 1/m})$$

where t' is the value exceeded with probability $\alpha/2$ by t_{n+m-2}, and $s^2 = \mathrm{SS}_{res}/(n + m - 2)$.

5.13 $1_n \in \mathcal{C}(X)$.

5.17 Denote by $P \, y, P_1 \, y, P_2 \, y$ the projections of y onto the linear spaces spanned by X, x_1, x_2, respectively. Moreover, denote by $\bar{P}_1 \, y, \bar{P}_2 \, y$ the two components of Py in directions x_1, x_2, so that $P \, y = \bar{P}_1 \, y + \bar{P}_2 \, y$. Then the following relationships hold:

$$\|\bar{P}_1 y\| + r\|\bar{P}_2 y\| = \|P_1 y\|,$$
$$\|\bar{P}_2 y\| + r\|\bar{P}_1 y\| = \|P_2 y\|,$$

where r is the cosine of the angle formed by x_1 and x_2, i e

$$r = \frac{\pm\|P_1 x_2\|}{\|x_2\|} = \frac{x_1^\top x_2}{\sqrt{(x_1^\top x_1)(x_2^\top x_2)}}.$$

commonly called the *sample correlation* between x_1 and x_2. By writing the length of the projections as the length of the vectors multiplied by suitable regression coefficients, obtain

$$\|x_1\| \, \hat{\beta}_1 + r \, \|x_2\| \, \hat{\beta}_2 = \|x_1\| \, \hat{\beta}_1^*,$$
$$\|x_2\| \, \hat{\beta}_2 + r \, \|x_1\| \, \hat{\beta}_1 = \|x_2\| \, \hat{\beta}_2^*,$$

giving

$$\hat{\beta}_1 = \frac{\hat{\beta}_1^* - \hat{\beta}_2^* r R}{1 - r^2}, \quad \hat{\beta}_2 = \frac{\hat{\beta}_2^* - \hat{\beta}_1^* r / R}{1 - r^2},$$

where $R = \|x_2\|/\|x_1\|$.

5.18 $\hat{\beta} = (X^\top W^{-1} X)^{-1} X^\top W^{-1} y$, $\mathrm{var}\left[\hat{\beta}\right] = (X^\top W^{-1} X)^{-1}$.

5.19 Since $V_n = (X^\top X)^{-1}$ is already available, then $V_{n+1} = (X^\top X + \tilde{x}_{n+1} \tilde{x}_{n+1}^\top)^{-1}$ can be readily computed using (A.25), which gives

$$V_{n+1} = V_n - h \, V_n \tilde{x}_{n+1} \tilde{x}_{n+1}^\top V_n,$$

where $h = 1/(1 + \tilde{x}_{n+1}^{\mathsf{T}} V_n \tilde{x}_{n+1})$. Replacing this expression in the usual LSE formula, we obtain the new estimate

$$
\begin{aligned}
\hat{\beta}_{n+1} &= V_{n+1}(X^{\mathsf{T}}y + \tilde{x}_{n+1}y_{n+1}) \\
&= \hat{\beta}_n + h\,V_n \tilde{x}_{n+1}(y_{n+1} - \tilde{x}_{n+1}^{\mathsf{T}}\hat{\beta}_n)
\end{aligned}
$$

after some algebra. Notice that the new estimate is a linear combination of the old estimate and the *prediction error* $(y_{n+1} - \tilde{x}_{n+1}^{\mathsf{T}}\hat{\beta}_n)$.

A minor variation of this procedure can be used to *delete* one observation $\{y_s, x_s\}$ for some $s \in \{1, \ldots, n\}$, instead of adding one. A procedure of this kind can be used to remove a suspected outlier, or simply to examine the effect of inserting/deleting a suspected outlier.

6.2 Models with multiplicative gamma errors have $Y = (\mu/\omega)\varepsilon$ where ε is distributed as $Gamma(\omega, 1)$, and ω is known. The corresponding density function of Y is as in Example 6.2.6.

6.3 Consider a GLM with normal error distribution and link function $g(t) = r^{-1}(t)$, under the condition that r depends on $x_1, \ldots, x_p$ only through a linear combination of them, $t = x_1\beta_1 + \cdots + x_p\beta_p$ say.

6.4 Formulate a GLM with normal error distribution, identity link function, and weight w_i for the ith observation given by the m_i of Exercise 5.18.

6.5 It should be found that the two curves are almost identical. This is not uncommon: in a number of cases of regression models for binary data, it turns out that the choice of the link function is not critical.

A.4 (a) $c_r = 2\,r^{1/r-1}\,\Gamma(1/r)$

   (b) The standard Laplace, the $N(0,1)$, and the $U(-1,1)$ density function, respectively.

   (c)

$$
\mathrm{E}[\,Y^m\,] = r^{m/r}\,\frac{\Gamma((m+1)/r)}{\Gamma(1/r)}
$$

if m is even, 0 otherwise.

A.6 It is sufficient to prove the statement for a single off-diagonal entry of the matrix. Let $Y_j = (X_j - \mathrm{E}[\,X_j\,])$ for $j = 1, 2$. By the Schwarz inequality, $\{\mathrm{E}[\,Y_1 Y_2\,]\}^2 \le \mathrm{E}[\,Y_1^2\,]\,\mathrm{E}[\,Y_2^2\,]$, which is equivalent to the statement in the text.

A.7 (c)

$$\text{cov}[Y, Z] = -\int_{|t|<c} t^2 \phi(t)\, dt + \int_{|t|>c} t^2 \phi(t)\, dt$$

$$= 1 - 4 \int_0^c t^2 \phi(t)\, dt.$$

(d) This covariance is 0 if c is the solution to the equation

$$\int_{|t|<c} t^2 \phi(t)\, dt = \int_{|t|>c} t^2 \phi(t)\, dt,$$

i.e

$$\int_0^c u^{\frac{1}{2}} e^{-\frac{1}{2}u}\, du = \int_c^\infty u^{\frac{1}{2}} e^{-\frac{1}{2}u}\, du;$$

hence the solution is the median of the χ_3^2 distribution. The exercise shows that one can construct normal uncorrelated but *not* independent random variables; this phenomenon is related to the fact that the pair (Y, Z) is not jointly normal.

A.8 The equation of the ellipse is

$$y_1^2 - 2\rho y_1 y_2 + y_2^2 = -2(1 - \rho^2) \ln \alpha,$$

obtained using Theorem A.5.4 and expression (A.28).

A.9 The joint distribution of (X, Y) is multivariate normal, namely

$$\begin{pmatrix} X \\ Y \end{pmatrix} \sim N_{2k}\left(\begin{pmatrix} \mu \\ \mu \end{pmatrix}, \begin{pmatrix} \Omega_X & \Omega_X \\ \Omega_X & \Omega_X + \Omega_Z \end{pmatrix} \right).$$

Hence, the conditional distribution of X given $Y = y$ is also normal, and its parameters can be computed using (A.29), obtaining

$$\text{E}[X|y] = \mu + \Omega_X(\Omega_Z + \Omega_X)^{-1}(y - \mu),$$
$$\text{var}[X|y] = \Omega_X + \Omega_X(\Omega_Z + \Omega_X)^{-1}\Omega_X.$$

Although, strictly speaking, these expressions provide the required answer, the expressions commonly used for the parameters are different, namely

$$\text{E}[X|y] = (\Omega_X^{-1} + \Omega_Z^{-1})^{-1}(\Omega_X^{-1}\mu + \Omega_Z^{-1}y),$$
$$\text{var}[X|y] = (\Omega_X^{-1} + \Omega_Z^{-1})^{-1},$$

which can be obtained from the previous ones by transforming them by means of (A.25).

Essential bibliography

Agresti, A (1990). *Categorical Data Analysis*. Wiley, New York.

Basawa, I. V. and Prakasa Rao, B. L. S. (1980). *Statistical Inference for Stochastic Processes*. Academic Press, London and New York.

Cox, D. R. and Hinkley, D. V. (1974). *Theoretical Statistics*. Chapman & Hall, London.

Cramér, H. (1946). *Mathematical Methods of Statistics*. Princeton University Press, Princeton, NJ.

Johnson, N. L. and Kotz, S. (1972). *Distributions in Statistics*, 4 volumes. Wiley, New York.

Kendall's Advanced Theory of Statistics.
  I. Stuart, A. and Ord, J. K. (1987). *Distribution Theory*. Edward Arnold, London.
  II. Stuart, A. and Ord, J. K. (1991). *Classical Inference and Relationships*. Edward Arnold, London.
  IIb. O'Hagan, A. (1994). *Bayesian Inference*. Edward Arnold, London.

Kotz, S., Johnson, N. L. and Read, C. B., (eds) (1982–1988). *Encyclopedia of Statistical Sciences*, 9 volumes. Wiley, New York.

Lehmann, E. L (1983). *The Theory of Point Estimation*. Wiley, New York.

Lehmann, E. L. (1986). *Testing Statistical Hypotheses*, 2nd edition. Wiley, New York.

Mardia, K. V., Kent, J. T. and Bibby, J. M. (1979). *Multivariate Analysis*. Academic Press, London and New York.

Priestley, M. B. (1981). *Spectral Analysis and Time Series*, 2 volumes. Academic Press, London.

Rao, C. R. (1973). *Linear Statistical Inference*, 2nd edition. Wiley, New York.

Scheffé, H. (1959). *The Analysis of Variance*. Wiley, New York.

Zacks, S. (1971). *The Theory of Statistical Inference*. Wiley, New York.

References

Abramowitz, M. and Stegun, I. A. (1965). *Handbook of Mathematical Functions*. Dover Publications, New York.

Agresti, A. (1990). *Categorical Data Analysis*. Wiley, New York.

Amari, S.-I. (1985). *Differential-Geometric Methods in Statistics*. Lecture Notes in Statistics 28. Springer-Verlag, Heidelberg.

Anderson, C. W. and Loynes, R. M. (1987). *The Teaching of Practical Statistics*. Wiley, New York.

Anderson, T. W. (1971). *The Statistical Analysis of Time Series*. Wiley, New York.

Andrews, D. F. and Herzberg, A. M. (1985). *Data. A Collection of Problems from Many Fields for the Student and Research Worker*. Springer-Verlag, New York.

Atkinson, A. C. (1985). *Plots, Transformations and Regression*. Clarendon Press, Oxford.

Barndorff-Nielsen, O. E. (1978). *Information and Exponential Families*. Wiley, New York.

Barndorff-Nielsen, O. E. and Cox, D. R. (1994). *Inference and Asymptotics*. Chapman & Hall, London.

Basawa, I. V and Prakasa Rao, B. L. S. (1980). *Statistical Inference for Stochastic Processes*. Academic Press, London and New York.

Basu, D. (1975). Statistical information and likelihood (with discussion). *Sankhyā A* **37**, 1–71.

Berkson, J. (1980). Minimum chi-square, not maximum likelihood! (with discussion). *Ann. Statist.* **8**, 457–487.

Bissel, A. F. (1972). A negative binomial model with varying element sizes. *Biometrika* **59**, 435–441.

Bliss, C. I. (1935). The calculation of the dosage–mortality curve. *Ann Appl. Biol.* **22**, 134-67.

Box, G. E. P. and Cox, D. R. (1964). The analysis of transformations (with discussion). *J. Roy. Statist. Soc. B* **26**, 211–252.

Brown, L. D. (1986). *Fundamentals of Statistical Exponential Families, with Applications in Statistical Decision Theory*. (Lecture Notes – Monograph Series). Institute of Mathematical Statistics, Hayward, CA.

Chatfield, C. (1988). *Problem Solving: A Statistician's Guide*. Chapman

& Hall, London.

Cleveland, W. S. (1995). *Visualizing Data*. Hobart Press, Summit, NJ.

Cochran, W. G. and Cox, G. M. (1950). *Experimental Designs*. Wiley, New York.

Cook, R. D. and Weisberg, S. (1982) *Residuals and Influence in Regression*. Chapman & Hall, London.

Cook, R. D. and Weisberg, S. (1994)   *An Introduction to Regression Graphics*. Wiley, New York.

Cox, D. R. (1958). Some problems connected with statistical inference. *Ann. Math. Statist.* **29**, 357–372.

Cox, D. R. (1977). The role of significance tests. *Scand. J. Statist.* **4**, 49–70.

Cox, D. R. and Snell, E. J. (1981). *Applied Statistics*. Chapman & Hall, London.

Cox, D. R. and Snell, E. J. (1989). *Analysis of Binary Data*, 2nd edition. Chapman & Hall, London.

Cramér, H. (1946). *Mathematical Methods of Statistics*. Princeton University Press, Princeton, NJ.

Cressie, N. and Read, T. R. C. (1989).   Pearson's X^2 and the log-likelihood ratio statistic G^2: A comparative review. *Int. Statist. Rev.* **57**, 19–43.

Edwards, A. W. F. (1972).   *Likelihood*. Cambridge University Press, Cambridge

Edwards, A. W. F. (1974). The history of likelihood. *Int. Statist. Rev.* **42**, 9–15.

Efron, B. (1975). Defining the curvature of a statistical problem (with applications to second order efficiency). *Ann. Statist.* **3**, 1189–1242.

Efron, B. and Hinkley, D. V. (1978). Assessing the accuracy of the maximum likelihood estimator: Observed versus expected Fisher information. *Biometrika* **65**, 457–487.

Feller, W. (1968). *An Introduction to Probability Theory and its Applications, Volume I*, 3rd edition. Wiley, New York.

Fisher, R. A. (1922). On the mathematical foundations of theoretical statistics. *Philos. Trans. Roy. Soc. London, Ser. A* **222**, 309–368.

Fisher, R. A. (1925). Theory of statistical estimation. *Proc. Cambridge Philos. Soc* **22**, Pt. 5, 309–368.

Fisher, R. A. (1958). *Statistical Methods for Research Workers*, 13th edition. Oliver and Boyd, Edinburgh.

Fuller, W. A. (1987). *Measurement Error Models*. Wiley, New York.

Hald, A. (1952). *Statistical Theory with Engineering Applications*. Wiley, New York.

Hampel, F. R., Ronchetti, R. M., Rousseeuw, P. J. and Stahel, W. A. (1986). *Robust Statistics*. Wiley, New York.

Härdle, W. (1990). *Applied Nonparametric Regression*. Cambridge University Press, London.

Jackson, D. (1921). A note on the median of a set of numbers. *Bull. Amer. Math. Soc.* **27**, 160–164.

Kalbfleisch, J. D. and Prentice, R. L. (1980). *The Statistical Analysis of Failure Time Data* Wiley, New York.

Lehmann, E. L. (1983). *The Theory of Point Estimation*. Wiley, New York.

Lehmann, E. L. (1986). *Testing Statistical Hypotheses*, 2nd edition. Wiley, New York.

Mäkeläinen, T., Schmidt, K. and Styan, P. H. (1981). On the existence and uniqueness of the maximum likelihood estimate of a vector-valued parameter in fixed-size samples. *Ann. Statist.* **9**, 758–767.

Mann, H. B. and Wald, A. (1943). On stochastic limit and order relationships. *Ann. Math Statist* **14**, 217–226.

McCullagh, P. and Nelder, J. A. (1989) *Generalized Linear Models*, 2nd edition. Chapman & Hall, London.

Mills, F. C. (1965). *Statistical Methods*. Pitman, London.

Naddeo, A. (1963). Statistica. In *Enciclopedia della Scienza e della Tecnica*. Vol. IX, pp. 580–587. Mondadori, Milan.

Nelder, J. A. and Wedderburn, R. W. M. (1972). Generalized linear models. *J. Roy. Statist. Soc., Ser. A* **135**, 370–384.

Page, E. (1977). Approximations to the cumulative normal function and its inverse for use on a pocket calculator. *Appl. Statist.* **26**, 75–76.

Patil, G. P. (1962). Maximum likelihood estimation for generalized power series distributions and its application to a truncated binomial distribution. *Biometrika* **49**, 227–237.

Pearson, E. S. and Hartley, H. O. (1970–72). *Biometrika Tables for Statisticians*, 2 volumes. Cambridge University Press, Cambridge.

Priestley, M. B. (1981). *Spectral Analysis and Time Series*, 2 volumes. Academic Press, London.

Priestley, M. B. (1988). *Non-linear and Non-stationary Time Series Analysis*. Academic Press, London

Ramsey, F. P. (1931). *The Foundations of Mathematics and Other Logical Essays*. Kegan Paul, Trench, Trubner & Co. Ltd, London.

Rao, C. R. (1961). Asymptotic efficiency and information. *Proc. Fourth Berkeley Symp. Math. Statist. Probab.* **1**, 531–545.

Ryan, B. F., Joiner, B. L and Ryan, T. A., Jr. (1985). *Minitab Handbook*, 2nd edition. PWS-Kent Publishing Company, Boston.

Sbr, A. M., Owen, R. D and Edgar, R. S. (1965). *General Genetics*, 2nd edition. Freeman & Co., San Francisco.

Scheffé, H. (1959). *The Analysis of Variance*. Wiley, New York.

Scholz, F. W. (1980). Towards a unified definition of maximum likelihood. *Canad. J. Statist* **8**, 193–203.

Seber, G. A. F. (1973). *The Estimation of Animal Abundance and Related Parameters*, 2nd edition. Griffin, London.

Serfling, R. J. (1980). *Approximation Theorems in Mathematical Statistics*. Wiley, New York.

Seshadri, V. (1993). *The Inverse Gaussian Distribution, A Case Study in Exponential Families*. Oxford University Press, Oxford.

Silvapulle, M. J. (1981). On the existence of maximum likelihood estimators for the binomial response model. *J Roy. Statist. Soc., Ser. B* **43**, 310–313.

Silverman, B. (1986). *Nonparametric Density Estimation*. Chapman & Hall, London.

Thisted, R. A. (1988). *Elements of Statistical Computing*. Chapman & Hall, London.

Wald, A. (1949). Note on the consistency of the maximum likelihood estimate. *Ann. Math. Statist* **20**, 595–601.

Weisberg, S. (1985). *Applied Linear Regression*. Wiley, New York.

Zacks, S. (1971). *The Theory of Statistical Inference*. Wiley, New York

Zehna, P. W. (1966). Invariance of maximum likelihood estimators. *Ann. Math. Statist.* **37**, 744.

Author index

Subject index